MW01051366

WILDLIFE TOXICOLOGY

Emerging Contaminant and Biodiversity Issues

WILDLIFE TOXICOLOGY

Emerging Contaminant and Biodiversity Issues

Edited by

Ronald J. Kendall

Thomas E. Lacher

George P. Cobb

Stephen B. Cox

With a foreword by Thomas Lovejoy

CRC Press is an imprint of the
Taylor & Francis Group, an **informa** business

CRC Press
Taylor & Francis Group
6000 Broken Sound Parkway NW, Suite 300
Boca Raton, FL 33487-2742

Printed in the United States of America on acid-free paper
10 9 8 7 6 5 4 3 2 1

International Standard Book Number: 978-1-4398-1794-0 (Hardback)

Library of Congress Cataloging-in-Publication Data

Wildlife toxicology : emerging contaminant and biodiversity issues / editors, Ronald J. Kendall ... [et al.].
p. cm.
Includes bibliographical references and index.
ISBN 978-1-4398-1794-0 (alk. paper)
1. Environmental toxicology. 2. Pesticides and wildlife. 3. Biodiversity. 4. Animals--Effect of pollution on. I. Kendall, Ronald J.

RA1226.W55 2010
363.17'92--dc22 2010005713

Visit the Taylor & Francis Web site at
http://www.taylorandfrancis.com

and the CRC Press Web site at
http://www.crcpress.com

Contents

Foreword

As the issues of global climate change and loss of biodiversity challenge the future environmental stability of planet Earth, there has never been a time more important than now to further assess the impacts of environmental contamination. Hence, the publication of the book *Wildlife Toxicology: Emerging Contaminant and Biodiversity Issues* comes at a time to stimulate and challenge the academic and research communities as well as environmental public policy and decision makers to think more globally about environmental contaminants and their potential impacts on biodiversity and environmental degradation. I compliment one of the outstanding wildlife toxicology research teams in the world for bringing together a robust and visionary database in a book that will advance the science of wildlife toxicology and better improve our ability to preserve biodiversity and enhance environmental protection to sustain us into the future.

Thomas Lovejoy

Preface

Wildlife Toxicology: Emerging Contaminant and Biodiversity Issues builds on the previous successful book *Wildlife Toxicology and Population Modeling: Integrated Studies of Agroecosystems*, published in 1994 and edited by Kendall and Lacher, now the co-editors of the current book. Because our first wildlife toxicology book was so successful and is now out of print, the current publisher, Taylor & Francis/CRC Press, contacted us to develop a second edition. The second edition evolved into a new book as we recognized a need to address emerging contaminant as well as biodiversity issues. Therefore, we have expanded the editorship and the scope of *Wildlife Toxicology: Emerging Contaminant and Biodiversity Issues*. Emerging issues of particular interest are the role of global climate change and atmospheric contaminants, losses of biodiversity at scales ranging from local to global, emerging diseases, agricultural trends and biofuels, and the widespread use of munitions and explosives from military- and industrial-related activities. Therefore, the goal of the present book is not only to address development within the field of wildlife toxicology, but also to integrate the broader issues of declining biodiversity and global climate change in a way that allows for better assessment of wildlife exposures to environmental contaminants.

The editors of *Wildlife Toxicology: Emerging Contaminant and Biodiversity Issues* acknowledge financial support for both the research and development of this book through the Strategic Environmental Research and Development Program (SERDP). Over the last decade, SERDP has provided funding to various organizations, including The Institute of Environmental and Human Health (TIEHH) at Texas Tech University, to conduct research on perchlorate and high-energy explosives residues in the environment. Building on our very successful textbook *Perchlorate Ecotoxicology*, edited by Kendall and Smith (2008), SERDP supported our interest in developing a wildlife toxicology textbook that not only includes research that we had conducted for SERDP but also expands the scope of the book to include previously mentioned issues as well as an integration of these issues with contaminant exposure and risk assessment. Research and development recommendations for the future are also discussed in the present book. We appreciate SERDP's role in advancing environmental toxicology, particularly wildlife toxicology and the integration of toxicological aspects of contaminants with other environmental issues, including global climate change.

The process of developing *Wildlife Toxicology: Emerging Contaminant and Biodiversity Issues* was a collegial and very professional process, with the editors working with an exceptional group of authors and co-authors to advance the field of wildlife toxicology. In particular, we, the editors of the present book— Kendall, Lacher, Cobb, and Cox—convened to develop and invite authorship for the various chapters that we selected as being important for this book. In a period of over a year, extensive communication occurred both electronically and via conference calls to develop outlines and a scope and direction of the chapters, as well as to develop drafts that evolved into a single draft of the book before we convened in a final, multi-day meeting held in Beaver Creek, Colorado. During the meeting, the text was refined, the risk-assessment chapter and the research recommendations were developed, and the book was brought together in a scientifically sound and professional manner to enhance this communication to the scientific community. We acknowledge the participants in the Beaver Creek meeting, including the editors, senior authors of chapters, and co-authors of chapters. Although the co-authors did not participate in the Beaver Creek Meeting, they did provide very useful and important scientific information that made the final chapters state-of-the-art in advancing the field of wildlife toxicology.

We had excellent support in the development of this textbook in the planning process, communication in early drafts of the book, and facilitation at the Beaver Creek meeting. These people include Ms. Lori Gibler, information technology specialist; Mr. Ryan Bounds, administrative support; and Ms. Tammy Henricks, secretarial and technical support for the book development. We offer this book as a major new advancement in the field of wildlife toxicology and we appreciate having the opportunity to make this contribution through the science of environmental toxicology.

Ronald J. Kendall, Ph.D.
Professor and Chair, Department of Environmental Toxicology
Director, The Institute of Environmental and Human Health (TIEHH)
Texas Tech University

Editors

Ronald J. Kendall, Ph.D., serves as the founding director of The Institute of Environmental and Human Health (TIEHH), a joint venture between Texas Tech University and Texas Tech University Health Sciences Center at Lubbock. He is a professor and the chairman of the Department of Environmental Toxicology at Texas Tech University. He graduated from the University of South Carolina and received his M.S. from Clemson University and his Ph.D. from Virginia Polytechnic Institute and State University. He received a United States Environmental Protection Agency post-doctoral traineeship at the Massachusetts Institute of Technology.

Dr. Kendall was a member of the United States Environmental Protection Agency's Science Advisory Panel from June 1995 to December 2002, serving as chairman from January 1999 to December 2002. He has served on many other boards, including the Society of Environmental Toxicology and Chemistry, as past president; on the board of directors of the SETAC Foundation for Environmental Education; on the Endocrine Disrupters Screening and Testing Advisory Committee of the United States Environmental Protection Agency; and on many panels of the National Research Council.

Dr. Kendall currently serves as editor for terrestrial toxicology for the journal *Environmental Toxicology & Chemistry*. In addition, he has authored more than 200 refereed journal and technical articles and has published or edited many books. He has made more than 170 public and scientific presentations in the field of wildlife and environmental toxicology and has successfully won 138 research grants from industries, foundations, and federal, state, and foreign governments. He has graduated 32 students at the graduate levels, including M.S. and Ph.D. degrees, and has authored 10 courses in environmental toxicology and wildlife toxicology. He has received numerous awards, addressed the United Nations Committee on Sustainable Development, and consulted with many foreign countries on environmental issues. Dr. Kendall was awarded a Fulbright Fellowship in 1991.

Thomas E. Lacher, Jr., Ph.D., is currently a full professor and the department head in Wildlife and Fisheries Sciences at Texas A&M University. He received his Ph.D. in Biological Sciences, Section of Ecology, Evolution and Systematics from the University of Pittsburgh in 1980, where he worked on the comparative social behavior of caviid rodents in northeastern Brazil. He has held positions at the University of Brasilia, Brazil; Western Washington University; Clemson University, where he was the executive director of the research consortium of the Archbold Tropical Research Center and the Springfield Field Station on the island of Dominica in the Windward Islands; and Texas A&M University, where he was a professor and Caesar Kleberg Chair in Wildlife Ecology in the Department of Wildlife and Fisheries Sciences. From 2002 to 2007 he was at Conservation International, where he was the founding director of the Tropical Ecology, Assessment, and Monitoring (TEAM) Initiative and then senior vice-president and executive director of the Center for Applied Biodiversity Science.

Dr. Lacher has been working in the neotropics for 30 years, with experience in Dominica, Costa Rica, Panama, Guyana, Suriname, Peru, and Brazil. His prior research has focused on the ecology and behavior of mammals, the applications of GIS technology to wildlife conservation, the effects of environmental contaminants on wildlife populations, and the integration of ecological and economic principles in conservation. He has served on the editorial boards of *Mastozoología Neotropical*, *Environmental Toxicology and Chemistry*, and *Sustainability: Science, Policy and Practice*. He has also served on numerous review panels for the NSF and EPA and was a member and the chair of the Area Advisory Committee for Latin America for the Council for International Exchange of Scholars (Fulbright Commission). He has been major advisor of 12 M.S. and 12 Ph.D. students and several postdoctoral fellows.

Dr. Lacher's current research is focused on the assessment of conservation status in mammals and the analysis and monitoring of large-scale patterns and trends in biodiversity, primarily in the tropics. At Texas A&M he is collaborating with faculty in the Department of Mathematics on applications of mathematics to questions in ecology and conservation. Current graduate student research includes morphological variation and river barriers in Amazonian marmosets, spatial and temporal patterns of space and resource use of macaws in Peru, and cultural values and conservation in the Rupununi savannas region of Guyana.

George P. Cobb, Ph.D., is a professor of environmental toxicology at Texas Tech University. Dr. Cobb received a B.S. in chemistry from the College of Charleston (1982) and a Ph.D. in chemistry from the University of South Florida (1989). Both degrees emphasized environmental chemistry and the potential adverse effects of toxicants. Dr. Cobb's research interests include sensor development for rapid trace analyses, toxicant bioaccumulation, non-lethal sampling techniques for exposure assessment, and evaluation of the effectiveness of ecological restoration procedures. His current research projects include steroid transport in air, mutagenic effects of nitramine explosives, age-dependent toxicity of metal nanoparticles, toxicant distributions left in the wake of Hurricane Katrina, and contaminant movement through trophic levels of ecosystems.

Dr. Cobb is a long-standing member of the American Chemical Society (ACS) and the Society of Environmental Toxicology and Chemistry (SETAC). Dr. Cobb is vice-president of SETAC North America and holds a seat on the SETAC World Council. He was program chair for the 2003 SETAC Annual Meeting in Austin, Texas. Within ACS, Dr. Cobb is treasurer and alternate councilor for the Environmental Chemistry Division.

Dr. Cobb has published over 90 peer-reviewed articles, and is a frequent participant in the U.S. EPA Science Advisory panels regarding terrestrial ecological risk assessment and product registration procedures. Dr. Cobb has served on the editorial board of *Environmental Toxicology & Chemistry* (1994 to 1996) and continues to review for numerous well-respected journals.

Stephen B. Cox, Ph.D., is an associate professor at The Institute of Environmental and Human Health (TIEHH) and Department of Environmental Toxicology, Texas Tech University. Dr. Cox received his B.S. in mathematics and Ph.D. in biology (1999), both from Texas Tech University. After conducting post-doctoral research at the National Center for Ecological Analysis and Synthesis at the University of California, Santa Barbara, he joined the faculty at Texas Tech in 2002.

Dr. Cox's research interests focus on the development and application of advanced quantitative techniques for understanding basic and applied questions in ecology, environmental biology, ecotoxicology, and human health. He views systems modeling and statistics as complementary tools for understanding the effects of toxicants on complex ecological systems. He is particularly interested in the interactive effects of natural and anthropogenic stressors and disturbance on ecological systems.

Authors

Todd A. Anderson, Ph.D.
Professor
Texas Tech University

Stephen B. Cox, Ph.D.
Associate Professor
Texas Tech University

Lee Hannah, Ph.D.
Senior Fellow
Conservation International

Ronald J. Kendall, Ph.D.
Professor/Director/Chair
Texas Tech University

Thomas E. Lacher, Jr., Ph.D.
Professor
Texas A&M University

Spencer R. Mortensen, Ph.D.
Monsanto Company

Steven M. Presley, Ph.D.
Associate Professor
Texas Tech University

Christopher J. Salice, Ph.D.
Assistant Professor
Texas Tech University

Philip N. Smith, Ph.D.
Associate Professor
Texas Tech University

Contributing Authors

Mohamad Afzal, Ph.D.
Kuwait University

Redha Al-Hasan, Ph.D.
Kuwait University

Galen P. Austin, Ph.D.
Texas Tech University

John W. Bickham, Ph.D.
Purdue University

Henk Bouwman, Ph.D.
North-West University, South Africa

Luisa E. Castillo, Ph.D.
Universidad Nacional, Costa Rica

George P. Cobb, Ph.D.
Professor
Texas Tech University

C. Brad Dabbert, Ph.D.
Texas Tech University

Michael H. Depledge, Ph.D.
Universities of Exeter and Plymouth, U.K.

Venugopal Dhananjayan, Ph.D.
Salim Ali Centre for Ornithology and Natural History, India

Cristina Fossi, Ph.D.
University of Siena, Italy

Claude Gascon, Ph.D.
Conservation International

Celine Godard-Codding, Ph.D.
Texas Tech University

Rhys E. Green, Ph.D.
Cambridge University, U.K.

Malsha Kitulagodage
University of Wollongong, Australia

Henrik Kylin, Ph.D.
Swedish University of Agricultural Sciences, Sweden

Robin Law
The Centre for Environment, U.K.

Letizia Marsili, Ph.D.
University of Siena, Italy

Robin D. Moore
Conservation International

Miguel Mora, Ph.D.
Texas A&M University

Thomas E. Nickson, Ph.D.
Monsanto Company

Todd O'Hara, Ph.D.
University of Alaska at Fairbanks

Manuel Spinola
Institute for Wildlife Conservation and Management
Universidad Nacional, Costa Rica

Paul Story
Australian Plague Locust Commission

Muralidharan Subramanian, Ph.D.
Salim Ali Centre for Ornithology and Natural History, India

1 Introduction and Overview

Ronald J. Kendall

CONTENTS

1.1 INTRODUCTION

The development of the book *Wildlife Toxicology: Emerging Contaminant and Biodiversity Issues* intends to build on the earlier successful book *Wildlife Toxicology and Population Modeling: Integrated Studies of Agroecosystems* (Kendall and Lacher 1994), which was widely distributed and used around the world. Since our first wildlife toxicology book was so successful, we were contacted by the publisher to develop a second edition; however, the current book has really evolved into a new work in itself, since we needed to address emerging contaminant as well as biodiversity issues. Although the field of wildlife toxicology has grown dramatically over the past quarter century (Kendall 1982) and many outstanding scientists are contributing to the development of the database on the response of wildlife to environmental contaminant exposures, things are really now more complex than ever, hence the reason for the expanded editorship to develop the book *Wildlife Toxicology: Emerging Contaminant and Biodiversity Issues.* Of particular interest has been the role of global climate change and atmospheric contaminants, impacts on biodiversity from a local to a global perspective, emerging diseases, agricultural trends and biofuels, and the widespread use of munitions and explosives for military-related activities and consequences of release into the environment. Therefore, the goal of this book will be to address not only developments within the field of wildlife toxicology but also issues such as declining biodiversity and global climate change, while integrating this information in a way in which we may better assess concomitant exposures to environmental contaminants.

1.2 SOME PERSPECTIVES

Quantifying the impacts of contaminants on wildlife populations remains as difficult a task as it was almost 30 years ago, when the field began to emerge as a major area of research (Kendall 1982, 1992; Kendall and Smith 2003). In fact, issues such as global climate change, deforestation, and desertification, among other emerging planet Earth environmental challenges, make the field of wildlife toxicology even more complicated today. For example, how does one determine the tolerable number or percentage of a wildlife species that may be killed outright, made more susceptible to disease or predation, or suffer reproductive impairment from exposure to toxic chemicals? At the same time, along with contaminant exposure, other emerging environmental factors that can cause increased stress or impaired reproduction in various wildlife species have created challenging opportunities for scientific research. In this chapter, I will discuss how the rapid expansion of environmental toxicology has produced a growing number of wildlife toxicologists to find answers to these questions. These toxicologists are developing and using ecological and related acute and chronic toxicological information to study wildlife potentially affected by environmental contaminants.

1.3 WHAT IS WILDLIFE TOXICOLOGY?

Wildlife toxicology has been reviewed and defined as the study of the effects of environmental contaminants on wildlife species as related to animals' well-being, general health, and reproduction (Kendall 1982). Wildlife toxicology has been expanded on by Hoffman et al. (1990), who noted three principal strategies for understanding xenobiotic effects on wildlife: chemical analysis, field ecology, and various controlled field studies. To further expand on this observation, I propose that wildlife toxicology requires an interdisciplinary approach of three major emphasis areas. Initially, environmental chemistry and analytical toxicology are extremely critical in identifying the toxic substances of concern and their concentrations both in the environment and *in vivo*, respectively. Then, biochemical toxicology, encompassing the physiological and biochemical disturbances of contaminants and their toxicokinetics, provides information as to the "mechanism of action" and dynamics of certain toxic substances that may gain entry into the bodies of wildlife. Finally, we must consider ecotoxicology, or the field effects of chemicals on both the aquatic and terrestrial environment as concerns impacts on wildlife and their populations. Through ecotoxicological testing methods in wildlife toxicology, we may conduct semi-controlled field studies, pen studies, and full-scale field studies, discussed later in this chapter. Thus, we are integrating analytical chemistry, biochemical toxicology, and toxicological testing methods with field ecotoxicological assessments. From these data sets, we are able to construct information on the fate and effects of toxic chemicals, so that through our studies of wildlife toxicology we may be able to conduct ecological risk assessments (Bascietto et al. 1990).

As we consider wildlife toxicology and the well-being of wildlife species, we can consider behavioral effects, such as whether there is a significant increase in the probability of being preyed upon, or an aberration in migratory behavior. Good general health implies that the organism or population exists in a sustainable, homeostatic condition with its environment and, therefore, can respond to various environmental situations, such as global climate change. Because the reproductive process is often very sensitive to chemical influences, these studies are a high priority in wildlife toxicology. Some researchers question what is so special about wildlife toxicology that it should be identified as a particular branch of toxicology. All toxicologists have in common a primary objective: to study the mechanism and processes by which toxic substances produce adverse effects in biological systems. Such studies are based largely on the use of certain laboratory animals, animal-derived tissue, or cell models regardless of whether they are conducted to assess hazards to humans or other species. It is in risk assessment that a certain diversification among toxicologists becomes noticeable. Those whose primary concern is assessing chemical hazards to humans should be able to judge whether animal data are relevant to humans. They must understand the toxic substance, the comparative physiology of the test species to humans, and the potential for human exposure to the chemicals in question. Likewise, wildlife toxicologists must understand the toxic substance, biological systems, its response, and the potential for exposure (Kendall 1992); however, in this case it is related to populations in the wild, where there is far less experimental control and much more uncertainty.

Although there are no clear taxonomic guidelines for determining what *wildlife* is, generally this pertains to vertebrate animals living in a natural, undomesticated state. In the early years, wildlife toxicology research focused on wildlife with economic benefits in terms of hunting and fishing, sources of food, enjoyment of nature, subjects of photography, or for aesthetic appreciation. This emphasis derived from the rationale that the most-studied species should be either beneficial or detrimental to human society (Giles 1978). Although "value-added" species such as birds and mammals are still quite important, this chapter includes issues related to amphibians and reptiles. These taxa have been receiving more attention in terms of their susceptibility to contaminant exposure.

Wildlife toxicology is truly interdisciplinary in nature, as already mentioned. It draws on several sub-disciplines, among which are analytical toxicology, biochemical toxicology, and ecotoxicology, which concerns the effects of contaminants on the ecology of wildlife species, including the effects on a species' behavior and foraging strategies (Kendall 1992). The field of wildlife toxicology has benefited from scientists whose diverse research interests helped assess exposure to and effects on wildlife from environmental contaminants. Initially, biologists, chemists, veterinarians, and pharmacologists who shared an interest in the effects of chemicals on wildlife and humans were involved in the study of wildlife toxicology, but they lacked a unified perspective. Academic programs specifically devoted to the interdisciplinary training of wildlife toxicology professionals were not available until the early 1980s.

1.4 WILDLIFE TOXICOLOGY — THE EARLY DAYS

The discovery that organochlorine pesticides could reduce eggshell thickness in raptorial species is perhaps the most well-known and extensively documented event in wildlife toxicology. During the 1940s and 1950s, Derrick Ratcliff of the British Nature Conservancy noted a decline in peregrine falcons (*Falco peregrinus*) across Europe. Soon thereafter, correlations between eggshell thickness and reproductive failure in these falcons, other raptors, and piscivorous (fish-eating) avian species were discovered (Ratcliff 1967). These findings, and evidence of exposure among humans, ultimately led to the ban on DDT (1,1,1-trichloro-2,2-bis (p-chlorophenyl) ethane) in the United States in 1973.

In addition, a number of events raised society's awareness and prompted the public's interest in environmental issues and, thus, wildlife toxicology. Rachel Carson's book *Silent Spring* (1962) fueled the debate on environmental contaminant effects on humans and wildlife and brought these important issues to the widespread attention of the American public. Many credit *Silent Spring* with spurring the ensuing environmental movement. Carson's cautionary words on the potential impacts of anthropogenic substances in the environment have inspired many environmental scientists to bring public awareness to this issue even to this day.

In 1979, the Society of Environmental Toxicology and Chemistry (SETAC) was founded to fill the need for a professional organization dedicated to the research on issues of environmental contamination. Among its early membership, SETAC included many of the first wildlife toxicologists, and today it remains an important organization for the dissemination of wildlife toxicology research (Kendall 2008). SETAC remains the largest and most influential organization for environmental and wildlife toxicology professionals and has become global in stature and reach.

The early days of wildlife toxicology saw extensive use of relatively simple experimental methods, including the generation of LD_{50} (lethal dose for 50% of a test group) and LC_{50} (lethal air or water concentration for 50% of test subjects) values. Generally speaking, overt lethality was the most common endpoint assessed by wildlife toxicologists. These methods were adapted from traditional toxicology. A new era of intensive synthetic chemical usage in the mid-1900s unfortunately resulted in many instances in which acute toxicity levels of contaminant levels created "die offs" among wildlife. Environmental regulations evolved, slowly at times, to address the booming industrial capacity, including those of the agricultural chemical industries, to produce and disseminate toxic substances into the environment.

Discoveries of wildlife mortality in the field led wildlife toxicologists to initiate laboratory dosing experiments using wild, inbred, or domestic animal models to establish benchmark data on acute and chronic toxicity of environmental contaminants, including pesticides, petroleum, and industrial waste products. For example, Aulerich and Ringer used PCB-laden fish to evaluate lethality and reproductive impairment of mink, which had begun to decline around the Great Lakes in the late 1960s and early 1970s (Aulerich and Ringer 1977). However, to generate statistically meaningful datasets in light of inherent variability of wildlife species, many of the early wildlife toxicity studies ironically resulted in the overkill of both laboratory animals and wildlife. Efforts to develop and validate sublethal indicators of exposure and effect for sometimes-critical wildlife populations were intensified.

Many of the well-known environmental contaminants, such as DDT, PCBs, and various metals, have been studied extensively for toxicity among wildlife. Although there is still much to be learned about the effects of these chemicals alone, and especially in mixtures, much progress has been made over the past approximately 30 years, the time frame of the most rapid development of the field of wildlife toxicology. Today, sublethal chronic testing of single and multiple contaminants focusing on alterations in wildlife physiological processes, reproductive success, and fitness have become more common than lethality tests, which had provided the earlier benchmark toxicity values. With the increased capabilities in genetic toxicology, we are also able to better understand the influence of environmental contaminants on wildlife genetics, which may ultimately manifest itself toxicologically with impairment of reproduction, health, or well-being.

Although it is sometimes possible in wildlife toxicology to use the actual species of interest in laboratory studies, generally data from one species must be extrapolated to another using comparative physiology. When risk assessments are conducted, wildlife toxicologists must consider hazards not only to one particular species, but also to a variety of species and their ecosystem. This requires sufficient understanding of the environmental chemistry of toxic chemicals (fate, activation, and degradation) as well as knowledge of the life cycles of wildlife, which may influence exposure.

In this area of risk assessment, wildlife toxicologists have one major advantage over human toxicologists: opportunities sometimes arise to conduct field studies to echo the impact of the chemical on certain species or groups of species (Balk and Koeman 1984). In most instances with humans, the effects of hazardous exposures to chemicals can be evaluated only indirectly.

The design of field experiments in wildlife toxicology requires the combination of disciplines previously mentioned: environmental chemistry and analytical toxicology, biochemical toxicology, biostatistics, and wildlife biology and field ecotoxicology. Consequently, wildlife toxicology should be seen as a rather broad area in which scientists with different educational backgrounds and training can and should collaborate.

1.5 WILDLIFE TOXICOLOGY — THE MOVEMENT FORWARD

1.5.1 Laws and Regulations

Although human health continues to generate a high level of concern among regulators, understanding the effects of chemicals in wildlife has increased the development of regulations governing the manufacture and use of chemical substances. In 1942, the U.S. Congress passed the Federal Insecticide, Fungicide and Rodenticide Act (FIFRA) to deal with pesticide registration, efficacy, and use under jurisdiction of the United States Department of Agriculture. FIFRA became the Federal Environmental Pesticide Control Act in 1972, charging the U.S. EPA with determining whether pesticides represented "unreasonable risk" to people, the environment, and wildlife. Other Congressional acts, including the Toxic Substance Control Act of 1976 and the Comprehensive Environmental Response, Compensation and Liability Act (Superfund) in 1980, have added strength to governmentally controlled chemicals and their release into the environment with possible impact on wildlife populations. These acts directly or indirectly led to increased funding for research in wildlife toxicology in the 1980s and early 1990s, fueled to a large degree by the need to evaluate the risk of agrochemicals in wildlife (Kendall 1992). Often, large complex field studies were combined with extensive laboratory testing on fish, birds, and mammals to provide data for registering and labeling a chemical or pesticide. These studies often included carcass searches in treated and untreated fields as well as chemical analysis of soil, vegetation, excrement, foot washes, and biological tissues. Bioassays ranging from cholinesterase measurements to induction of metabolizing enzyme evaluations aided in assessing physiological responses to chemical exposure, thus allowing spatial analysis of exposure and effects. Moreover, population-level indices of contaminant effects, such as reproductive success and survival, help assess contaminant effects beyond the individual level. However, in the 1990s, changes in EPA policy reduced requirements for extensive field studies. In the 2000s, very few field studies were required by the EPA, and thus few were conducted.

The U.S. federal government demonstrated early concern for wildlife interactions with environmental contaminants. The U.S. Fish and Wildlife Service's (USFWS) environmental contaminant program began in the 1950s. It became intimately involved with environmental quality issues similar to ones addressed by the EPA, but its mission focused on the health and well-being of fish and wildlife. Today, the USFWS has approximately 75 locations throughout the United States, with contaminant specialists within the USFWS focusing on pollution prevention, contaminant identification and risk assessment, cleanup, and technical support for other USFWS biologists.

In 1992, the EPA released an Ecological Risk Assessment Framework in which the evaluation of wildlife health, reproduction, and survival were established as criteria for registering chemicals for commercial use and contaminants found at hazardous waste sites (U.S. EPA 1992). This document provided the necessary outline for examining environmental issues facing wildlife populations and incorporating exposure and effects assessments into a quantitative process that accounts for associated uncertainties in regulatory decision making.

In 1998, this document was replaced with the EPA's *Guidelines for Ecological Risk Assessment*, which was designed to improve the quality and consistency of the risk assessment process (U.S. EPA 1998). This framework allows for the assessment of toxic chemical impacts and other stressors on ecological systems. In the formulation phase, a conceptual model is usually developed that describes routes of exposure, biota of concern, and anticipated effect endpoints. The actual risk of chemicals to wildlife is then determined using exposure and effects data for the chemicals of interest. Toxicity data for species of concern at either the individual or population level may be incorporated in the risk-characterization phase (Kendall and Ackerman 1992). Exposure and effects data accumulated in the analysis phase are combined and the risk potential is characterized.

1.6 ECOLOGICAL RISK ASSESSMENT

The EPA has published a good example of an ecological risk assessment involving the exposure of wildlife to the insecticide carbofuran (2,3-dihydro-2,2-dimethyl-7-benzofuranol methylcarbamate) (Houseknecht 1993). This study documented widespread and repeated mortality events, particularly when birds ingested carbofuran granules in agricultural systems. According to legislation promulgated by FIFRA, chemicals in the environment could not pose unreasonable adverse effects to birds or other wildlife populations. In addition, the Migratory Bird Treaty Act protects species internationally by prohibiting the killing of songbirds or waterfowl with a pesticide. New tools have emerged in conducting ecological risk assessments using probabilistic risk assessment with both fish and wildlife that will improve our ability to understand the implications of toxic chemical impacts on fish and wildlife resources (Giddings et al. 2005; Solomon et al. 2001).

1.7 ENDOCRINE DISRUPTION — AN EMERGING ISSUE

The 1990s presented several issues with direct consequences to wildlife toxicology, but endocrine-disrupting properties topped the list. With the passage of the Food Quality Protection Act of 1996, the U.S. Congress charged the U.S. EPA to address scientific questions regarding the potential for chemicals to cause endocrine disruption in both humans and wildlife. With the hypothesis that these compounds impact reproduction of wildlife, Theo Colborn's book *Our Stolen Future* stimulated an intense debate (Colborn et al. 1996). Guillette et al. (1994) had previously reported alterations in alligator (*Alligator mississippiensis*) sex hormones and gonadal development related to environmental contaminants in Florida. The combination of legislation and other findings generated intense scientific debate, workshops, and ultimately books addressing the potential influence of contaminants on endocrine function (Kendall et al. 1998). In response, the EPA established the Endocrine-Disruptor Screening and Testing Advisory Committee (EDSTAC), on which I participated, to examine this issue and make recommendations on testing and regulation of endocrine-disrupting chemicals.

Endocrine disruptors remain an issue of concern as new chemicals with these types of properties are identified. Recently, perchlorate, a thyroid hormone inhibitor, gained national attention because of its widespread distribution in groundwater and surface water supplies. Perchlorate detected in numerous plant, fish, and wildlife species (Smith et al. 2001) can effectively inhibit metamorphosis and shift sex ratios in amphibians (Goleman et al. 2002). As endocrine-altering chemicals continue to emerge, the demand for studies to examine their effects on wildlife will increase. In fact, even at the time of the preparation of this book in 2010, industry is preparing to extensively increase testing of the effects of endocrine-disrupting compounds on wildlife.

1.8 AMPHIBIANS — A NEW ERA IN WILDLIFE TOXICITY TESTING

Because of declining populations and discoveries of malformations in amphibians and reptiles, they have received much attention from the general public and biologists in the past decade (Campbell and Campbell 2002). Numerous studies have indicated dramatic reductions in amphibian populations worldwide, with explanations such as climate change, parasitic, bacterial, fungal, and viral infections, and increased predation. In 1998, the U.S. federal government established the Interagency Task Force on Amphibian Declines and Deformities to examine the changes seen in amphibian populations (USFWS 1999). Although amphibians have become a major focus, reptiles remain relatively underrepresented in the wildlife toxicology literature (Sparling et al. 2000). As new analytical tools and methods improve our abilities to evaluate subtle changes in environmental quality and wildlife health, additional challenges are sure to emerge.

1.9 ATRAZINE — THE GREAT DEBATE

The herbicide atrazine (6-chloro-N-ethyl-N′-isopropyl-1,3,5-triazine-2,4-diamine) is widely used in agriculture in the production of corn and other crops. After application, atrazine is sometimes found in surface waters, which can be inhabited by aquatic organisms, particularly amphibians. Some studies reported adverse effects of atrazine on sexual development in frogs (Hayes et al. 2002; Hayes et al. 2003). The suggestion that atrazine induced hermaphroditism at 0.1 ppb in American leopard frogs (*Rana pipiens*) created intense debate in the scientific community as well as at the U.S. EPA (Solomon et al. 2008). To assess whether atrazine causes adverse effects in frogs through mechanisms mediated by endocrine and other pathways, several hypotheses were tested in laboratory and field studies using guidelines for the identification of causative agents of disease. Extensive research of more than a decade involving laboratory and field studies with a variety of species of frogs on several continents revealed that, based on the weight of the evidence and analysis of all the data, the central theory that environmentally relevant concentrations of atrazine affect reproduction and/or reproductive development in amphibians is not supported by the vast majority of observations (Solomon et al. 2008). The same conclusions also hold for the supporting theories, such as induction of aromatase, the enzyme that converts testosterone to estradiol. For other responses, such as immune function, stress endocrinology, and parasitism of population level effects, there is no indication of effects, or there is such a paucity of good data that definitive conclusions cannot be made (Solomon et al. 2008).

1.10 WILDLIFE TOXICITY — SOME EXAMPLES AND STRATEGIES FOR ASSESSMENT

Laboratory tests generally required for the development of data for risk assessment in wildlife are limited, providing only basic toxicological and biochemical characteristics of the test product in free-ranging wildlife. If preliminary laboratory testing, when available, indicates that a substance may harm wildlife, then field-testing may be required for an ecological risk assessment. Prior to the initiation of such testing, a chemical such as a pesticide is evaluated to characterize potential impacts, such as mortality, behavioral effects, and biochemical or physiological disturbances.

Initial testing generates acute toxicity data through LD_{50} and/or LC_{50} determinations in addition to laboratory reproductive toxicity data. Also, representative species resident in a geographic area of planned field tests can be exposed to the compound under laboratory conditions to obtain lethality and toxicity data. These studies maximize the use of the test animals not simply by quantifying mortality but also by evaluating behavioral and biochemical impacts of the chemical or toxic substance. In the case of organophosphate pesticides, for example, which have been relatively well studied in relation to wildlife toxicology, cholinesterase activities are measured in both blood and brain tissues.

The use of plasma-cholinesterase activities in live-trapped wild animals has several benefits compared to conventional brain-cholinesterase analysis. First, it allows multiple captures and samplings from radio-tagged or otherwise marked animals, so that biochemical toxicity in individuals can be evaluated over time (Kendall and Lacher 1994). Depression of plasma cholinesterase, occurring with exposures well below lethal levels, best provides a very sensitive non-lethal means of establishing low-level OP exposure in wildlife, particularly birds that have been well studied.

Sublethal effects of pesticides on behavior of wildlife also have been examined. For instance, northern bobwhite (*Colinus virginianus*) have been assessed for predator-evasion responses subsequent to being exposed to the organophosphate pesticide methyl parathion (phosphorothioic acid, O,O-dimethyl O-(4-nitrophenyl) ester). The organophosphate-treated birds were less capable of escaping predators than were controls in simulated environments (Galindo et al. 1985). The activity and mortality of radio-tagged northern bobwhite exposed to non-lethal doses of methyl parathion and released back into their natural habitat have been evaluated (Buerger et al. 1991). Multiyear tests have shown that methyl-parathion-exposed bobwhites are more susceptible to predation, primarily from raptors. The evidence also suggests that exposure to organophosphates may influence the integrity of northern bobwhite coveys. Individuals suffering from acute toxicity may have difficulty maintaining covey affiliation, a behavior important for winter survival and predator detection in this species. These several examples provide insight as to how disturbed behavior can ultimately manifest itself in the impairment or death of exposed wildlife.

In terms of reproductive toxicity, which is often difficult to measure with environmental contaminants, studies with European starlings (*Sturnus vulgaris*) have used enhanced local populations attracted into test fields with artificial nest boxes. These tests have led to greater insight into the reproductive success of a passerine species exposed, for instance, to organophosphate pesticides (Kendall et al. 1989).

The starling is an excellent species for use in environmental monitoring. It is geographically widespread and utilizes many habitat types. As an introduced species in North America, it has acquired the status of a pest species, thus increasing its acceptability as a wild test organism. Nest boxes placed in study fields are readily utilized, providing a large synchronistic breeding population of passerines on the treatment site. The starlings' diet during the breeding season consists of primarily terrestrial invertebrates that live in direct contact with pesticides present in the soil. An additional important benefit is the starlings' tolerance of handling and monitoring methods used to quantify breeding success. This provides an example of the necessary strategy for identifying the wildlife to sample and assess for contaminant impacts.

The enhanced avian population model entails the establishment of starling nest boxes on study sites. One or more of the sites are treated with a chemical of interest, such as a pesticide. In field tests with the organophosphate methyl parathion, controlled birds had 48% successful nesting (fledging at least one bird from the nest) while those in a field treated with 1 kilogram of methyl parathion per hectare had only a 19% success rate (Robinson et al. 1988).

Nest boxes placed in agroecosystems not recently treated with pesticides have provided healthy nestlings that have been evaluated in the nest for their responses to increasing levels of the organophosphate pesticide diazinon (phosphorothioic acid O,O diethyl-O-[6-methyl-2-(1-methylethyl-4-pyrimidinyl] ester). The compound was administered orally in a corn oil vehicle to evaluate the differential sensitivity between nesting starlings of different ages. Newly hatched young were nearly 20 times more sensitive to an acute dose of diazinon when compared with fledglings (Hooper et al. 1990).

As previously mentioned, in the 1980s wildlife toxicology saw a significant increase in the number of field studies conducted to rebut the assumption of risk, particularly of pesticides to wildlife through the EPA Pesticide Registration Process. Field studies conducted in the early to mid-1980s on the effects of pesticides in birds generally included environmental chemistry data and response of birds (including mortality) to applied pesticides (Brewer et al. 1988). The EPA subsequently evolved a set of criteria upon which pesticide field studies could be conducted at either Level 1 or Level 2 (Fite et al. 1988).

A Level 1 field study generally involved mortality as a key endpoint for a variety of sites treated with pesticides and assessed for the response of birds and other wildlife possibly exposed to a chemical. Some residue analyses as well as behavioral observations were employed. These Level 1 field studies were generally known as "screening trials" and were a more qualitative assessment of the potential hazard of the pesticide or other toxic chemical to birds and other wildlife that may utilize habitats receiving such agents, for instance. Data acquired from Level 1 field trials then made it possible for an assessment for higher levels of field study, or a Level 2 study. A Level 2 field study sought to quantify the response of wildlife populations to chemicals such as agricultural pesticides. This quantification can take the form of extensive chemical analyses, radio telemetric monitoring techniques, and starling nest-box studies. Conducting a field experiment generally includes a number of sites for sufficient replication and a variety of observations that allow quantification of the response of wildlife to chemicals in terms of their reproduction, health, and well-being (Kendall and Lacher 1994). A Level 2 field study conducted in Iowa to assess the response of wildlife to the corn rootworm insecticide Counter 15-G (terbufos; phosphorodithioic acid S-[(1,1-dimethylethyl) thio] O,O diethyl) was submitted to the EPA to meet requirements of FIFRA guidelines (Kendall and Ackerman 1992; Kendall 1992). This is one of only a few known studies that have ever been submitted to the EPA to address Level 2 requirements. Data generated from Level 1 and particularly Level 2 studies provided useful information for the development of ecological models related to the effects of pesticides in avian and other wildlife species. Although very few pesticide-related field studies are conducted currently in the field of wildlife toxicology, their need is still extremely important when trying to fully understand the response of wildlife in their natural habitats to toxic chemical exposures.

1.11 THE FUTURE OF WILDLIFE TOXICOLOGY

Wildlife toxicology has much more sophisticated tools now than 30 years ago, when the field really began to emerge as a dynamic area of toxicology research. New analytical equipment and more economical technologies, such as enzyme-linked immunosorbent assays, passive sampling devices, and accelerated solvent extractors, have improved detection capabilities and removed many of the restrictions on measuring contaminants in various environmental matrices. For example, until the 1990s, no reliable method was available to detect perchlorate in water at the parts-per-billion concentrations now commonly detected in environmental samples. Scientists at the California EPA developed a new method in the 1990s sensitive enough to reveal widespread contamination of ground and surface waters in that state and elsewhere. The analytical method was evaluated and refined for detection of perchlorate in soil, sediment, and plants (Ellington and Evans 2000). A method for tissue was developed at a later time (Anderson 2002).

Tools for assessing physiological changes in wildlife related to environmental contaminants have also become increasingly more sophisticated. Polymerase chain reaction, DNA fingerprinting, cDNA microarrays, and other molecular techniques now provide more detailed information on the impacts of chemicals beyond individual and cellular levels. Thus, studies of contaminant effects of wildlife today may include measurement endpoints on all levels, from molecular, cellular, organ system, individual, and population to entire ecosystems.

Clearly the health of the environment influences the viability of people and wildlife. Current risk assessments for chemicals in the environment that establish protective limits for humans often rely on wildlife exposure data. Therefore, wildlife toxicologists have an opportunity to participate in regulatory processes aimed at protecting environmental and human health.

The field of wildlife toxicology continues to evolve while maintaining its original interdisciplinary nature by enlisting diverse specialists to help understand complexities of contaminant movement, fate, bioavailability, and physiological, population, and ecosystem effects. In the future, wildlife toxicologists could contribute significantly to assessing the global threats of chemical contamination, and they also have the expertise in environmental assessments to understand and perhaps counter potential terrorist attacks that use chemical and biological agents. Attacks with biological and chemical weapons intended for humans or livestock could have simultaneously devastating effects on wildlife resources and diversity (Dudley and Woodford 2002; Kendall et al. 2008).

REFERENCES

Anderson, T.A. and T.H. Wu. 2002. *Extraction, clean-up, and analysis of the perchlorate anion in tissue samples.* Bulletin of Environmental Contamination and Toxicology 68:684–691.

Aulerich, R.J. and R.K. Ringer. 1977. *Current status of PCB toxicity to mink, and effect on their reproduction.* Archives of Environmental Contamination and Toxicology 6:279–292.

Balk, F. and J.H. Koeman. 1984. *Future hazards from pesticide use.* Environmentalist 4:1–100.

Bascietto, J., D. Hinckley, J. Plafkin, and M. Slimak. 1990. *Ecotoxicity and ecological risk assessment.* Environmental Science & Technology 24:10–15.

Brewer, L.W., C.J. Driver, R.J. Kendall, J.C. Galindo and G.W. Dickson. 1998. *Avian response to a turf application of Triumph4E.* Environmental Toxicology and Chemistry 7(5):391–401.

Buerger, T.T., R.J. Kendall, B. Mueller, T. DeVos, and B.A. Williams. 1991. *Effects of methyl parathion on northern bobwhite survivability.* Environmental Toxicology and Chemistry 10(4):527–532.

Campbell, K.R. and T.S. Campbell. 2002. *A logical starting point for developing priorities for lizard and snake ecotoxicology: a review of available date.* Environmental Toxicology and Chemistry 21:894–898.

Carson, R. 1962. *Silent Spring*, Houghton Mifflin, Boston.

Colborn, T., D. Dumanoski, and J.P. Myers. 1996. *Our Stolen Future*, Plume: New York.

Dudley, J.P. and M.H. Woodford. 2002. *Bioweapons, biodiversity, and ecocide: potential effects of biological weapons on biological diversity.* Bioscience 52:583–592.

Ellington, J.J. and J.J. Evans. 2000. *Determination of perchlorate at parts-per-billion levels in plants by ion chromatography.* Journal of Chromatography A 898:193–199.

Fite, E.C., L.W. Turner, N.J. Cook, and C. Stunkard. 1988. *Guidance Document for Conducting Terrestrial Field Studies.* U.S. EPA, Washington, DC.

Galindo, J., R.J. Kendall, C.J. Driver, and T.E. Lacher, Jr. 1985. *The effect of methyl parathion on the susceptibility of bobwhite quail (Colinus virginianus) to domestic cat predation.* Behavioral and Neural Biology 43:21–36.

Giddings, J.M., T.A. Anderson, L.W. Hall, Jr., A.J. Hosmer, R.J. Kendall, R.P. Richards, R.R. Solomon, and M.W. Williams. 2005. *Atrazine in North American Surface Waters: A Probabilistic Aquatic Ecological Risk Assessment.* Society of Environmental Toxicology and Chemistry (SETAC) Press, 432 pp.

Giles, Jr., R.H. 1978. *Wildlife Management*, W.H. Freeman, San Francisco.

Goleman, W.L., L.J. Urquidi, T.A. Anderson, R.J. Kendall, R.J., E.E. Smith, and J.A. Carr. 2002. *Environmentally relevant concentrations of ammonium perchlorate inhibit development and metamorphosis in Xenopus laevis.* Environmental Toxicology and Chemistry 21:424–430.

Guillette, L.J., Jr., T.S. Gross, G.R. Masson, J.M. Matter, H.F. Percival, and A.R. Woodward. 1994. Environmental Health Perspective 102:680–688.

Hayes, T.B., A. Collins, M. Mendoza, N. Noriega, A.A. Stuart, and A. Vonk. 2002. *Hermaphroditic, demasculinized frogs exposed to the herbicide atrazine at low ecologically relevant doses.* Proceedings of the National Academy of Sciences USA 99:5476–5480.

Hayes, T.B., K. Haston, M. Tsui, A. Hoang, C. Haeffele, and A. Vonk. 2003. *Atrazine-induced hermaphroditism at 0.1 ppb in American leopard frogs (Rana pipiens): laboratory and field evidence.* Environmental Health Perspective 111:568–575.

Hoffman, D.J., B.A. Rattner, and R.J. Hall. 1990. *Wildlife Toxicology.* Environmental Science and Technology 24:276–283.

Hooper, M.J., L.W. Brewer, G.P. Cobb, and R.J. Kendall. 1990. *An integrated laboratory and field approach for assessing hazards of pesticide exposure to wildlife.* In: L. Somerville and C.H. Walker, Eds. Pesticide Effects on Terrestrial Wildlife; Taylor and Francis, New York.

Houseknecht, C.R. 1993. *A Review of Ecological Assessment Case Studies from a Risk Perspective.* EPA/630/R-92-005 Vol. 3. U.S. Environmental Protection Agency; U.S. Government Printing Office, Washington, DC.

Kendall, R.J. and J. Akerman. 1992. *Terrestrial wildlife exposed to agrochemicals: an ecological risk assessment perspective.* Environmental Toxicology and Chemistry 11(12):1727–1749.

Kendall, R.J., L.W. Brewer, T.E. Lacher, B.T. Marden, and M.L. Whitten. 1989. *The Use of Starling Nest Boxes for Field Reproductive Studies; Provisional Guidance Document and Technical Support Document:* U.S. EPA, Washington, DC, EPA/600/8-89/056, [NTIS publication number: PB89 195 028/AS].

Kendall, R.J., R.L. Dickerson, J.P. Giesy, and W.A. Suk. 1998. *Principles and Processes for Evaluating Endocrine Disruption in Wildlife*; Society of Environmental Toxicology and Chemistry (SETAC) Press, Pensacola, FL, 491pp.

Kendall, R.J. and P.N. Smith. 2003. *Wildlife Toxicology Revisited,* Environmental Science & Technology 37(9):178A–183A.

Kendall, R.J. 1992. *Farming with agrochemicals: the response of wildlife.* Environmental Science & Technology 26(2):238–245.

Kendall, R.J. 2008. *Wildlife Toxicology: Integration of Ecological and Toxicological Research Strategies.* Society of Environmental Toxicology and Chemistry 29th Annual Meeting, Tampa, FL., November 16–20, 2008.

Kendall, R.J. and T.E. Lacher, Jr. 1994. *Wildlife Toxicology and Population Modeling: Integrated Studies of Agroecosystems;* Lewis Publishers/CRC Press, 576pp.

Kendall R.J., S.M. Presley, G.P. Austin, and P.N. Smith, editors. 2008. *Advances in Biological and Chemical Terrorism Countermeasures*; Taylor & Francis/CRC Press, 280pp.

Ratcliffe, D.A. 1967. *Decrease in eggshell weight in certain birds of prey.* Nature 215:201–210.

Robinson, S.C., R.J. Kendall, C.J. Driver, T.E. Lacher, Jr., and R. Robinson. 1988. *Effects of agricultural spraying of methyl parathion on cholinesterase activity and reproductive success in wild starlings (Sturnus vulgaris).* Environmental Toxicology and Chemistry 7(5):343–349.

Smith, P.N., C.W. Theodorakis, T.A. Anderson, and R.J. Kendall. 2001. *Preliminary assessment of perchlorate in ecological receptors at the Longhorn Army Ammunition Plant (LHAAP), Karnack, Texas.* Ecotoxicology 10:305–313.

Solomon, K.R., J.P. Giesy, R.J. Kendall, L.B. Best, J.R. Coats, K.R. Dixon, M.J. Hooper, E.E. Kenaga, and S.T. McMurry. 2001. Chlorpyrifos: ecotoxicological risk assessment for birds and mammals in corn agroecosystems. Human and Ecological Risk Assessment 7(3):497–632.

Solomon, K.R., J.A. Carr, L.H. Du Preez, J.P. Geisy, R.J. Kendall, E.E. Smith, and G.J. Van Der Kraak. 2008. *Effects of atrazine on fish, amphibians, and aquatic reptiles: a critical review.* Critical Reviews in Toxicology 38:721–772.

Sparling, D., G. Linder, and C. Bishop, Eds. 2000. *Ecotoxicology of amphibians and reptiles.* Society of Environmental Toxicology and Chemistry, Pensacola, FL.

United States Environmental Protection Agency. 1992. *Framework for Ecological Risk Assessment,* U.S. EPA/630/R-92/001; U.S. Government Printing Office, Washington, DC.

United States Environmental Protection Agency. 1998. *Guidelines for Ecological Risk Assessment,* EPA/630/R-95/002F, Risk Assessment Forum, Washington, DC.

United States Fish and Wildlife Service. 1999. Department of Environmental Quality, Amphibian Declines and Deformities, http://contaminants.fws.gov/Issues/Amphibians.cfm.

2 Environmental Toxicology of Munitions-Related Compounds

Nitroaromatics and Nitramines

Todd A. Anderson

CONTENTS

2.1 INTRODUCTION

Interactions between environmental contaminants and ecological receptors are important for assessing risk, developing management objectives that protect the environment, and maintaining the health status of organisms. Despite numerous strategies to contain or minimize the release of fugitive chemical residues into the environment, organisms continue to be exposed to these potentially hazardous substances. As such, there continues to be a need for environmental and wildlife toxicologists to determine the effects of these chemical residues in water and soil. As basic toxicological information on a chemical is developed, scientists often shift their attention toward (1) more subtle endpoints beyond death, and/or (2) potential effects of the primary degradation products of the parent chemical. Such has largely been the case with munitions-related compounds (Talmage et al. 1999).

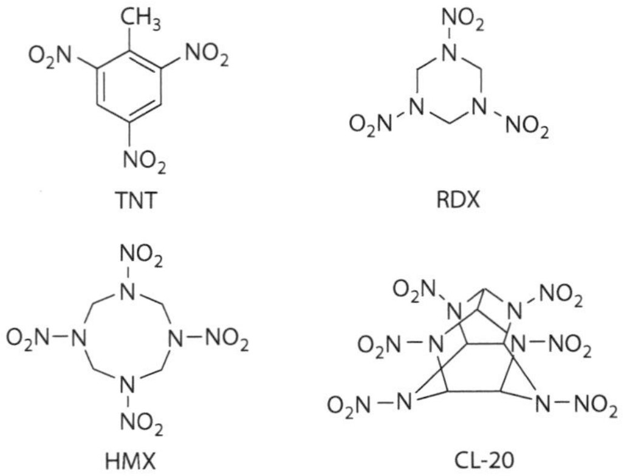

FIGURE 2.1 Structures of some nitroaromatic and nitramine explosives. TNT = 2,4,6-trinitrotoluene; RDX = hexahydro-1,3,5-trinitro-1,3,5-triazine; HMX = octahydro-1,3,5,7-tetranitro-1,3,5,7-tetrazocine; CL-20 = hexanitrohexaazaisowurtzitane.

Explosives and munitions are a class of energetic materials used by both the military and commercial sectors. Explosives as a group include mixtures and chemically pure compounds. Nitroaromatics, such as 2,4,6-trinitrotoluene (TNT), are some of the most common explosives of the last half century. In addition, the chemically related nitramine explosives, such as the well-known (and well-characterized) hexahydro-1,3,5-trinitro-1,3,5-triazine (RDX), octahydro-1,3,5,7-tetranitro-1,3,5,7-tetrazocine (HMX), and the recently developed hexanitrohexaazaisowurtzitane (CL-20), are also common (Figure 2.1).

The U.S. military is required to maintain a delicate balance between mission readiness through training and environmental protection (Boice 2006). The production, testing, use, and disposal of explosives indicate that they will directly or indirectly enter the environment. In addition, it is likely that their combustion and other degradation products will also be found in various environmental compartments. In fact, the Department of Defense (DoD) has an estimated 12,000 sites (Zhang et al. 2006) that require some form of remediation as a result of the presence of explosives. Coupled with this is the fact that the federal government owns more than 264 million hectares of land within the United States; it is the largest landowner and is thus responsible for land management (Stein et al. 2008). Because of their remoteness and limited access, DoD facilities (live-firing ranges, etc.) often harbor an abundance of indigenous wildlife; in some instances these organisms include threatened or endangered species (Figure 2.2). DoD lands contain a disproportionately greater number of species listed as threatened or endangered (by the Endangered Species Act [ESA] or other regulations)—more species than in lands managed by any other federal agency (Stein et al. 2008). This trend has held for more than 10 years (the last comprehensive analysis) and has provided the impetus for the DoD to continue to manage its facilities in ways that help sustain biodiversity and minimize habitat loss (Efroymson et al. 2009). The DoD, through the Strategic Environmental Research and Development Program

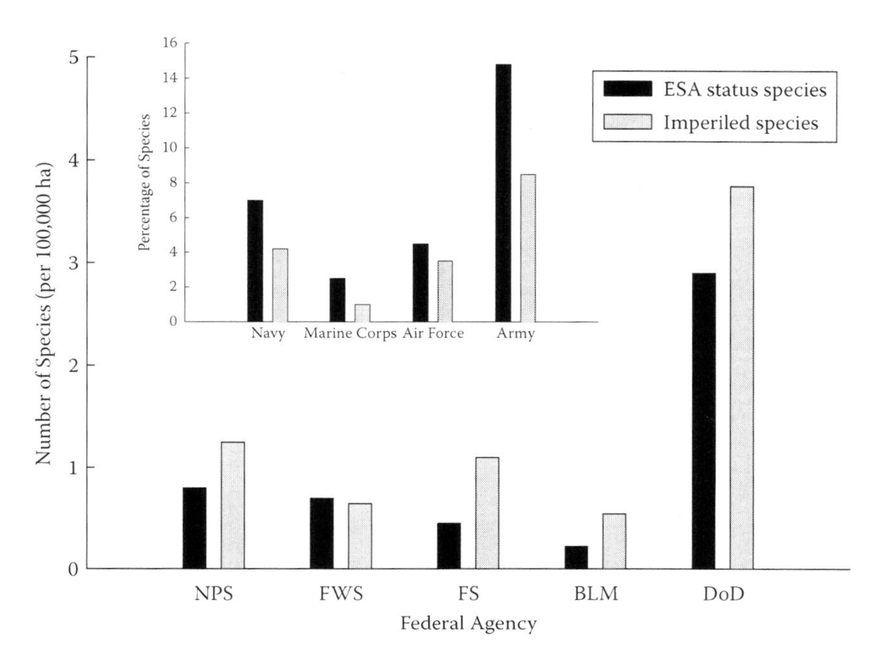

FIGURE 2.2 Density of endangered and imperiled species on federal land. Inset indicates the distribution of endangered and imperiled species within the Department of Defense. NPS = National Park Service, FWS = U.S. Fish & Wildlife Service, FS = Forest Service, BLM = Bureau of Land Management, DoD = Department of Defense. (Adapted from Stein et al. 2008 *BioScience*. 58:339–347.)

(SERDP), has taken an active role in supporting internal and external research to determine the potential effects of munitions-related compounds on a variety of important organisms.

Explosives can be placed in such categories as "first generation" and "second generation" based on the oxygen/nitrogen balance of the material. First-generation explosives have a low oxygen balance while more modern (second-generation) explosives have higher or even positive oxygen balances. Most of the well-known (first-generation) explosives (such as nitroaromatic TNT) have been studied in detail; the environmental fate and the toxicity of these materials have been reviewed, previously by Talmage et al. (1999) and recently by Sunahara et al. (2009). However, in comparison, relatively little toxicological data exist for the degradation products of these well-known nitroaromatic and nitramine explosives or some of the more recently developed (second-generation) explosives. This chapter summarizes the environmental effects of selected nitroaromatic and nitramine compounds on various broad taxonomic groups (aquatic and terrestrial invertebrates, reptiles and amphibians, fish, and small mammals and birds). The compounds of interest will include the degradation metabolites of TNT (first used in artillery shells in 1902) and RDX (widely used during World War II), as well as HMX (first synthesized in 1930) and CL-20 (originally synthesized circa 1987).

2.2 ENVIRONMENTAL FATE

An assessment of the risks associated with soil or water contamination of nitroaromatic and nitramine explosives would be less meaningful without information on the fate and transport of these compounds. Their transport, transformation, and uptake potential will depend on their individual physicochemical properties as well as the environment in which they reside. In general, the environmental fates of many nitroaromatics and nitramines, including TNT, RDX, and HMX, have been well studied. However, much less information is available on CL-20. An excellent summary on the availability and uptake of nitroaromatic and nitramine energetic compounds has recently been conducted (Johnson et al. 2009).

Nitroaromatic and nitramine compounds as a group have a wide range of physicochemical properties and corresponding environmental fates. Reported water solubilities in the literature vary somewhat with source, but in general range from the slightly soluble CL-20 (4.8 mg/L) to the readily soluble 2,4-DNT (270 mg/L). HMX and RDX are expected to be relatively mobile in the environment based on the available soil-sorption (K_{OC}) data. TNT and CL-20 are considerably less mobile in soil based on the low water solubility and the measured K_{OC} > 1000 mL/g, respectively. Volatilization from water and/or soil does not appear to be an important environmental fate process for most nitroaromatic and nitramine compounds. In addition, the data available in the literature suggest that bioconcentration in aquatic and terrestrial organisms should not occur, based on the relatively low octanol-water partition coefficients (K_{OW}) for this group of compounds.

The uptake of nitroaromatic and nitramine explosives into terrestrial plants appears to be a significant fate process. There are several studies in the literature that have documented the reproductive effects in deer mice (Schneider et al. 1996). More recently, studies have also been conducted on RDX (Harvey et al. 1991), HMX (Groom et al. 2002), and even CL-20 (Rocheleau et al. 2008). Data indicate that RDX and HMX are translocated to aboveground plant parts, while CL-20 is primarily sequestered in the roots (Rocheleau et al. 2008). Bioaccumulation factors for these compounds generally range from 10 to 15.

Photolytic degradation plays a significant role in the environmental fates of nitroaromatic and nitramine explosives (Hawari et al. 2004), especially in the aquatic environment; photolytic half-lives of a few days have been observed. Nitroaromatic and nitramine explosives on the soil surface are also susceptible to photodegradation. Available data for TNT indicate that the rates are slightly slower than those in water, but the degradation products (trinitrobenzaldehyde and trinitrobenzene, for example) are consistent with those produced in water.

The biological degradation of nitroaromatic and nitramine explosives in soils/sediments, water, and in some cases plants, has been well characterized (Boopathy 2001; McCormick et al. 1981; McCormick et al. 1976; Karakaya et al. 2009). Overall, there is little biological degradation of these compounds under aerobic conditions. In contrast, a number of studies have indicated that under reduced (anaerobic) conditions, TNT, RDX, and HMX are readily transformed by microorganisms. Further, the transformation products of TNT (Figure 2.3) and RDX (Figure 2.4) reduction

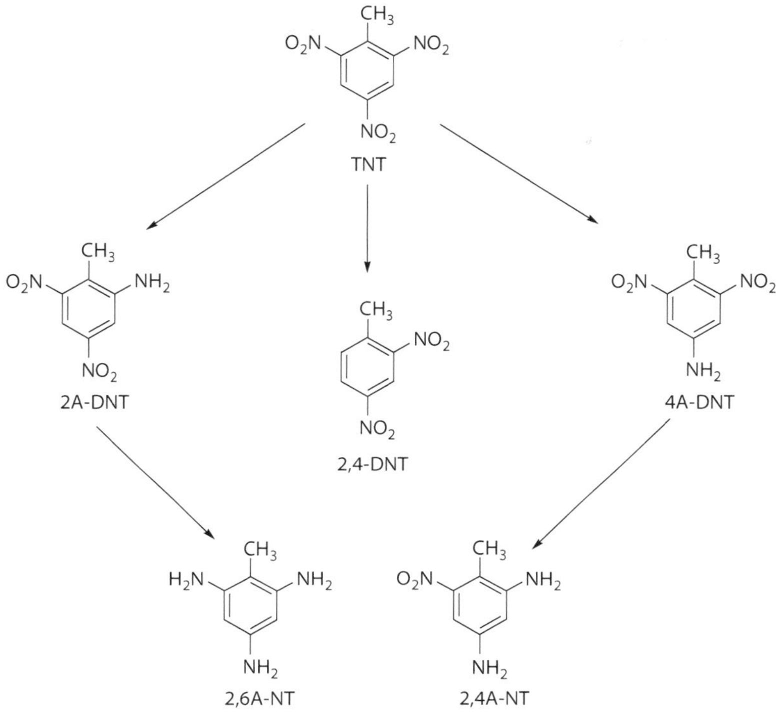

FIGURE 2.3 TNT and some of its biotic and abiotic metabolites. TNT = 2,4,6-trinitrotoluene; 2,4-DNT = 2,4-dinitrotoluene; 2A-DNT = 2-amino-4,6-dinitrotoluene; 4A-DNT = 4-amino-2,6-dinitrotoluene; 2,6A-NT = 2,6-diamino-4-nitrotoluene; 2,4A-NT = 2,4-diamino-6-nitrotoluene. The arrows do not necessarily imply a one-step reaction.

have been characterized; overall, less is known about the degradation products of HMX (Radtke 2005).

Under anaerobic conditions, TNT is transformed (Figure 2.3) initially to 2-aminodinitrotoluene derivatives: 2-amino-4,6-dinitroltoluene (2A-DNT) and 4-amino-2,6-dinitrotoluene (4A-DNT). These primary degradates can be further metabolized to the corresponding 2,6-diamino-4-nitrotoluene and 2,4-diamino-6-nitrotoluene. There is some controversy as to whether 2,4-dinitrotoluene (2,4-DNT) is a metabolite of TNT or simply an impurity of TNT. There are authors who claim that 2,4-DNT and

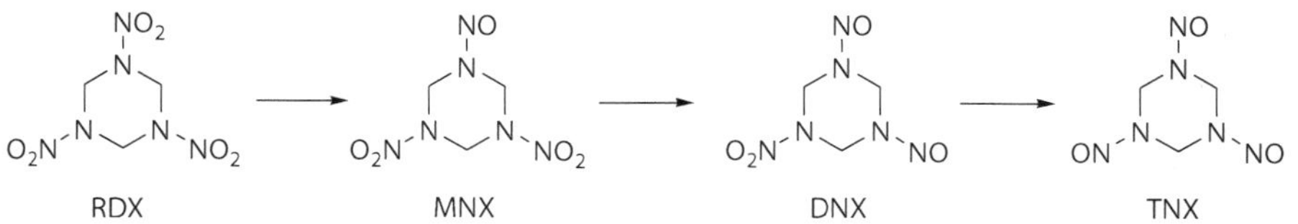

FIGURE 2.4 RDX and its anaerobic degradation products. RDX = hexahydro-1,3,5-trinitro-1,3,5-triazine; MNX = hexahydro-1-nitroso-3,5-dinitro-1,3,5-triazine; DNX = hexahydro-1,3-dinitroso-5-nitro-1,3,5-triazine; TNX = hexahydro-1,3,5-trinitroso-1,3,5-triazine.

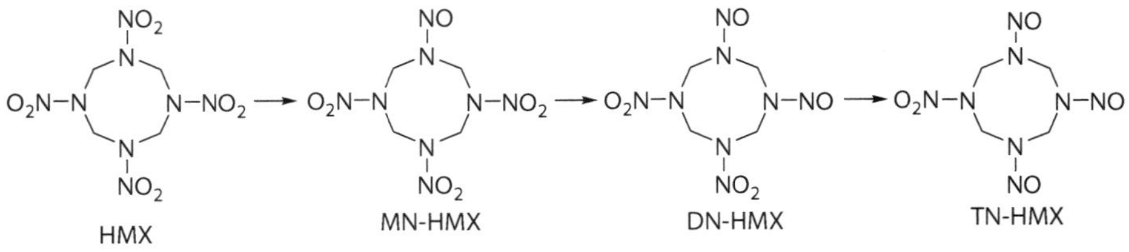

FIGURE 2.5 HMX and some of its known anaerobic degradation products. HMX = octahydro-1,3,5,7-tetranitro-1,3,5,7-tetrazocine; MN-HMX = mononitroso; DN-HMX = dinitroso; TN-HMX = trinitroso.

2,6-DNT are simply impurities of TNT (for example Hughes et al. 1999). However, there are also reports that TNT can be metabolized to both 2,6-DNT (Nefso et al. 2005) and 2,4-DNT (Esteve-Nunez et al. 2001 and references therein).

The nitramine explosive RDX also undergoes reductive transformation in water and soil (Figure 2.4). The degradation metabolites of RDX are a series of N-nitroso compounds: hexahydro-1-nitroso-3,5-dinitro-1,3,5-triazine (MNX); hexahydro-1,3-dinitroso-5-nitro-1,3,5-triazine (DNX); and hexahydro-1,3,5-trinitroso-1,3,5-triazine (TNX). There is also some evidence that these metabolites are formed in the gut of deer mice (Pan et al. 2007).

HMX can be transformed by the white-rot fungus *Phanerochaete chrysosporium* (Fournier et al. 2004), as well as anaerobically in sludge (Hawari et al. 2001). The transformation rate of HMX under these conditions has been described as slower than RDX and TNT transformation (Radtke 2005). Some degradation metabolites have been identified; these include nitroso compounds (Figure 2.5) similar to anaerobic RDX degradation: mononitroso-HMX (MN-HMX), dinotroso-HMX (DN-HMX), and trinitroso-HMX (TN-HMX).

2.3 EFFECTS OF EXPLOSIVES ON INVERTEBRATES

Invertebrate organisms in soil and water represent a diverse group of animals. They play critical roles in the ecological function of terrestrial and aquatic environments, including the conversion of detritus and the cycling of nutrients. In addition, invertebrates often serve as food sources for vertebrate organisms; thus they represent a pathway through which environmental contaminants can move from soil or water to higher animals. Assessments of contaminated sites rely on measurements of residues in invertebrates to determine exposure potential to higher organisms. Further, information about the acute and chronic effects of contaminants of concern on invertebrates helps maintain the ecological integrity of soil and aquatic environments. Additional information on the potential effects of explosives on soil organisms (including invertebrates) has recently been compiled (Kuperman et al. 2009).

2.3.1 CL-20

To date, few studies have been conducted to evaluate the effects of CL-20 on aquatic invertebrates (Table 2.1). We were able to locate data on a single-celled alga

TABLE 2.1
Laboratory Studies on the Toxic Effects of Hexanitrohexaazaisowurtzitane (CL-20) on Various Aquatic Organisms

Test Species	Test	Test System	Results	Reference
Green algae				(Haley et al. 2007)
Selenastrum	96 h	U.S. EPA Method	IC_{50} = 116 mg/L	
capricornutum	10 d	U.S. EPA Method	IC_{50} = 86 mg/L	
Daphnia				(Haley et al. 2007)
Ceriodaphnia dubia	7 d reproduction	U.S. EPA Method	EC_{50} = 1.9 mg/L	
Fathead minnow				(Haley et al. 2007)
Pimephales	7 d growth	U.S. EPA Method	EC_{50} = 2.7 mg/L	
promelas	7 d survival	U.S. EPA Method	LC_{50} = 2.0 mg/L	

(*Selenastrum capricornutum*) and a daphnid (*Ceriodaphnia dubia*). The 96-hour and 10-day IC_{50} values for CL-20 to *S. capricornutum* were 116 mg/L and 86 mg/L, respectively. CL-20 had an IC_{50} = 1.9 mg/L in a 7-day *C. dubia* reproduction study.

We were able to locate several additional studies concerning CL-20 toxicity on terrestrial invertebrates (Table 2.2), including data for two earthworm species (*Eisenia andrei* and *Eisenia fetida*) and two potworm species (*Enchytraeus crypticus* and *Enchytraeus albidus*). In *E. andrei*, CL-20 affected survival, growth, and reproduction; toxicity was influenced by soil type, particularly organic matter content. Growth and reproductive endpoints were at least as sensitive and order of magnitude more sensitive than survival. Reproductive effects (cocoon production, cocoon hatching, juvenile survival) were observed at CL-20 concentrations far below 1 mg/kg in sandy loam soil. Similar results were observed for both potworm species. Namely, CL-20 affected reproduction at soil concentrations < 1 mg/kg. In contrast, however, CL-20 was lethal to both species (LC_{50} < 1 mg/kg), particularly in soil low in organic carbon.

2.3.2 HMX

The acute (24- and 48-hour) toxicity of HMX to aquatic invertebrates has been studied in *Daphnia magna* and the midge *Chironomus tentans*. At 24 and 48 hours, no mortality was observed for either species at the highest HMX concentration tested (32 mg/L). Additional studies on aquatic organisms are summarized in Table 2.3.

HMX has been tested on the same terrestrial invertebrates as was CL-20: two earthworm species (*Eisenia andrei* and *E. fetida*) and two potworm species (*Enchytraeus crypticus* and *E. albidus*). Studies of HMX effects on earthworms have emphasized reproductive endpoints. In artificial soil, HMX reduced fecundity of *Eisenia andrei* (LOEC = 280 mg/kg) but did not cause mortality. In sandy forest soil, *E. andrei* fecundity was reduced at HMX concentrations > 15 mg/kg. In a sandy loam soil, freshly spiked HMX negatively impacted reproduction of *E. fetida* (EC_{20} for cocoon production = 2.7 mg/kg; EC_{20} for juvenile production = 0.4 mg/kg), but did not affect either reproduction endpoint in HMX-weathered (aged) soil.

TABLE 2.2
Laboratory Studies on the Toxic Effects of Hexanitrohexaazaisowurtzitane (CL-20) on Various Terrestrial Organisms

Test Species	Test	Test System	Results	Reference
Earthworm				(Robidoux et al. 2004)
Eisenia andrei	28 d survival	Sandy loam soil	LC_{50} = 53.4 mg/kg	
		Forest soil	LC_{50} = >125 mg/kg	
	28 d growth	Sandy loam soil	EC_{50} = 0.04 mg/kg	
		Forest soil	EC_{50} = <0.01 mg/kg	
	56 d reproduction	Sandy loam soil	EC_{50} = 0.09 mg/kg cocoon production	
		Sandy loam soil	EC_{50} = 0.07 mg/kg cocoon hatching	
		Sandy loam soil	EC_{50} = 0.05 mg/kg juvenile survival	
		Forest soil	EC_{50} = 1.5 mg/kg cocoon production	
		Forest soil	EC_{50} = 1.45 mg/kg cocoon hatching	
		Forest soil	EC_{50} = 1.30 mg/kg juvenile survival	
Earthworm				(Gong et al. 2007)
Eisenia fetida	14 d contact	0–2.15 μg/cm²	Retardation, stiffness, and body shrink at concentrations as low as 0.02 μg/cm². Neurotoxicity was reversible. CL-20 more potent than RDX.	
Potworm				(Dodard et al. 2005)
Enchytraeus crypticus	14 d survival	Sandy loam soil	LC_{50} = 0.4 mg/kg	
	28 d reproduction	Sandy loam soil	EC_{50} = 0.12 mg/kg	
	14 d survival	Ag soil	LC_{50} = 0.1 mg/kg	
	28 d reproduction	Ag soil	EC_{50} = 0.08 mg/kg	
Enchytraeus albidus	21 d survival	Sandy loam soil	LC_{50} = 0.2 mg/kg	
	42 d reproduction	Sandy loam soil	EC_{50} = not determined	
	21 d survival	Ag soil	LC_{50} = >1.0 mg/kg	
	42 d reproduction	Ag soil	EC_{50} = not determined	
Japanese quail				(Bardai et al. 2005)
Coturnix coturnix	14 d subacute 42 d subchronic	0–5304 mg/kg 0–1085 mg/kg (feeding)	No overt toxicity. Weight loss with dose. No overt toxicity. Decrease eggs laid and embryo weights. Developmental deformities.	

TABLE 2.3
Laboratory Studies on the Toxic Effects of Octahydro-1,3,5,7-Tetranitro-1,3,5,7-Tetrazocine (HMX) on Various Aquatic Organisms

Test Species	Test	Test System	Results	Reference
Daphnia				(Bentley et al. 1977)
Daphnia magna	24 h	Static	LC_{50} = > 32 mg/L	
	48 h	Static	LC_{50} = > 32 mg/L	
Midge				(Bentley et al. 1977)
Chironomus tentans	24 h	Static	LC_{50} = > 32 mg/L	
	48 h	Static	LC_{50} = > 32 mg/L	
Mediterranean mussel				(Rosen and Lotufo 2007)
Mytilus galloprovincialis	48 h; 96 h	Survival	No effect; highest concentration (1.9 mg/L)	
	48 h; 96 h	Byssal thread formation	No effect; highest concentration (1.9 mg/L)	
	48 h; 96 h	Larval development	No effect; highest concentration (1.9 mg/L)	
Scud				(Bentley et al. 1977)
Gammarus fasciatus	24 h	Static	LC_{50} = > 32 mg/L	
	48 h	Static	LC_{50} = > 32 mg/L	
Fathead minnow				(Bentley et al. 1977)
Pimephales promelas	96 h	Static; 7-d old fry	LC_{50} = 15 mg/L	
Bluegill				(Bentley et al. 1977)
Lepomis macrochirus	24 h	Static	LC_{50} = > 32 mg/L	
	48 h	Static	LC_{50} = > 32 mg/L	
	96 h	Static	LC_{50} = > 32 mg/L	
Channel catfish				(Bentley et al. 1977)
Ictalurus punctatus	24 h	Static	LC_{50} = > 32 mg/L	
	48 h	Static	LC_{50} = > 32 mg/L	
	96 h	Static	LC_{50} = > 32 mg/L	
Rainbow trout				(Bentley et al. 1977)
Oncorhynchus mykiss	24 h	Static	LC_{50} = > 32 mg/L	
	48 h	Static	LC_{50} = > 32 mg/L	
	96 h	Static	LC_{50} = > 32 mg/L	

Potworms appear to be insensitive to HMX in soil as determined by both survival and reproduction endpoints. In agricultural soil with extremely high (23%) organic matter, there was no effect of HMX (up to 918 mg/kg) on survival or reproduction in *Enchytraeus crypticus* or *E. albidus*. In another study on *E. crypticus* in a sandy loam soil, an NOEC of 21,750 mg/kg was observed for both 14-day survival and 28-day reproduction. Additional studies on HMX effects on terrestrial organisms are summarized in Table 2.4.

2.3.3 RDX Metabolites

The toxicity of RDX metabolites (MNX and TNX) has been tested on two species of terrestrial invertebrates, the earthworm *Eisenia fetida* and the cricket *Acheta domesticus* (Table 2.5). MNX and TNX toxicity to earthworms were determined for two soil types, a sandy loam (1.3% organic carbon) and a silt loam (2.5% organic carbon). In general, MNX toxicity was not affected by soil type in 7-day and 14-day bioassays. TNX appeared to be more toxic in the sandy loam, based on the 7-day survival data; however, the 14-day survival data were essentially equivalent in both soil types. Reproductive toxicity assays for MNX and TNX in sandy loam soil indicated that these compounds have equal effects on *E. fetida* cocoon production (EC_{50} = 27.0 mg/kg and EC_{50} = 27.3 mg/kg, respectively) and juvenile production (EC_{50} = 28.5 mg/kg and EC_{50} = 28.4 mg/kg, respectively). MNX was slightly more toxic than TNX with respect to cocoon hatching in sandy loam soil (EC_{50} = 9.5 mg/kg versus EC_{50} = 14.6 mg/kg).

TNX was slightly more toxic to cricket (*A. domesticus*) eggs than MNX. Zhang and co-workers (2006) observed TNX EC_{20} and EC_{50} values of 12 mg/kg and 48 mg/kg, respectively. MNX had EC_{20} and EC_{50} values of 29 mg/kg and 52 mg/kg, respectively. Interestingly, both compounds were more toxic to cricket eggs in sand than in topical exposures.

2.3.4 TNT Metabolites

With respect to aquatic invertebrates, TNT metabolites have been tested almost exclusively on green algae (*Selenostrum capricornutum*) (Sunahara et al. 1999) (Table 2.6). Two exceptions are that the toxicity of 2,4A-NT has also been tested on *Hyalella azteca* (LC_{50} = 19 mmol/kg) (Steevens et al. 2002), and 2A-DNT and 2,4A-NT have been tested on the midge *Chironomus tentans* (Lotufo and Farrar 2005). Green algae data (96-hour growth) indicate that the order of toxicity from greatest to least is 2A-DNT (EC_{50} = 13 μM), 4A-DNT (EC_{50} = 59.4 μM), 2,6A-NT (EC_{50} = 211 μM), and 2,4A-NT (EC_{50} = 293 μM). 2A-DNT and 2,4A-NT were found to be equally toxic (LC_{50} = 307 μmol/kg) to the midge in a 10-day survival study in sediment.

TNT metabolites have been tested on one terrestrial invertebrate, the earthworm *Eisenia andrei*. Earthworm data (14-day survival in a forest soil) indicate that 4A-DNT (LC_{50} = 105 mg/kg) is slightly more toxic than 2A-DNT (LC_{50} = 215 mg/kg). In addition, 2,4A-NT (NOEC = 482 μmol/kg) and 2,6A-NT (NOEC = 520 μmol/kg) had minimal effects on survival at the concentrations tested (LaChance et al. 2004).

TABLE 2.4
Laboratory Studies on the Toxic Effects of Octahydro-1,3,5,7-Tetranitro-1,3,5,7-Tetrazocine (HMX) on Various Terrestrial Organisms

Test Species	Test	Test System	Results	Reference
Earthworm *Eisenia andrei*	28 d reproduction	Artificial soil	Reduced fecundity. LOEC = 280 mg/kg. Growth of adult worms reduced, but no mortality.	(Robidoux et al. 2001)
Earthworm *Eisenia andrei*	28 d reproduction	Sandy forest soil	Reduced fecundity at concentrations > 15.6 mg/kg. Adult growth and survival were not reduced up to 711 mg/kg.	(Robidoux et al. 2002)
Earthworm *Eisenia fetida*	Reproduction	Sandy loam soil	EC_{20} = 2.7 mg/kg cocoon production EC_{20} = 0.4 mg/kg juvenile production	(Simini et al. 2004)
		Sandy loam soil	Cocoon production and juvenile production in weathered/aged HMX-treated soils were not different from control soils.	
Potworm *Enchytraeus crypticus*	14 d survival	Ag soil (23% OM)	No effect up to 918 mg/kg.	(Dodard et al. 2005)
	28 d reproduction	Ag soil (23% OM)	No effect up to 918 mg/kg.	
Enchytraeus albidus	21 d survival	Ag soil (23% OM)	No effect up to 918 mg/kg.	
	42 d reproduction	Ag soil (23% OM)	No effect up to 918 mg/kg.	
Potworm *Enchytraeus crypticus*	14 d survival	Sandy loam soil	NOEC = 21,750 mg/kg; LOEC = >21,750.	(Kuperman et al. 2004)
	28 d reproduction	Sandy loam soil	NOEC = 21,750 mg/kg; LOEC = >21,750.	
Green anole *Anolis carolinensis*	Contaminated food	Up-and-Down	LD_{50} = >2000 mg/kg. HMX accumulated in adults and eggs.	(Jones 2007)
Quail *Colinus virginianus*	Contaminated food	Exposed adults to 0, 12.5, 50, and 125 mg/kg	HMX was deposited into eggs. Metabolic rate of eggs at 5, 9, and 21 days was not affected by HMX.	(Liu et al. 2008)
Quail *Colinus virginianus*	Contaminated food	Exposed adults to 0, 12.5, 50, 125, and 250 mg/kg	Apparent food aversion (and weight loss) at 125 and 250 mg/kg. Alteration in body mass in 12.5 and 50 mg/kg dose groups. No HMX-related effects in hatchling survival and growth.	(Brunjes et al. 2007)

TABLE 2.5
Laboratory Studies on the Toxic Effects of RDX Metabolites on Various Terrestrial Organisms

Test Species	Test	Test System	Results	Reference
MNX				
Earthworm				(Zhang, Kendall, et al. 2006)
Eisenia fetida	7 d survival	Sandy loam soil	NOEC = 102.8 mg/kg	
			LC_{20} = 114.3 mg/kg	
			LC_{50} = 262.1 mg/kg	
	14 d survival	Sandy loam soil	NOEC = 102.8 mg/kg	
			LC_{20} = 114.3 mg.kg	
			LC_{50} = 262.1 mg/kg	
	7 d survival	Silt loam soil	NOEC = 211.7 mg/kg	
			LC_{20} = 233.8 mg/kg	
			LC_{50} = 389.9 mg/kg	
	14 d survival	Silt loam soil	NOEC = 94.4 mg/kg	
			LC_{20} = 112.7 mg/kg	
			LC_{50} = 244.6 mg/kg	
Earthworm				(Zhang et al. 2008)
Eisenia fetida	Cocoon production	Sandy loam soil	NOEC = 10 mg/kg	
			EC_{50} = 27 mg/kg	
	Juvenile production	Sandy loam soil	NOEC = 10 mg/kg	
			EC_{50} = 27.3 mg/kg	
	Cocoon hatching	Sandy loam soil	NOEC = 0.1 mg/kg	
			EC_{20} = 3.1 mg/kg	
			EC_{50} = 9.5 mg/kg	

Cricket				(Zhang et al. 2006)
Acheta domesticus	Egg hatching	Sand	EC_{20} = 29 mg/kg	
			EC_{50} = 52 mg/kg	
	Egg hatching	Topical exposure	EC_{20} = 65 mg/kg	
			EC_{50} = 140 mg/kg	
Deer mice				(Smith et al. 2007)
Peromyscus maniculatus	21 d old mice	Up-and-down method	LD_{50} = 181 mg/kg	
	50 d old mice		LD_{50} = 575 mg/kg	
	200 d old mice		LD_{50} = 542 mg/kg	
TNX				(Zhang, Kendall, et al. 2006)
Earthworm				
Eisenia fetida	7 d survival	Sandy loam soil	NOEC = 43.9 mg/kg	
			LC_{20} = 113.7 mg/kg	
			LC_{50} = 253.6 mg/kg	
	14 d survival	Sandy loam soil	NOEC = 43.9 mg.kg	
			LC_{20} = 111.5 mg.kg	
			LC_{50} = 251.3 mg/kg	
	7 d survival	Silt loam soil	NOEC = 87.9 mg/kg	
			LC_{20} = 200.1 mg/kg	
			LC_{50} = 362.1 mg/kg	
	14 d survival	Silt loam soil	NOEC = 87.9 mg/kg	
			LC_{20} = 96.7 mg/kg	
			LC_{50} = 216.3 mg/kg	
Earthworm				(Zhang et al. 2008)
Eisenia fetida	Cocoon production	Sandy loam soil	NOEC = 10 mg/kg	
			EC_{50} = 28.5 mg/kg	
	Juvenile production	Sandy loam soil	NOEC = 10 mg/kg	
			EC_{50} = 28.4 mg/kg	

(*Continued*)

TABLE 2.5 (CONTINUED)
Laboratory Studies on the Toxic Effects of RDX Metabolites on Various Terrestrial Organisms

Test Species	Test	Test System	Results	Reference
	Cocoon hatching	Sandy loam soil	NOEC = 1 mg/kg EC_{20} = 4.7 mg/kg EC_{50} = 14.6 mg/kg	
Cricket				(Zhang et al. 2006)
Acheta domesticus	Egg hatching	Sand	EC_{20} = 12 mg/kg EC_{50} = 48 mg/kg	
	Egg hatching	Topical exposure	EC_{20} = 47 mg/kg EC_{50} = 128 mg/kg	
Deer mice				(Smith et al. 2007)
Peromyscus maniculatus	21 d old mice 50 d old mice 200 d old mice	Up-and-down method	LD_{50} = 338 mg/kg LD_{50} = 338 mg/kg LD_{50} = 999 mg/kg	
Deer mice				(Smith et al. 2006)
Peromyscus maniculatus	Reproduction	0, 1, 10, 100 µg/L	Bioaccumulation in the liver. Exposures associated with postpartum mortality. Dose-dependent decrease in body weight from birth to weaning.	
Deer mice				(Smith et al. 2009)
Peromyscus maniculatus	Multi-generational	0, 10, 100, 1000 µg/L	Decreased litter size at TNX exposure of 1000 µg/L. Increased mortality of offspring born to parents exposed to 10 and 1000 µg/L. Little effect on body and organ weights of survivors.	

TABLE 2.6
Laboratory Studies on the Toxic Effects of TNT Metabolites on Various Aquatic Organisms

Test Species	Test	Test System	Results	Reference
2A-DNT				
Green algae				(Sunahara et al. 1999)
Selenastrum capricornutum	96 h growth	Static	EC_{50} = 13 μM	
Midge				(Lotufo and Farrar 2005)
Chironomus tentans	10 d survival	Sediment	LC_{50} = 307 μmol/kg	
Sheepshead minnow				(Lotufo and Lydy 2005)
Cyprinodon variegatus	Toxicokinetic study	Water	Low bioaccumulation potential and rapid elimination. Higher BCF than TNT, but unlikely to pose risk to fish.	
4A-DNT				
Green algae				(Sunahara et al. 1999)
Selenastrum capricornutum	96 h growth	Static	EC_{50} = 59.4 μM	
2,4A-NT				
Green algae				(Sunahara et al. 1999)
Selenastrum capricornutum	96 h growth	Static	EC_{50} = 293 μM	
Amphipod				(Steevens et al. 2002)
Hyalella azteca	10 d survival	Static	LC_{50} = 19 mmol/kg	
Midge				(Lotufo and Farrar 2005)
Chironomus tentans	10 d survival	Sediment	LC_{50} = 307 μmol/kg	
Sheepshead minnow				(Lotufo and Lydy 2005)
Cyprinodon variegatus	Toxicokinetic study	Water	Low bioaccumulation potential and rapid elimination. Lower BCF than TNT, and unlikely to pose risk to fish.	
2,6A-NT				
Green algae				(Sunahara et al. 1999)
Selenastrum capricornutum	96 h growth	Static	EC_{50} = 211 μM	

2.4 EFFECTS OF EXPLOSIVES ON REPTILES AND AMPHIBIANS

Reptiles and amphibians are both cold-blooded vertebrates that occupy both aquatic and terrestrial environments. Reptiles are air-breathing, while amphibians metamorphose from water-breathing juvenile forms to air-breathing adult forms. Most reptiles are carnivores and overall have slow digestive processes. Amphibians have varied diets (Henry 2000).

Toxicity data for the new nitramine explosives or the metabolites of older nitroaromatic and nitramine explosives on amphibian and reptiles were nearly non-existent (Johnson and Salice 2009). We were able to locate two studies in the literature. Jones (2007) measured the LD_{50} for HMX in the green anole (*Anolis carolinensis*) using the up-and-down method with exposure through contaminated food. HMX did accumulate in both adults and eggs; however, it was not toxic to anoles at the highest concentrations tested (the LD_{50} was estimated as >2000 mg/kg) (Table 2.4).

Johnson and co-workers (Johnson, Holladay et al. 2000; Johnson, Vodela et al. 2000) evaluated the acute toxicity of 2A-DNT and 4A-DNT to the tiger salamander (*Ambystoma tigrinum*) in soil and through a feeding trial. Initial soil concentrations of 2A-DNT and 4A-DNT were 39 mg/kg and 62 mg/kg, respectively. Following a 14-day exposure, there were no adverse effects for any of the endpoints evaluated (immunological indicators, blood parameters, and histopathology of the liver and kidney). In the feeding trial, salamanders fed contaminated earthworms (2A-DNT = 2.1–2.6 μg/g; 4A-DNT = 2.1–2.5 μg/g) for 14 days also showed no adverse effects (Table 2.7).

2.5 EFFECTS OF EXPLOSIVES ON FISH

Fish are ectothermic vertebrates that are abundant in both freshwater and saltwater environments. They are used as a food source by other aquatic and terrestrial vertebrates, including humans, as well as some avian species (piscivorous birds). As such, they represent a potential pathway through which contaminants could bioaccumulate in aquatic environments or move from aquatic to terrestrial environments. They are also the focus of toxicity assessments for contaminants of concern because of their sensitivity and ecological importance. Finally, adequate background information concerning the life history of various fish species exists, making it easier to interpret and compare toxicity data.

In general, a variety of fish species have been tested for effects in response to nitroaromatic and nitramine explosives. However, there are fewer studies on newer explosives or explosive metabolites. In addition, monitoring data have indicated the presence of energetic materials in marine environments, but a limited amount of toxicological data are available for marine and estuarine species (Nipper et al. 2009).

Only one fish species has been evaluated for toxic effects upon exposure to CL-20 (Table 2.1). Fathead minnows (*Pimephales promelas*) have been tested in both 7-day growth and 7-day survival assays using EPA test protocols (Haley et al. 2007). CL-20 was lethal to fathead minnows (LC_{50} = 2.0 mg/L) and also affected growth (EC_{50} = 2.7 mg/L), with lethality being the more sensitive endpoint.

In contrast, a larger variety of fish species have been tested with HMX, including the scud (*Gammarus fasciatus*), fathead minnow (*P. promelas*), bluegill

TABLE 2.7
Laboratory Studies on the Toxic Effects of TNT Metabolites on Various Terrestrial Organisms

Test Species	Test	Test System	Results	Reference
2A-DNT				
Earthworm				(LaChance et al. 2004)
Eisenia andrei	14 d survival	Forest soil	LC_{50} = 215 mg/kg	
Tiger salamanders				(Johnson, Holladay et al. 2000)
Ambystoma tigrinum	14 d survival	Soil	Initial soil concentration of 39 mg/kg did not produce any adverse effect for the endpoints evaluated.	
		Feeding study	Salamanders fed earthworms containing 2.1–2.6 μg/g showed no adverse effects.	
4A-DNT				
Earthworm				(LaChance et al. 2004)
Eisenia andrei	14 d survival	Forest soil	LC_{50} = 105 mg/kg	
Tiger salamanders				(Johnson, Holladay et al. 2000)
Ambystoma tigrinum	14 d survival	Soil	Initial soil concentration of 62 mg/kg did not produce any adverse effect for the endpoints evaluated.	
		Feeding study	Salamanders fed earthworms containing 2.1–2.5 μg/g showed no adverse effects.	
2,4A-NT				
Earthworm				(LaChance et al. 2004)
Eisenia andrei	14 d survival	Forest soil	NOEC = 482 μmol/kg	
2,6A-NT				
Earthworm				(LaChance et al. 2004)
Eisenia andrei	14 d survival	Forest soil	NOEC = 520 μmol/kg	

(continued)

TABLE 2.7 (CONTINUED)
Laboratory Studies on the Toxic Effects of TNT Metabolites on Various Terrestrial Organisms

Test Species	Test	Test System	Results	Reference
2,4-DNT				
Western fence lizard				(Suski et al. 2008)
Sceloporus occidentalis	14 d survival	Oral dose	LD_{50} = 380 mg/kg for males and 577 mg/kg for females.	
	60 d subchronic	Oral dose	Dose-dependent mortality at 25, 42, and 70 mg/kg/d. Changes in body weight, kidney weight, food consumption, and blood chemistry at doses above 9 mg/kg/d.	
White-footed mouse				(USACHPPM 1996)
Peromyscus leucopus	14 d survival	Feeding study	NOAEL = 158 mg/kg/d in males and 74 mg/kg/d in females. LOAEL = 286 mg/kg/d in males and 158 mg/kg/d in females.	
Northern bobwhite				(Johnson et al. 2005)
Colinus virginianus	14 d survival	Oral dose	LD50 = 55 mg/kg. NOAEL = 15 mg/kg/d. LOAEL = 35 mg/kg/d.	
	60 d subchronic	Oral dose	NOAEL = 1 mg/kg/d. LOAEL = 5 mg/kg/d.	
2,6-DNT				
White-footed mouse				(USACHPPM 1996)
Peromyscus leucopus	14 d survival	Feeding study	NOAEL = 103 mg/kg/d in males and 44 mg/kg/d in females. LOAEL = 238 mg/kg/d in males and 103 mg/kg/d in females.	

(*Lepomis macrochirus*), channel catfish (*Ictalurus punctatus*), and rainbow trout (*Oncorhynchus mykiss*) (Table 2.3). Similar to the results for aquatic invertebrates, HMX was rarely toxic to any of these fish species at 24, 48, or 96 hours (LC_{50}s > 32 mg/L) (Bentley et al. 1977). The lone exception was the fathead minnow, in which a 96-hour HMX LC_{50} of 15 mg/L was observed.

Lotufo and Lydy (2005) tested the toxicokinetics of 2A-DNT and 2,4A-NT in the sheepshead minnow (*Cyprinodon variegatus*) (Table 2.6). Both compounds were characterized as having a low bioaccumulation potential with rapid elimination. 2A-DNT had a higher bioconcentration factor (BCF) than TNT (13.1 versus 9.6), while 2,4A-NT had the lowest BCF (0.5) of the compounds examined. The authors concluded that neither of these TNT metabolites was likely to pose unacceptable risks to fish.

2.6 EFFECTS OF EXPLOSIVES ON SMALL MAMMALS AND BIRDS

Small mammals and birds are endothermic vertebrates that primarily occupy terrestrial environments. They are carnivores, herbivores, or omnivores. In order to maintain a relatively high and constant body temperature, small mammals will forage and eat frequently. Several small mammals are fossorial. Birds have a high metabolic rate and often satisfy their water needs through the food they consume. Both small mammals and birds make excellent models for assessing effects of contaminants (Kendall and Dickerson 1996). Additional information beyond that presented below on the potential effects of explosives on small mammals and birds has recently been compiled (Johnson and Salice 2009).

We were able to obtain one study on the effects of CL-20 on birds (Table 2.2). Bardai and coworkers (2005) conducted subacute (14-day) and subchronic (42-day) studies using Japanese quail (*Coturnix coturnix japonica*). In the subacute study, there was no overt toxicity, but weight loss was observed with increasing dose (0–5304 mg/kg). In the subchronic feeding study (0–1085 mg/kg), CL-20 did not produce overt toxicity in quail, but did decrease the number of eggs laid as well as the weight of embryos. CL-20 also produced developmental deformities.

HMX as well has been tested on a quail species, the northern bobwhite (*Colinus virginianus*) (Table 2.4). Quail were exposed to HMX through contaminated food. There was food aversion and concomitant weight loss in adults at the two highest HMX concentrations tested (125 and 250 mg/kg) (Brunjes et al. 2007). In a companion study, HMX was detected in eggs laid following exposure; however, the metabolic rate of the eggs was not affected by HMX (Liu et al. 2008). In addition, there were no HMX-related effects in hatchling survival and growth.

RDX metabolites (MNX and TNX) have been tested in one small mammal species, the deer mouse *Peromyscus maniculatus* (Table 2.5). Using the up-and-down method and mice in three age groups (21-day, 50-day, and 200-day-old mice), the LD_{50}s were 181 mg/kg, 575 mg/kg, and 542 mg/kg for MNX and 338 mg/kg, 338 mg/kg, and 999 mg/kg for TNX (Smith et al. 2007). In addition, TNX was also tested for reproductive effects in deer mice. Smith and et al. (2006) observed bioaccumulation of TNX in the liver as well as postpartum mortality associated with TNX exposure. Pups experienced a dose-dependent decrease in body weight from birth to weaning. Multigenerational effects were also seen when deer mice were exposed to aqueous TNX (Smith et al. 2009). TNX exposure decreased litter size and increased postpartum mortality of offspring in cohorts receiving 1000 μg/L.

2.7 CONCLUSIONS

Nitroaromatic and nitramine explosives or munitions are energetic materials used by both military and commercial sectors. Recently, much research effort has been focused on filling data gaps related to the fate and toxicity of these compounds in soil and water. While much toxicity information typically exists for the parent explosives (especially nitroaromatics), much less data generally describe the degradation metabolites of these explosives. Products of the biotic and abiotic degradation of these compounds may also pose some toxicological risk to terrestrial and aquatic organisms. Further, some newer parent explosives are relatively understudied.

As the largest landowner in the United States, the federal government, in particular the DoD, has taken an active role in managing its facilities in ways that help sustain biodiversity and in supporting research on the potential effects of munitions-related compounds. DoD lands contain a disproportionately greater number of species listed as threatened or endangered. Potential habitat loss from range construction and management, as well as species sensitivities to energetic materials, are of particular importance.

From an exposure assessment perspective, we generally understand most of the fate processes for nitroaromatic and nitramine explosives. However, the fates of some newer explosives such as CL-20 and relatively understudied older explosives such as HMX are much less understood. This is particularly true for the degradation metabolites of some of the newer energetic materials. As a group, nitroaromatic and nitramine compounds are readily susceptible to photodegradation, which can limit their persistence in the environment. However, numerous studies have indicated that these compounds can be taken up by vegetation, which could represent a pathway of exposure from soil to higher organisms.

In this chapter, we have attempted to summarize the literature currently available on the environmental effects of selected nitroaromatic and nitramine compounds on broad taxonomic groups (aquatic and terrestrial invertebrates, reptiles and amphibians, fish, and small mammals and birds). Given that most of the first-generation explosives (such as TNT) have been studied in detail and the toxicity of these materials has been reviewed as recent as 2009, we focused our efforts on the degradation products of these well-known nitroaromatic and nitramine explosives and some of the more recently developed explosives where little toxicological data exist. Most of the toxicological information for this select group of nitroaromatic and nitramine compounds is on terrestrial and aquatic invertebrates. Importantly, the data available suggest that in several instances the degradation metabolites can have a toxicity similar to that of the parent explosives. Much less information (if any) is available on the toxicological effects of these compounds on fish, reptiles, amphibians, small mammals, and birds. Collectively, this represents a significant data gap in our understanding of the potential impacts of energetic compounds on wildlife species.

REFERENCES

Bardai, G., G.I. Sunahara, P.A. Spear, M. Martel, P. Gong, and J. Hawari. 2005. Effects of dietary administration of CL-20 on Japanese quail, *Coturnix coturnix japonica*. *Archives of Environmental Contamination and Toxicology* 49:215–222.

Bentley, R.E., G.A. LeBlanc, T.A. Hollister, and B.H. Sleight. 1977. Acute toxicity of 1,3,5,7-tetranitrooctahydro-1,3,5,7-tetrazocine (HMX) to aquatic organisms. Washington, DC: U.S. Army Medical Research and Development Command.

Boice, L.P. 2006. Defense and conservation: Compatible missions. *Endangered Species Bulletin* 31:4–7.

Boopathy, R. 2001. Enhanced biodegradation of cyclotetramethylenetetranitramine (HMX) under mixed electron-acceptor condition. *Bioresource Technology* 76 (3): 241–244.

Brunjes, K.J., S.A. Severt, J. Liu, X. Pan, J. Brausch, S.B. Cox, G.P. Cobb, S.T. McMurry, R.J. Kendall, and P.N. Smith. 2007. Effects of octahydro-1,3,5,7-tetranitro-1,3,5,7-tetrazocine (HMX) exposure on reproduction and hatchling development in northern bobwhite quail. *Journal of Toxicology and Environmental Health* 70:682–687.

Dodard, S.G., G.I. Sunahara, R.G. Kuperman, M. Sarrazin, P. Gong, G. Ampleman, S. Thiboutot, and J. Hawari. 2005. Survival and reproduction of enchytraeid worms, *Oligochaeta*, in different soil types amended with energetic cyclic nitramines. *Environmental Toxicology & Chemistry* 24:2579–2587.

Efroymson, R.A., V. Morrill, V.H. Dale, T.F. Jenkins, and N.R. Giffen. 2009. Habitat disturbance at explosives-contaminated ranges. In *Ecotoxicology of Explosives*, edited by G.I. Sunahara, G. Lotufo, R.G. Kuperman, and J. Hawari. Boca Raton, FL: CRC Press.

Esteve-Nunez, A., A. Caballero, and J. L. Ramos. 2001. Biological degradation of 2,4,6-trinitrotoluene. *Microbiology and Molecular Biology Reviews* 65 (3): 335–352.

Fournier, D., A. Halasz, S. Thiboutot, G. Ampleman, D. Manno, and J. Hawari. 2004. Biodegradation of octahydro-1,3,5,7-tetranitro-1,3,5,7-tetrazocine (HMX) by *Phanerochaete chrysosporium*: New insight into the degradation pathway. *Environmental Science and Technology* 38:4130–4133.

Gong, P., L.S. Inouye, and E.J. Perkins. 2007. Comparative neurotoxicity of two energetic compounds, hexanitrohexaazaisowurtzitane and hexahydro-1,3,5-trinitro-1,3,5-triazine, in the earthworm *Eisenia fetida*. *Environmental Toxicology & Chemistry* 26:954–959.

Groom, C.A., A. Halasz, L. Paquet, N. Morris, L. Olivier, C. Dubois, and J. Hawari. 2002. Accumulation of HMX (octahydro-1,3,5,7-tetranitro-1,3,5,7-tetrazocine) in indigenous and agricultural plants grown in HMX contaminated anti-tank range soil. *Environmental Science and Technology* 36:112–118.

Haley, M.V., J.S. Anthony, E.A. Davis, C.W. Kurnas, R.G. Kuperman, and R.T. Checkai. 2007. Toxicity of the cyclic nitramine energetic material CL-20 to aquatic receptors. Edgewood Chemical Biological Center. Aberdeen Proving Ground, MD: U.S. Army Research, Development and Engineering Command.

Harvey, S.D., R.J. Fellows, D.A. Cataldo, and R.M. Bean. 1991. Fate of the explosive hexahydro-1,3,5-trinitro-1,3,5-triazine (RDX) in soil and bioaccumulation in bush bean hydroponic plants. *Environmental Toxicology & Chemistry* 10:845–855.

Hawari, J., A. Halasz, S. Beaudet, L. Paquet, G. Ampleman, and S. Thiboutot. 2001. Biotransformation routes of octahydro-1,3,5,7-tetranitro-1,3,5,7-tetrazocine by municipal anaerobic sludge. *Environmental Science and Technology* 35:70–75.

Hawari, J., S. Deschamps, C. Beaulieu, L. Paquet, and A. Halasz. 2004. Photodegradation of CL-20: Insights into the mechanisms of initial reactions and environmental fate. *Water Research* 38 (19): 4055–4064.

Henry, P.F.P. 2000. Aspects of amphibian anatomy and physiology. In *Ecotoxicology of Amphibians and Reptiles*, edited by D.W. Sparling, G. Linder, and C.A. Bishop. Pensacola, FL: SETAC Press.

Hughes, J.B., C.Y. Wang, and C. Zhang. 1999. Anaerobic biotransformation of 2,4-dinitrotoluene and 2,6-dinitrotoluene by *Clostridium acetobutylicum*: A pathway through dihydroxylamine intermediates. *Environmental Science and Technology* 33:1065–1070.

Johnson, M.S., S.D. Holladay, K.S. Lippenholz, J.L. Jenkins, and W.B. McCain. 2000. Effects of 2,4,6-trinitrotoluene in a holistic environmental exposure regime to a terrestrial salamander: *Ambystoma tigrinum. Toxicological Pathology* 28:334–341.

Johnson, M.S., J.K. Vodela, G. Reddy, and S.D. Holladay. 2000. Fate and the biochemical effects of 2,4,6-trinitrotoluene exposure to tiger salamanders (*Ambystoma tigrinum*). *Ecotoxicology and Environmental Safety* 46:186–191.

Johnson, M.S., M.W. Michie, M.A. Bazar, and R.M. Gogal. 2005. Influence of oral 2,4-dinitrotoluene exposure to the northern bobwhite (*Colinus virginianus*). *International Journal of Toxicology* 24:265–274.

Johnson, M.S., and C.J. Salice. 2009. Toxicity of energetic compounds to wildlife species. In *Ecotoxicology of Explosives,* edited by G.I. Sunahara, G. Lotufo, R.G. Kuperman, and J. Hawari. Boca Raton, FL: CRC Press.

Johnson, M.S., C.J. Salice, B.E. Sample, and P.Y. Robidoux. 2009. Bioconcentration, bioaccumulation, and biomagnification of nitroaromatic and nitramine explosives in terrestrial systems. In *Ecotoxicology of Explosives,* edited by G.I. Sunahara, G. Lotufo, R.G. Kuperman and J. Hawari. Boca Raton, FL: CRC Press.

Jones, L.E. 2007. Accumulation and effects of HMX in the green anole (*Anolis carolinensis*). Thesis in Environmental Toxicology, Texas Tech University, Lubbock.

Karakaya, P., C. Christodoulatos, A. Koutsospyros, W. Balas, S. Nicolich, and M. Sidhoum. 2009. Biodegradation of the high explosive hexnitrohexaazaiso-wurtzitane (CL-20). *International Journal of Environmental Research and Public Health* 6:1371–1392.

Kendall, R.J., and R.L. Dickerson. 1996. Principles and processes for evaluating endocrine disruption in wildlife. *Environmental Toxicology & Chemistry* 15:1253–1254.

Kuperman, R.G., R.T. Checkai, M. Simini, C.T. Phillips, J.E. Kolakowski, C.W. Kurnas, and G.I. Sunahara. 2004. Survival and reproduction of *Enchytraeus crypticus* (Oligochaeta, Enchytraeidae) in a natural sandy loam soil amended with the nitro-heterocyclic explosives RDX and HMX. *Pedobiologia* 47:651–656.

Kuperman, R.G., M. Simini, S. Siciliano, and P. Gong. 2009. Effects of energetic materials on soil organisms. In *Ecotoxicology of Explosives*, edited by G.I. Sunahara, G. Lotufo, R.G. Kuperman, and J. Hawari. Boca Raton, FL: CRC Press.

LaChance, B., A.Y. Renoux, M. Sarrazin, J. Hawari, and G.I. Sunahara. 2004. Toxicity and bioaccumulation of reduced TNT metabolites in the earthworm *Eisenia andrei* exposed to amended forest soil. *Chemosphere* 55:1339–1348.

Liu, J., S.B. Cox, B. Beall, K.J. Brunjes, X. Pan, R.J. Kendall, T. A. Anderson, S.T. McMurry, and P.N. Smith. 2008. Effects of HMX exposure upon metabolic rate of northern bobwhite quail (*Colinus virginianus*) *in ovo. Chemosphere* 71:1945–1959.

Lotufo, G.R., and J.D. Farrar. 2005. Comparative and mixture sediment toxicity of trinotrotoluene and its major transformation products to a freshwater midge. *Archives of Environmental Contamination and Toxicology* 49:333–342.

Lotufo, G.R., and M.J. Lydy. 2005. Comparative toxicokinetics of explosive compounds in sheepshead minnows. *Archives of Environmental Contamination and Toxicology* 49 (2): 206–214.

McCormick, N.G., F.E. Feeherry, and H.S. Levinson. 1976. Microbial transformation of 2,4,6-trinitrotoluene and other nitroaromatic compounds. *Applied and Environmental Microbiology* 31 (6): 949–958.

McCormick, N.G., J.H. Cornell, and A.M. Kaplan. 1981. Biodegradation of hexahydro-1,3,5-trinitro-1,3,5-triazine. *Applied and Environmental Microbiology* 42:817–823.

Nefso, E.K., S.E. Burns, and C.J. McGrath. 2005. Degradation kinetics of TNT in the presence of six mineral surfaces and ferrous iron. *Journal of Hazardous Materials B* 123:79–88.

Nipper, M., R.S. Carr, and G. Lotufo. 2009. Aquatic toxicology of explosives. In *Ecotoxicology of Explosives*, edited by G. I. Sunahara, G. Lotufo, R. G. Kuperman, and J. Hawari. Boca Raton, FL: CRC Press.

Pan, X., B. Zhang, J.N. Smith, M. San Francisco, T. A. Anderson, and G.P. Cobb. 2007. N-nitroso compounds produced in deer mice (*Peromyscus maniculatus*) GI tract following hexahydro-1,3,5-trinitro-1,3,5-triazine (RDX) exposure. *Chemosphere* 67:1164–1170.

Radtke, C.W. 2005. Laboratory investigation of explosives degradation in vadose zone soil using carbon source additions. Department of Environmental Toxicology, Texas Tech University, Lubbock.

Robidoux, P.Y., J. Hawari, S. Thiboutot, G. Ampleman, and G.I. Sunahara. 2001. Chronic toxicity of octahydro-1,3,5,7-tetranitro-1,3,5,7-tetrazocine (HMX) in soil determined using the earthworm (*Eisenia andrei*) reproduction test. *Environmental Pollution* 111:283–292.

Robidoux, P.Y., G.I. Sunahara, K. Savard, Y. Berthelot, S. Dodard, M. Martel, P. Gong, and J. Hawari. 2004. Acute and chronic toxicity of the new explosive CL-20 to the earthworm (*Eisenia andrei*) exposed to amended natural soils. *Environmental Toxicology & Chemistry* 23:1026–1034.

Robidoux, P.Y., J. Hawari, G. Bardai, L. Paquet, G. Ampleman, S. Thiboutot, and G.I. Sunahara. 2002. TNT, RDX, and HMX decrease earthworm (*Eisenia andrei*) life-cycle responses in a spiked natural forest soil. *Archives of Environmental Contamination and Toxicology* 43:379–388.

Rocheleau, S., B. Lachance, R.G. Kuperman, J. Hawari, S. Thiboutot, G. Ampleman, and G.I. Sunahara. 2008. Toxicity and uptake of cyclic nitramine explosives in ryegrass, *Lolium perenne*. *Environmental Pollution* 156:199–206.

Rosen, G., and G.R. Lotufo. 2007. Toxicity of explosive compounds to the marine mussel, *Mytilus galloprovincialis*, in aqueous exposures. *Ecotoxicology and Environmental Safety* 68:228–236.

Schneider, K., J. Oltmanns, T. Radenberg, T. Schneider, and D. Pauly-Mundegar. 1996. Uptake of nitroaromatic compounds in plants: Implications for risk assessment of ammunition sites. *Environmental Science and Pollution Research* 3:135–138.

Simini, M., R.T. Checkai, R.G. Kuperman, C.T. Phillips, J.E. Kolakowski, C.W. Kurnas, and G.I. Smith, J.N., et al. 2006. Reproductive effects of hyxahydro-1,3,5-trinitroso-1,3,5-triazine in deer mice (*Peromyscus maniculatus*) during a controlled exposure study. *Environmental Toxicology & Chemistry* 25:446–451.

Smith, Jordan N., Jun Liu, Marina A. Espino, and George P. Cobb. 2007. Age dependent acute oral toxicity of hexahydro-1,3,5-trinitro-1,3,5-triazine (RDX) and two anaerobic N-nitroso metabolites in deer mice (*Peromyscus maniculatus*). *Chemosphere* 67 (11): 2267–2273.

Smith, J.N., M.A. Espino, J. Liu, N.A. Romero, S.B. Cox, and G.P. Cobb. 2009. Multigenerational effects in deer mice (*Peromyscus maniculatus*) exposed to hexahydro-1,3,5-trinitroso-1,3,5-triazine (TNX). *Chemosphere* (in press).

Steevens, J.A., B.M. Duke, G.R. Lotufo, and T.S. Bridges. 2002. Toxicity of the explosives 2,4,6-trinitrotoluene, hexahydro-1,3,5-trinitro-1,3,5-triazine, and octahydro-1,3,5,7-tetranitro-1,3,5,7-tetrazocine in sediments to *Chironomus tentans* and *Hyalella azteca*: Low-dose hormesis and high-dose mortality. *Environmental Toxicology & Chemistry* 21:1475–1482.

Stein, B.A., C. Scott, and N. Benton. 2008. Federal lands and endangered species: The role of military and other federal lands in sustaining biodiversity. *BioScience* 58:339–347.

Sunahara, G.I., S. Dodard, M. Sarrazin, L. Paquet, J. Hawari, C.W. Greer, G. Ampleman, S. Thiboutot, and A.Y. Renoux. 1999. Ecotoxicological characterization of energetic substances using a soil extraction procedure. *Ecotoxicology and Environmental Safety* 43 (2): 138–148.

Sunahara. 2004. Reproduction and survival of *Eisenia fetida* in a sandy loam soil amended with the nitro-heterocyclic explosives RDX and HMX. *Pedobiologia* 47:657–662.

Sunahara, G.I., G.R. Lotufo, R.G. Kuperman, and J. Hawari, eds. 2009. *Ecotoxicology of Explosives*. Boca Raton, FL: CRC Press.

Suski, J.G., C.J. Salice, J.T. Houpt, M.A. Bazar, and L.G. Talent. 2008. Dose-related effects following oral exposure of 2,4-dinitrotoluene on the western fence lizard, *Sceloporus occidentalis*. *Environmental Toxicology & Chemistry* 27:352–359.

Talmage, S.S., D.M. Opresko, C.J. Maxwell, C.J.E. Welsh, F.M. Cretella, P.H. Reno, and F.B. Daniel. 1999. Nitroaromatic munition compounds: Environmental effects and screening values. *Reviews of Environmental Contamination and Toxicology* 161:1–156.

USACHPPM. 1996. Toxicological Study No. 85-4152-96. 14-day feeding study of 2,6-dinitrotoluene in the white-footed mouse, *Peromyscus leucopus*, 24 June–12 July 1996. Cincinnati, OH.

———. 1996. Toxicological Study No. 4152-31-96-06-14. 14-day feeding study of 12,4-dinitrotoluene in the white-footed mouse, *Peromyscus leucopus*, 1 July–19 July 1996. Cincinnati, OH.

Zhang, B., C.M. Freitag, J.E. Canas, Q. Cheng, and T.A. Anderson. 2006. Effects of hexahydro-1,3,5-trinitro-1,3,5-triazine (RDX) metabolites on cricket (*Acheta domesticus*) survival and reproductive success. *Environmental Pollution* 144:540–544.

Zhang, B., R.J. Kendall, and T.A. Anderson. 2006. Toxicity of the explosive metabolites hexahydro-1,3,5-trinitroso-1,3,5-triazine (TNX) and hexahydro-1-nitroso-3,5-dinitro-1,3,5-triazine (MNX) to the earthworm, *Eisenia fetida*. *Chemosphere* 64:86–95.

Zhang, B., P.N. Smith, and T.A. Anderson. 2006. Evaluating the bioavailability of explosive metabolites (MNX and TNX) in soils using passive sampling devices. *Journal of Chromatography A* 1101:38–45.

Zhang, B., S.B. Cox, S.T. McMurry, W.A. Jackson, G.P. Cobb, and T.A. Anderson. 2008. Effect of two major N-nitroso hexahydro-1,3,5-trinitro-1,3,5-triazine (RDX) metabolites on earthworm reproductive success. *Environmental Pollution* 153:658–667.

3 Agriculture
Pesticides, Plant Genetics, and Biofuels

Spencer R. Mortensen, Thomas E. Nickson, and George P. Cobb

CONTENTS

3.1 INTRODUCTION

The relationships between pesticides, plant genetics, and biofuels comprise an important component of agricultural land use, development, and sustainability. Each of these aspects has the potential to alter wildlife habitat quality and quantity.

As of this writing, the world population is estimated to be over 6.8 billion people. That number is projected to grow to between 8 and 10.5 billion by 2050. Global food demand is estimated to at least double in the next 50 years (Yu 2008). Fiber and fuel demand will also need to increase substantially to meet the needs of our growing world population. How do we continue to grow enough food and fiber and produce enough fuel to sustain the growing population? Perhaps more important, how do we do so while protecting our environment and the wildlife it supports? While there are no easy answers to these questions, it is certain that pesticides, plant genetics, and biofuels will play major roles, and each will pose its corresponding risks to wildlife.

This chapter will address our current understanding of pesticides, plant genetics, and biofuels in the context of potential ecological effects. Given the more complete understanding of pesticide use and distribution, these materials will be discussed in the context of risk assessment. As part of this evaluation, we will explore the changing paradigm of pesticides and cropping practices with the advent of more selective chemical products and plant genetics that incorporate additional tools to help address pest resistance issues. The chapter will offer insights into agricultural practices that are likely to decrease ecological risks as well as data collection and evaluation processes that hold promise for improving future risk assessment paradigms in light of these changing practices. The transition to biofuel production will also be discussed in the context of alterations in agricultural practices, nutrient cycling, and contaminants. Potential ecological and toxicological effects from these changes will be explored.

3.2 PESTICIDES

Pesticides have been used, to one degree or another, for improving the quantity and quality of crops for the production of food and fiber (and more recently fuels) almost as long as humans have been involved in agriculture. The United States Environmental Protection Agency (U.S. EPA) defines a pesticide (or a crop-protection product) as "any substance or mixture of substances intended for preventing, destroying, repelling, or mitigating any pest" (U.S. EPA 2009). Pesticides encompass a number of physical, chemical, or biologic agents that are more specifically classified by their use pattern and type of pest they control (Ecobichon 2001). The most common types of pesticides (and target pests) are insecticides (insects), herbicides (weeds), fungicides (fungi), and rodenticides (rodents), but there are also aracnicides or miticides (mites), molluscicides (snails and mollusks), larvicides (insect larvae), and pediculocides (lice). Additionally, attractants (pheromones), defoliants, desiccants, plant growth regulators, and repellants also are considered to be pesticides, especially from a regulatory standpoint (Costa 2008; Ecobichon 2001). Furthermore, within each broad pesticide class (e.g., insecticides), there are a number of subclasses (e.g., chlorinated hydrocarbons, organophosphates, carbamates, pyrethroids, botanical insecticides, insect growth regulators, neonicitinoids, formamides, microbials, fumigants, inorganics, amidinohydrazones, pyrazoles, oxadiazines, sulfonamides, pyridazinones, nereistoxin analogs, pyridine azomethines, pyrimidinamines, nicotinimides, benzene dicarboximides, dichloropropenyl ethers, tetronic acids, tetramic acids, anathranilic diamides, and pesticidal oils and soaps). Even within subclasses, important chemical and toxicological differences can exist, as is the case, for example, of the organophosphate insecticides (e.g., parathion, methyl parathion, and malathion). Thus, a comprehensive understanding of a pesticide's physical and chemical properties, fate and metabolism, toxicological effects, and temporal- and spatial-use profile is essential for the proper evaluation of its risk to non-target organisms, including wildlife.

It is not the intent nor is it possible in this chapter to provide a thorough review of the chemistry nomenclature, biotransformation, degradation, environmental effects, toxicity in target and non-target organisms, or mode of action of all pesticides. The reader is referred to the following articles for additional information: Matsumura (1985), Dikshith (1991), Hayes and Laws (1991), Krieger (2001), Senseman and Armbrust (2007), Costa (2008), and Yu (2008).

3.2.1 The Need for Pesticides

Pesticides continue to have a clear role in helping provide a sustainable supply of food, fiber, and biofuels for a growing human population, while protecting us from vector-borne diseases and organisms that threaten to damage or destroy the buildings in which we live and work.

3.2.2 Food Production

In parts of the world, excessive loss of food crops to pests contributes directly or indirectly to starvation, and in those places the use of pesticides has an obvious risk-benefit relationship. Today more than a billion people in the world are hungry. The primary cause of hunger is flawed policies, but wars, revolutions, and natural disasters, including pests, compounded by climate change, are also key factors. Of pests, the data clearly show that weeds are the major cause of food production losses. It is estimated that weeds cause nearly $100 billion a year in loss of global food production, which translates to approximately 3.63 million metric tons of wheat, or more than half of the world production expected in 2009 (FAO 2009). Even with pesticide use, about a third of the world's food crops are destroyed by pests during growth, harvesting, and storage (Ware and Whitacre 2004). Losses are greater in developing countries. For example, in Latin America, approximately 40% of crops are lost because of pests. It has been estimated that 50% of cotton production in developing countries would be destroyed without the use of insecticides. In the United States alone, current crop losses due to pests are estimated to be 30%. It has been estimated that removing pesticides from U.S. agriculture alone would cause a decline of crop production by as much as 50%, depending on the crop species (NRC 2000). Additionally, there would be an expected proportional decrease in farm exports with a concomitant increase in food prices to the consumer, ultimately leading to food price inflation. In the United States, it is estimated that we would spend three to five times more on food than we currently do. The National Research Council (NRC 2000) concluded that chemical pesticides should remain part of a larger toolbox of diverse pest management options for the foreseeable future.

3.2.3 Disease Control

Pesticides play a major role in the control of insects, mites, rodents, and other pests that are involved in the life cycle of vector-borne diseases such as malaria, yellow fever, typhus, and many other diseases that represent a major threat to the health of human populations. Several well-known diseases are transmitted to humans by invertebrates (Table 3.1); a detailed discussion of disease-related issues is presented in Chapter 4.

The World Health Organization (WHO) has documented that the use of synthetic insecticides can greatly reduce the risk of insect-borne diseases, especially in the case of malaria (Nauen 2007). The case of DDT (1,1,1-trichloro-2,2-bis (*p*-chlorophenyl) ethane) highlights the difficulty in striking a balance between the benefits of pesticide use with environmental risk. When introduced in 1942, DDT held immense promise of benefiting agricultural economics and protecting human health against vector-borne diseases. In fact, the public-health benefits of DDT were so great that Paul Müller was awarded the Nobel Prize in Medicine in 1948 for recognizing DDT's insecticidal properties. However, because of its propensity to bioaccumulate in the environment, particularly in human mother's milk, insect resistance, and concerns about its effects on avian reproduction, DDT was banned in most countries by the mid-1970s. In South Africa, DDT was not banned until 1996, and at the time fewer than 10,000 cases of malaria were registered in the country. By the year 2000, cases of malaria had reached approximately 62,000. DDT was subsequently reintroduced at the end of 2000, and cases afterward decreased to 12,500. There are still hundreds of millions of people in the world who are at risk from vector-borne diseases, particularly in Africa and some Asian countries. Human health protection and ecological risks posed by pesticides must be carefully weighed in these situations (Costa 2008).

TABLE 3.1
Some Common Diseases That Are Transmitted to Humans by Invertebrate Pests

Disease	Vector
African sleeping sickness, anthrax, dysentery, onchocerciasis, leishmaniasis, trypanosomiasis	Flies
Bubonic plague, endemic typhus	Fleas
Dengue fever, encephalitis, malaria, St. Louis encephalitis, West Nile virus, yellow fever	Mosquitoes
Epidemic typhus	Lice
(Hemorrhagic fevers), lyme disease, Q fever, relapsing fevers, Rocky Mountain spotted fever	Ticks (and mites)

Source: Adapted from Ware, G.W. and Whitacre, D.M., *The Pesticide Book*, MeisterPro Information Resources, Willoughby, OH, 2004. With permission.

3.2.4 Trends in Pesticide Use

As is mentioned in Chapter 9 of this book, risk is a function of exposure and toxicity. Understanding pesticide use, therefore, is very important for risk assessment purposes as it defines the exposure parameter of risk equations.

Quantitative and descriptive analyses of pesticide use generally are difficult and can be misleading because pesticide use patterns are driven by numerous factors: weather changes, pest population changes, introduction of new pesticides, regulatory changes, economic factors, changes in agronomic practices, introduction of non-chemical pest control products, voluntary cancellation of pesticides, and changes in crop preferences (Gianessi and Silvers 2000). Many if not all of these factors fluctuate annually, and with these fluctuations come changes in both amounts and types of pesticides used.

3.2.4.1 Pesticide Use Resources

Several good resources provide pesticide-use data that are inexpensive or free. The U.S. EPA periodically publishes a report titled "Pesticides Industry Sales and Usage" that can be accessed via the Internet at http://www.epa.gov/oppbead1/pestsales/ (Aspelin 1997; Aspelin and Grube 1999; Donaldson et al. 2002; Kiely et al. 2004). The National Center for Food and Agricultural Policy published a report titled "Trends in Crop Pesticide Use: Comparing 1992 and 1997" (Gianessi and Silvers 2000). The California Department of Pesticide Regulation (CA DPR) provides perhaps the most extensive pesticide use information in its annual "Summary of Pesticide Use Report Data" (see http://www.cdpr.ca.gov/docs/pur/purmain.htm) (CDPR 2008). The State of New York's "Pesticide Sales and Use Reporting" site (http://pmep.cce.cornell.edu/psur/) provides state-specific pesticide use data as well as links to a number of additional sources of important data, including two U.S. Department of Agriculture sites: the USDA Pesticide Data Program (http://www.ams.usda.gov/AMSv1.0/science) and the USDA National Agricultural Statistics Service (http://usda.mannlib.cornell.edu/MannUsda/viewDocumentInfo.do?documentID=1001).

As a whole, pesticide use increased steadily in the United States between 1950 and 1980, but has since plateaued (with moderate increases or decreases from year to year), which is thought to be attributed to the utilization of more efficacious active ingredients and formulations, introduction of integrated pest management programs, and organic farming in developed countries (Costa 2008). This is not, however, the trend for all classes of pesticides. For example, cholinesterase-inhibiting insecticides (organophosphates and carbamates) grew steadily from the 1950s to the mid-1970s, but have since declined steadily, with the exception of 1999, which showed an increase in the use of organophosphate insecticides as a result mainly of the increased amount of malathion used as part of the USDA-sponsored Boll Weevil Eradication Program (Kiely et al. 2004; Mortensen 2006). Interestingly, California saw nearly 10% decreases in all pesticides used in 2007 compared to 2006. More importantly, decreases were seen (both in pounds of active ingredients applied and in cumulative acres treated) in pesticides classified as "known to cause reproductive effects," "known to cause cancer," the cholinesterase-inhibiting pesticides, pesticides on the DPR's groundwater protection list, and pesticides on CADPR's toxic air contaminants list. There were also decreases in fumigants, oil pesticides (in pounds applied, but marginal increase in cumulative acres treated), and biopesticides (CADPR 2008).

Of the 2.3 million metric tons of pesticide active ingredient used globally in 2000 and 2001, 37% was herbicide, 24% was insecticide, 9% was fungicide, and 21% was other (nematicides, fumigants, rodenticides, molluscicides, etc.) (Kiely et al. 2004). It is noteworthy that of the 75 new active ingredients registered in the United States between 1999 and 2002, more than half were conventional chemical, reduced-risk products (Ware and Whitacre 2004).

Most pesticide-use summaries are not intended to serve as indicators of pesticide risk to the public or the environment (CADPR 2008). Pesticide-use data and trend analyses, however, can be useful in estimating exposure, which in turn may help risk assessors and risk managers make better informed decisions by providing estimates of typical applications and how pesticides are used.

3.2.5 Effects on Wildlife from Pesticide Use

When *Wildlife Toxicology and Population Modeling: Integrated Studies of Agroecosystems* (Kendall and T. E. Lacher 1994) was published, the major questions of wildlife risk from pesticide exposure included chemical occurrence, transport, and transformation, as well as wildlife encounters with pesticides and the various tolerances to toxicities among organisms. During the past sesquidecade, significant progress has been made in chemical modeling and understanding interspecies toxicities. However, site-specific chemical degradation, interspecies exposures, and interspecies sensitivities remain poorly understood in some situations (USEPA 2003). The U.S. Geological Survey maintains records of wildlife mortality in the United States (USGS 2008), and the incidence of poisonings has diminished over time (Figure 3.1). Even though the frequency of incidents and extent of mortality involving organophosphate (OP) and carbamate insecticides are at 20-year lows, poisonings still occur.

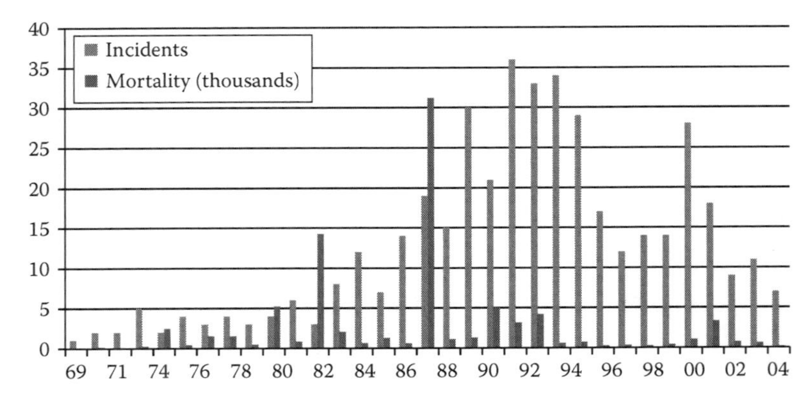

FIGURE 3.1 Avian mortality incidents in the United States reported to have involved organophosphorus or carbamate insecticides. (Data obtained from *Wildlife Mortality Information: Wildlife Organophosphate/Carbamate Poisoning (OP/CARB)*. Washington, DC: U.S. Geological Survey, 2008.)

The decrease in the incidence of wildlife mortality due to pesticide poisonings may be a function of the phase-out of certain chemical classes of pesticides as well as the high specificity, rapid degradation, and low bioaccumulation of recently registered synthetic pesticides. For example, in the United States, the European Union (EU), and many other parts of the world, most of the persistent and bioaccumulative chlorinated hydrocarbon insecticides have been phased out of use unless there has been a compelling benefit for its use (such as the use of DDT to control malaria in India and certain African countries) (Yu 2008). Additionally, there has been a relatively steady decline in the use of the more toxic cholinesterase-inhibiting (organophosphorus and carbamate) insecticides (Mortensen 2006), which as a class have been implicated in a number of wildlife mortalities (USGS 2008). While most of the cases of wildlife mortality were likely associated with product misuse, some clearly were not (Cobb et al. 2003; Tank et al. 1993). For example, Swainson's hawks (*Buteo swainsoni*) migrating from Canada and the United States to Argentina were killed by the hundreds when they gorged themselves on grasshoppers treated with monocrotophos (dimethyl (*E*)-1-methyl-2-(methylcarbamoyl) vinyl phosphate), a product whose use had been banned in the United States since 1977 (Hooper 2003). Because of this and other similar events, a very focused and collaborate effort involving a number of government agencies, university personnel, industry, and non-governmental organizations led to a withdrawal of this insecticide and eventual ban of its use in Argentina.

Pyrethroid insecticides are synthetic derivatives of pyrethrum, a solvent extract of dried flowers of *Chrysanthemum cinerariaefolium*. Pyrethroids are a widely used class of insecticides with the second-largest market share (approximately 20%) relative to the cholinesterase-inhibiting insecticides (OPs = 25% and carbamates = 11%). Like the cholinesterase-inhibiting insecticides, they are very effective at controlling a wide range of insect pests, but they are not as toxic to terrestrial organisms. However, pyrethroid insecticides are highly toxic to many aquatic organisms, with the lowest EC/LC_{50} values in the part-per-trillion range. They are nonpolar compounds that are strongly hydrophobic (K_{ow} or Log P values in the 5–7 range), and they tend to bind tightly to soils and sediments (K_{oc} values in the 40,000–300,000 range). Despite their high K_{ow}/Log P values, they do not tend to bioconcentrate in fish as much as would be expected (BCF values in the 400–6,000 range; many below 1,000). Recently, however, there have been concerns about the effects of some pyrethroid insecticides on benthic invertebrates that dwell in the sediments of creeks within residential neighborhoods (Westin et al. 2005), but not necessarily within agricultural environments, specifically water bodies near cotton-growing regions of the United States (Solomon et al. 2001; Giddings et al. 2001; Hendley et al. 2001; Travis and Hendley 2001; Maund et al. 2001).

A relatively new class of insecticide, the neonicitinoids (or chloronicotinyls), represents nearly 16% of the global market share of the major classes of insecticides (Nauen 2007). The neonicitinoids are analogs of nicotine, but unlike nicotine these insecticides are much less toxic to mammals, with LD_{50} values ranging from 400 mg/kg to >5,000 mg/kg (Yu 2008). This class of chemicals recently received some bad press when it was implicated as one of the suspected causes of honeybee colony collapse disorder, or CCD. Consequently, all of the uses of neonicitinoids were banned from being used in some countries in the EU, most notably France. However, the ban of neonicitinoids has not eliminated CCD, and recent research data suggest that the primary causes of CCD are likely biological (viruses and/or mites) and that other factors, including pesticides, may play a contributing role. It should be noted that samples of bee bread, pollen, and bees themselves indicate that the most likely pesticide contributors to CCD may be the very pesticides (acaricides) that have traditionally been used to control the mites that parasitize the bees. There was an unfortunate incident recently involving neonicitinoids and bees that occurred in Germany. Corn seeds being treated with the neonicitinoid clothianadin did not include enough binding agent used to ensure that the insecticide would adhere to the seed. The equipment used to drill the seeds (a pneumatic device) blew pesticide-laden dust from the corn seeds into the air rather than toward the ground. Prevailing winds blew the dust toward flowering canola fields that were filled with bees pollinating the crop. The result was the death of many colonies of bees. This incident was not related to CCD, but was instrumental in developing improved label language, including the employment of proper equipment under more favorable environmental conditions.

While habitat loss or alteration, introduction of nonindigenous species, overexploitation and collection, and, more recently, diseases such as the chytrid fungus rank among the greatest risks to amphibians (Solomon et al. 2008), there has been growing interest in the potential effects of pesticides on amphibians. A recent review on amphibians and agricultural chemicals (Mann et al. 2009) provides excellent information on this subject. Of particular interest is the purported effect of the herbicide atrazine (2-chloro-4-ethylamino-6-isopropylamino-*s*-triazine) on developing frogs at concentrations as low as 0.1 μg/L (Hayes et al. 2002; Hayes et al. 2003). However, data from research by a number of other investigators support the conclusion that "the weight of available evidence does not substantiate claims that atrazine is a reproductive toxicant that feminizes and demasculinizes male frogs" (Solomon et al. 2008). Based on a recent examination of all the studies and their results, the U.S. EPA concluded that atrazine does not adversely affect amphibian gonadal development when exposure falls within the range of 0.01 to 100 μg/L (USEPA 2007).

Effects of pesticide mixtures on wildlife in laboratory or semi-field studies have received significant attention recently (Relyea 2009). Interestingly, a USGS report (Gilliom et al. 2007) noted that "more than 6,000 unique 5-compound mixtures were found at least 2 percent of the time in agricultural streams," but "the number of unique 5-compound mixtures found in agricultural streams is less than 100 when only concentrations greater than 0.1 micrograms per liter (μg/L) are considered." The USGS noted that, "many mixtures do not occur very often at high concentrations, and the most frequently occurring mixtures are composed of relatively few pesticides. Furthermore, a study conducted by the Danish Environmental Protection Agency (Kudsk 2005) examined 22 pesticides in a total of 101 binary combinations of pesticide mixtures and test systems suggests that significant divergence from additivity was only observed in relatively few cases.

Given the potential for adverse effects by pesticides to non-target organisms, assessment of risks posed by pesticides is an essential part of sound management strategies. Where pesticides co-occur at environmentally relevant concentrations and for a biologically meaningful duration, they should be evaluated for potential synergistic effects and included in the risk assessment process.

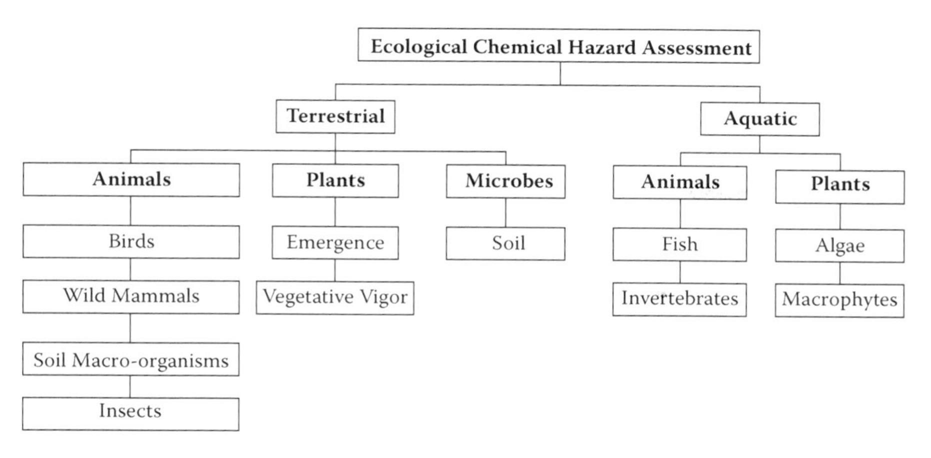

FIGURE 3.2 Organisms used in ecotox-testing programs with pesticides to determine hazard.

3.2.6 Ecological Effects Testing

Several wildlife toxicity studies are required as part of data requirements for pesticide registration. There certainly are differences in the requirements between regulatory bodies as to the organisms and guidelines used in the conduct of ecotoxicological studies, but a global pesticide registration generally requires toxicity studies with representative or surrogate species from the following major groups: terrestrial vertebrates and invertebrates, aquatic (freshwater and estuarine/marine) vertebrates and invertebrates, terrestrial and aquatic plants, and soil macro- and microorganisms (Figure 3.2).

The purpose of the first tier of studies is to develop acute (generally median lethal mortality or morbidity) and chronic (generally growth and development) endpoints to be used in simple hazard quotients that are compared to a specific level of concern. In addition to the first tier of studies required for registration, regulatory authorities can require pesticide registrants to conduct additional toxicity studies that may include higher-tiered studies (such as full life-cycle, mico/mesocosm, semi-field, field, and pen). The purpose of the higher-tiered studies is to refine early-tiered hazard or risk assessments by evaluating surrogate wildlife species that are exposed to pesticides under more environmentally realist scenarios. It is noteworthy that requirements are continuously reviewed and updated as necessary. Recently, both the United States and the EU have made revisions to ecotoxicology study requirements and guidelines. For example, the United States recently updated the avian acute oral toxicity test requirements by adding another species. Now, a pesticide registrant is required to test either an upland game species (northern bobwhite quail, *Colinus virginianus*), or a waterfowl (mallard duck, *Anas platyrhynchos*) along with a passerine species (CFR 2008).

3.2.7 Pesticides and Ecological Risk Assessment

The overall U.S. ecological risk assessment process is presented below (Figure 3.3) (USEPA 1998). Similar ecological risk assessment processes exist for most countries that have a formal regulatory program for pesticide registrations.

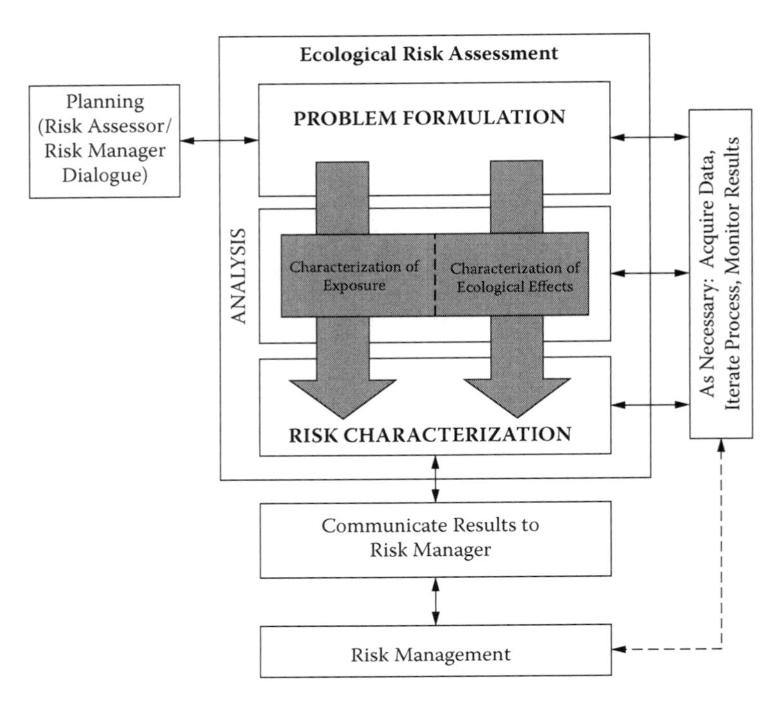

FIGURE 3.3 The ecological risk assessment process. (From U.S. EPA's *Guidelines for Ecological Risk Assessment*, April 1998.)

The first and arguably most important step in the process is the problem formulation phase, where the purpose for the assessment is articulated, the problem is defined, and a plan for analyzing and characterizing risk is determined. During the analysis phase, data are evaluated to determine how exposure to stressors is likely to occur, the magnitude of the exposure, and the potential and types of ecological effects that are expected given different exposure scenarios. The risk characterization phase includes a summary of assumptions and identification of uncertainties as well as strengths and limitations of the analyses.

It is important to understand that the ecological risk assessment process is considered to be an iterative process. Something learned during the analysis or characterization phase may lead to a reevaluation of the problem formulation or require the collection of additional data with subsequent analysis (USEPA 1998).

From a U.S. regulatory perspective, the primary purpose of the risk assessment process is to compare the appropriate exposure endpoint with the appropriate effects (toxicity) endpoint and calculate a risk or hazard quotient (see Chapter 9). The risk quotient is then compared to a predetermined level of concern (LOC). In the first tier, very conservative estimates of exposure and effects are assumed so that if the exposure/toxicity ratio is less than the LOC, the risk to the non-target organism from exposure to the pesticide under consideration is considered to be low. Conversely, if the exposure/toxicity ratio is greater than the LOC, there is a presumption of risk, and refinements to the risk assessment must be made if possible. If not, additional research is needed or mitigation measures are required. The end result is that the risk assessor provides the risk manager with a series of risk presumptions based on the comparison of hazard quotients with the respective LOC. This first-tier assessment is deterministic and does not comply with the true definition of risk, which implies that the probability of a particular adverse effect occurring under a given exposure scenario will be evaluated (see probabilistic approaches in Chapter 9 in this book).

In the United States, the ecological risk assessors send their assessments to the risk managers, who weigh the risks against their benefits. In a perfect world, communication between the risk assessors and risk managers would occur early in the process during problem formulation. However, this is not always followed in practice and can cause some confusion. For example, the risk assessor may determine that one or more risk quotients for a given pesticide are greater than their corresponding LOCs. The risk manager, however, may determine that the benefits of the pesticide outweigh the risks and so the pesticide may be registered. If the risk assessment but not the benefits analysis is made available to the public, it may not be clear to the public why the pesticide was registered. Clear and transparent risk communication between the risk assessors and risk managers and between risk managers and the public would likely establish more trust among all stakeholders.

3.3 GENETICS AND CROP PROTECTION

3.3.1 History of Genetics and Crop Development

Humans have been using genetics to improve agricultural production ever since agriculture began over 10,000 years ago. Targets of these efforts have been improved food and fiber quality as well as quantity, and the tool has been selection-based breeding (Knauft and Gardner 1998). More recently, science and technology have played a significant role in advancing goals for agriculture. Understanding the scientific basis of genetics was advanced when Gregor Mendel noted in the 1850s that the flower color of peas followed a predictable pattern. Within the last 60 years, our knowledge of the molecular details of genetic structure, function, and mechanisms of action has expanded greatly. As a result, many of the major commercial commodity crops have been highly domesticated through breeding that involves strong selection for characteristics desirable to growers and consumers. Recently, plant breeders have utilized the techniques of modern biotechnology to complement proven practices in crop improvement. Today, molecular breeding and genetic modification involving recombinant DNA are valuable means by which crop protection technologies are incorporated into the genetics of the plant. Consequently, yields and production efficiency have been increased to the point where much less land and fewer human resources are required to produce certain foods, feeds, and fibers (McCloud 1998).

Since 1996, biotechnology-derived crops, also known as genetically modified (GM) crops, have integrated crop protection technologies that are useful to growers. This has taken place in countries that have developed workable regulatory systems. Farmers in the major commodity-producing areas of the world have widely adopted GM corn (*Zea mays*) and cotton (*Gossypium* spp.) products that are both tolerant to

TABLE 3.2
Biotechnology-Derived Crop Protection Technologies Available in the U.S.

Crop	Trait	Gene(s)
Canola	Herbicide tolerant	*cp4 epsps* & *gox* (glyphosate tolerance) *bar* (glufosinate tolerance) *pat* (glufosinate tolerance)
Corn	Herbicide tolerant	*cp4 epsps* (glyphosate tolerance) *bar* (glufosinate tolerance)
	Insect protected	*cry1Ab, cry2Ab, cry1A.105, cry1F* (lepidopteran protected) *mcry3A, cry3Bb, cry34/35* (coleopteran protected)
Cotton	Herbicide tolerant	*cp4 epsps* (glyphosate tolerance) *bar* (glufosinate tolerance)
	Insect protected	*cry1Ab, cry1Ac, cry2Ab, cry1F* & *vip3A* (lepidopteran protected)
Papaya	Virus resistant	CMV-PRV (papaya ringspot virus coat protein)
Plum	Virus resistant	PPV-CP (plum pox virus coat protein)
Squash	Virus resistant	CMV/WMV-2/ZYMV (cucumber mosaic virus, watermelon mosaic virus 2, zucchini yellow mosaic virus coat proteins)
Sugarbeet	Herbicide tolerant	*cp4 epsps* (glyphosate tolerance)
Soybean	Herbicide tolerant	*cp4 epsps, gat4601* (glyphosate tolerance) *pat* (glufosinate tolerance) *gm-hra* (imidazolinone tolerance)

Note: Agbios keeps an excellent global database on GM crops at http://www.agbios.com/main.php.

broad-spectrum herbicides and able to resist key insect pests. Herbicide tolerance in corn, cotton, canola (*Brassica napus*), and soybeans (*Glycine max*) is a common, perhaps baseline, trait in efficient, conventional production systems, while herbicide-tolerant sugar beet (*Beta vulgaris*) is new to the market in the United States. Resistance to virus-induced diseases is commercially available in such crops as papaya (*Carica papaya*), plum (*Prunus domestic*), and yellow squash (*Cucurbita pepo*). Application of modern biotechnology to agriculture has increased the number of crop protection technologies available to growers (Table 3.2).

3.3.2 Current Status of GM Crops

Between 2007 and 2008, the area producing GM crops grew by 9.4% or 10.7 million hectares (26.43 million acres), to reach a total of 125 million hectares (309 million acres) (James 2008). According to James (2008), 13.3 million farmers in 25 countries are growing GM crops, and 90% (12.3 million) of these are resource-poor farmers in 15 developing countries (Table 3.3). Some of the reasons given for the rapid adoption of herbicide-tolerant canola, cotton, maize, and soybean have been ease of weed control, improved control, less crop injury, concordance with soil conservation practices, economic return, and use of herbicides that are less toxic to humans and degrade rapidly in the environment (Burnside 1996; Culpepper and

TABLE 3.3
Global Plantings of Biotech Crops

Rank	Country	Area (million hectares)	Biotech Crops
1	USA*	62.5	Soybean, maize, cotton, canola, squash, papaya, alfalfa, sugar beet
2	Argentina*	21.0	Soybean, maize, cotton
3	Brazil*	15.8	Soybean, maize, cotton
4	India*	7.6	Cotton
5	Canada*	7.6	Canola, maize, soybean, sugar beet
6	China*	3.8	Cotton, tomato, poplar, petunia, papaya, sweet pepper
7	Paraguay*	2.7	Soybean
8	South Africa*	1.8	Maize, soybean, cotton
9	Uruguay*	0.7	Soybean, maize
10	Bolivia*	0.6	Soybean
11	Philippines*	0.4	Maize
12	Australia*	0.2	Cotton, canola, carnation
13	Mexico*	0.1	Cotton, soybean
14	Spain*	0.1	Maize
15	Chile	<0.1	Maize, soybean, canola
16	Colombia	<0.1	Cotton, carnation
17	Honduras	<0.1	Maize
18	Burkina Faso	<0.1	Cotton
19	Czech Republic	<0.1	Maize
20	Romania	<0.1	Maize
21	Portugal	<0.1	Maize
22	Germany	<0.1	Maize
23	Poland	<0.1	Maize
24	Slovakia	<0.1	Maize
25	Egypt	<0.1	Maize

* Countries growing 50,000 hectares, or more, of biotech crops

York 1998; McKinley et al. 1999; Carpenter and Gianessi 2001; Bullock and Nitsi 2001; Kalaitzandonakes and Suntornpithug 2001; Fawcett and Towery 2002; Shelton et al. 2002; Runge and Ryan 2003). The economic and environmental impacts of biotech crops were investigated over a 10-year period, and researchers concluded that farmers have benefited economically (Brookes and Barfoot 2005, 2006). They estimated an overall reduction of 224 million kilograms of pesticides sprayed.

The current diversity of crop and gene combinations (Table 3.2) is rather narrow compared with the number of plants that can be transformed using recombinant DNA techniques (Dunwell 2000; Babu et al. 2003; silvers et al. 2003). Important food crops such as wheat (*Triticum aestivum*), rice (*Oryza sativa)*, sweet potato (*Ipomoea batatas*), and cassava (*Manihot esculenta*) have been successfully transformed and field-tested around the world. Their importance to food security and economic development makes these and other plants key targets for improvement using GM techniques. Similarly, a greater diversity of traits and trait combinations are being developed for eventual introduction into farming systems (Nickson 2005). Tolerance to abiotic stressors such as drought, salt, heat, and cold, as well as traits that give improved nutritional qualities, are being aggressively developed. Many of these products should be a priority for developing world agriculture (Thomas et al. 2003; Qaim and Zilberman 2003). However, their deployment into commercial food and feed production systems will be possible only after each product has completed appropriate regulatory reviews. A key step in this process is the development of an environmental risk assessment that provides evidence for an authority to conclude with sufficient certainty that the risks posed by the GM crop are acceptable compared to those posed by the conventional plant.

3.3.3 Environmental Risk Assessment (ERA) of GM Crops

Products developed through traditional breeding are largely not regulated from the perspective of food, feed, and environmental safety. Canada is the only country where a new crop/trait combination derived from traditional breeding may be subject to regulatory review for food, feed, and environmental safety depending on the "novelty" of the trait (see http://www.inspection.gc.ca/english/sci/biotech/reg/novnoue.shtml). Conversely, virtually all other countries in the world use the process of "modern biotechnology" as a trigger for regulatory review. Experience to date has been that all GM crops have undergone regulatory review; and food, feed and environmental risk assessments (ERA) have been integral components of every deregulation/approval.

The early basis for the development of the ERAs conducted for GM crops can be found in four publications (OECD 1986; OECD 1993; NRC 1989; Tiedje et al. 1989). Since the publication of these seminal documents, numerous publications have appeared intended to either describe the ERA process or to clarify the conceptual basis, or both (for example, UNEP 1995; Rissler and Mellon 1996; Kjellsson 1997, EI 1998; Kjaer et al. 1999; Nickson and McKee 2002; Hancock 2003; Wilkinson et al. 2003; Evaristo de Jesus et al. 2006; Craig et al. 2008). Certain principles have emerged that are common to all ERA frameworks for GM crops (Hill and Sendashonga 2003). ERAs should utilize multiple lines of evidence and consider all the available information, including both qualitative (expert judgment) and quantitative data. There is debate regarding the quantity of data needed for regulatory decision-making, and whether using laboratory data alone is sufficient to approve GM crops in situations where little information from field experiments is available regarding a product's effects on non-target organisms (Romeis et al. 2008; U.S. EPA 2006). In addition, ERAs should be sufficiently flexible to allow them to be recursive and comparative in nature. A fundamental principle for evaluating a GM plant is to compare it to an appropriate counterpart whose environmental risks are known to some degree. Information obtained in the course of evaluating the risks associated with the GM plant relative to its counterpart may need to be reconsidered in light of assumptions made earlier in the assessment or to re-evaluate earlier risk characterizations.

Recently, some authors have described the conceptual basis of the ERA for GM plants in a manner harmonized with processes used for chemical pesticides (Nickson and McKee 2002; Hill and Sendashonga 2003; Raybould 2006; Nickson 2008). The importance of problem formulation in ERA was highlighted by Raybould (2006), who noted that testing appropriate risk hypotheses was essential to avoid collection of irrelevant data. Many experts in this field have witnessed the proliferation of data requirements with little justification other than scientific curiosity. Raybould (2007) subsequently described that there is a distinct difference between an ecological and ecotoxicological approach to data collection. The former is appropriate for asking basic questions of science, while the latter is necessary to conduct efficient and rigorous ERAs for the purpose of regulation. The term *assessment endpoint* as described by the EPA (USEPA 1998) is infrequently encountered in the ERAs for GM crops. This highlights the problem that risk hypotheses are rarely properly described. In fact, many practitioners confuse individual elements of hazard and exposure as "risks." It is quite common to encounter such statements as "gene flow is a risk" and "toxicity of a *Bt* protein to butterflies is a risk" when, in fact, gene flow is a component of exposure and toxicity is an element of hazard. These experiences only exemplify the challenges that one can encounter when conducting an ERA for a GM crop.

Three very broad high-level assessment endpoints are typically the focus of an ERA for a GM crop. These are (1) abundance of pests (weeds, animals, and disease), (2) microbial processes involved in agricultural productivity, and (3) abundance of beneficial animals in agricultural fields. First, the ERA is typically concerned with the question of whether the GM plant becomes a pest of agriculture or other environments through gene flow (exposure). Likewise, risk assessors are concerned that the GM plant will create a worse pest—for example, by increasing the virulence of viral pathogens through some scientifically reasonable mechanism. Second, the ERA typically addresses the potential for the GM plant to affect microbial processes that are valuable to farm productivity and in-field biogeochemical process (e.g., hydrologic cycles). Finally, determining the potential for the GM plant to adversely affect beneficial organisms (non-target organisms) is an area that the ERA must specifically address. The potential for the GM plant to cause harm either directly or indirectly is examined for all three (broad) assessment endpoints. Of course, specific assessment endpoints and data needed for the ERA will vary depending on the regulatory authority and their legal requirements. Generation of specific risk hypotheses should be based on knowledge of the conventional crop that was modified, the nature of the trait that was introduced into the crop, the likely receiving environment, and the likely interactions among these in the context of the assessment endpoints. In this way, relevance to the ERA is ensured because the necessary linkage between the assessment endpoints, the characteristics of the specific GM plant, and the specific hypotheses tested is clear. In this way, the relevance of the data for regulatory decision-making (approval) is obvious.

There is no consensus on a prescriptive set of data necessary for every ERA. A principle of case-by-case is applied, but there has been a good amount of commonality with regard to the data used in the assessments conducted on the currently commercial GM crop production products. Figure 3.4 provides an outline of the information commonly submitted as part of an ERA for a herbicide-tolerant, insect- or virus-protected crop.

Decisions on exactly what data should be submitted should be guided by a concept developed about 20 years ago called *familiarity* (OECD 1993; Nickson and

- Product Characterization:
  - molecular analysis
- Identification of hazard potential: Plant
  - Phenotypic and environmental interaction information
  - Composition studies
- Identification of hazard potential: Trait
  - Mode of action of gene product
  - Toxicity potential/data
- Exposure assessment
  - Expression levels of the gene product
    - Temporal expression
    - Appropriate plant parts
  - Gene Flow
    - Pollen and/or seed
    - Sexually compatible wild relatives
  - Environmental Fate
    - Fate in soil
    - Fate in water

FIGURE 3.4 General description of data and information requirements for an ERA for a GM crop.

Horak 2006). Familiarity is useful to decision makers; it comes from preexisting knowledge—experimental results as well as expert opinion—and experience gained over time. Familiarity encompasses experience with the crop, the trait, the environment, and interactions. By definition, familiarity increases with time and experience, and thus it helps address uncertainty in the risk assessment and direct future information collection (e.g., monitoring). Familiarity is critically important in selecting only those risk hypotheses that are plausible or reasonable. It also then aids in selecting appropriate testing procedures and species. Importantly, familiarity is not a safety conclusion; instead it encompasses the information available at a given point in time, and it serves as a basis from which the risk assessment should proceed.

Underlying familiarity are two important assumptions: (1) the process of genetic engineering is not inherently more risky than conventional plant breeding, and introduced transgenes behave in essentially the same manner as any other gene within the plant genome; and (2) there is a significant history of introducing new traits into crop plants and of evaluating these new varieties in agriculture. In other words, the ERA should focus on the phenotype of the plant (not the genotype), and one's experience with a plant and familiarity with its phenotype guides the planning phase (generation of risk hypotheses and selection of specific test hypotheses). According to Hokanson et al. (2008), "Familiarity allows decision-makers to draw upon the vast experience with introduction of plants into the environment, and to compare genetically engineered plants to their non-engineered counterparts."

As described in Figure 3.4, a typical ERA for a GM crop protection product involves molecular and expression analyses as well as comparative compositional analysis and agronomic/phenotypic evaluations. This information is useful in characterizing the GM product and assessing its hazard potential. These analyses should be designed to detect differences that are relevant to the risk assessment. That is, the data should allow a risk assessor the ability to decide with reasonable certainty what, if anything, is sufficiently different such that it may cause an adverse effect to some environmental attribute that must be protected (see above). Experience with crops tells us that differences between genotypes will frequently be observable in the phenotype, such as maturity, flower color, and plant architecture. These differences, however, are almost always not meaningful in terms of environmental risks. As such, phenotypic and compositional studies must be designed appropriately to reveal differences that are relevant to the ERA. For the crop protection products in Table 3.2, the only meaningful differences observed in our assessments to date have been the presence of the introduced trait and the resulting phenotype.

Figure 3.4 also highlights that other factors related to hazard potential and exposure have been considered. Once we have high certainty that the only meaningful difference between the GM crop and its conventional counterpart is the intended difference, risk hypotheses are formulated for the introduced trait—for example, "The presence of trait x will have no increased adverse effects on the environmental components of concern." Testing these hypotheses is done in a tiered manner where the characteristics of the trait are carefully considered with regard to their potential to cause harm (Garcia-Alonzo et al. 2006; Romeis et al. 2008). Based on the properties of the expressed proteins conferring herbicide tolerance or virus resistance to the crops, it was concluded that there was no need to conduct further (effects) tests. In addition, the potential impacts associated with gene flow from these GM crops could be assessed without the need for any further data. It was sufficient to conclude that, based on the biology of the crop, if gene flow is possible, it will occur. Risk characterization for gene flow then focused on the potential for harm to occur based on the characteristics of the trait.

Conversely, based on what was known about the insecticidal properties of the proteins from *Bacillus thuringiensis* (Bt), a battery of tests were conducted against representative arthropods for products expressing insecticidal Bt proteins (Romeis et al. 2006). Subsequently, no adverse effects have been observed in numerous laboratory tests conducted on a wide diversity of organisms with the Bt proteins identified in Table 3.2 (those interested in more details on these tests are encouraged to obtain the Biopesticide Registration Action Documents available from the U.S. EPA: http://www.epa.gov/pesticides/biopesticides/ingredients/). One significant challenge in conducting laboratory tests is acquiring sufficient quantities of test material (protein). This is typically done using recombinant bacteria, e.g., *E. coli*, since expression in plants is usually very low. In these cases, an "equivalence" study is required to demonstrate that the protein produced in *E. coli* is an appropriate substitute for the protein produced in plants for use in toxicological tests.

The environmental fate of the GM crop protection products must also be considered in any ERA of these GMOs. The characteristics of the proteins introduced into herbicide-tolerant and virus-resistant crops are such that there is sufficient information to conclude that they have negligible likelihood to persist or accumulate. Thus, no reasonable risk hypotheses could be formulated with regard to potential harm to soil and soil processes. Experiments were designed, however, to examine the potential for Bt proteins to persist and accumulate. Much of the earlier interest in this question was generated by a publication (Tapp and Stotzky 1998) several years ago. Not surprisingly, many of the commentaries on this publication mischaracterized the result as a "risk." Since this time, much research has been done examining the environmental fate of Bt proteins and tissues expressing these protein (for example, see Margarit et al. 2008). Fate data have been collected either as a part of the ERA, a condition or registration for the U.S. EPA, or as basic academic research. Figure 3.4 simply makes note that environmental fate is an element of the exposure assessment with an ERA for a GM crop.

3.3.4 Summary

Several valuable crop protection technologies are now available to growers as the result of a natural evolution and adoption of technology in agriculture. The use of biotechnology has allowed us to take another step forward in developing new products that must meet societal standards of acceptability. Many GM crops demonstrate how chemistry and genetics can complement each other and integrate into more sustainable agricultural systems. Both chemical and GM products have had to meet rigorous regulatory requirements to ensure the users and general public that these products pose no unacceptable risks to the environment. In the future, many more trait, crop, and chemical combinations will become available to agriculture only when these products also are shown to meet regulatory and other standards of the agricultural systems.

3.4. BIOFUELS

Most experts agree that sustainable fuels are needed, and many consider biofuels a good alternative to fossil fuels (Groom et al. 2008). Along with the need for conversion to biofuels is the need to limit the release of contaminants into the environment. While this is a laudable goal, all participants in biofuel production must acknowledge that there is a concomitant increase in contaminant emission with production of any fuel and waste products. Biofuel production is no different. While the volumes and toxicities of biofuel-derived waste may be lower than for fossil fuels, there remains a concern that increased biofuel production will increase the release of contaminants from the feed stock, transport, production, and distribution processes associated with biofuel production. There are also significant concerns regarding the conversion of croplands and forested areas for use in producing biofuel feedstocks.

Reducing the emission of CO_2 into the atmosphere is one critical goal of conversion from fossil fuels to biofuels. Current estimates suggest that corn-derived ethanol reduces greenhouse gases (GHGs) by approximately 13% compared to petroleum-based fuels, but that lignocellulosic fuels can reduce GHGs by approximately 85% (Hoekman 2009). Unfortunately, the current total worldwide cropland use of 0.24 ha/person for all purposes stands in stark contrast to the 0.7 ha needed to produce sufficient sugarcane ethanol to displace the carbon emissions from one vehicle. And this comparison gets far worse for corn ethanol and biodiesel. Therefore, bioenergy policies must seriously consider the impact of these acreages on cropland and on natural habitats for wildlife. Any use of food crops for biofuels or conversion of arable land into biofuel crops such as grasses has a significant potential to impact food resources in developing countries (Ugarte and He 2007). Concern for wildlife and ecosystem health is seldom a major consideration in regions where human populations have insufficient food. Therefore, processes that do not use food crops or land that is needed for food crops are needed in developing countries, not only to benefit the human condition, but to allow inhabitants of these countries to consider the ecosystem services that accrue from protecting wildlife and natural resources.

3.4.1 General Considerations for Biofuel Production

One promising attribute of biofuel production is that available biofuel resources are more evenly distributed globally than are fossil fuel resources (Sagar and Kartha 2007). However, the most mature of these biofuel technologies, grain-based ethanol, is by far the least sustainable. Total energy conversion from wheat is approximately 1.09, and for corn this conversion efficiency is 1.2 to 1.4 (Dong et al. 2008). In fact, the use of row crops in temperate zones to produce ethanol is likely to consume energy rather than produce it.

Equatorial sugar cane (*Saccharum* spp.) is the only feedstock from which ethanol can be produced by fermentation with an energy conversion ratio of 1.6 or higher (Dong et al. 2008). It is possible for efficient use of sugarcane to produce 3.7 to 8.2 times the energy input, and cellulosic ethanol production can generate 6 to 11 times more energy than is input for its generation. Bio-diesel may achieve conversion efficiencies of 1.2 to 3 times input energy, depending on feedstock source (Sage et al. 2009). Although the cellulosic processes were not considered, research to compare CO_2 emissions, natural resource utilization, and contaminant emission suggests that few complete biofuel production, distribution, and consumption processes are less harmful than processes for fossil fuels (Zah et al. 2007; Scharlemann and Laurance 2008). Therefore, new approaches to biofuel production are needed (Groom et al. 2008). Standard measures of greenhouse gas emissions in the lifecycle of all fuels need to be evaluated as part of decision-making processes for fuel production and use (Liska and Cassman 2008). We believe that lifecycle assessments must consider all inputs, not only GHG and fuels, but all resources. Recent data suggest that only the most efficient biofuel production processes utilizing feedstocks that are non-destructive to existing ecosystems reduce environmental impacts compared to petroleum production (Zah et al. 2007).

In the United States, biofuels comprise approximately 0.7% of energy production, with direct combustion of wood waste accounting for another 3.3% (Hoekman 2009). The largest feed stocks for bioethanol and biodiesel are corn and soy, respectively. Biofuel production consumed 14% of the corn crop in 2006, demonstrating the need to increase other types of biofuel production, such as cellulosic ethanol (Bies 2006). Regulations in the United States have provided incentives to produce biofuels through cellulosic conversion of non-food crops. Biochemical and thermochemical methods for cellulosic conversion are being developed separately, but future production facilities are likely to incorporate both processes to convert a wider variety of feed stocks (Bies 2006).

There are limitations on the amount of biofuel production that can be envisioned from harvesting crops, grasses, or wood byproducts (Hoekman 2009). Issues surrounding wood include accessibility, transportation, and soil stability. Issues surrounding crops and grasses focus on needed improvements in biomass yield, water consumption, and nutrient inputs. Despite this, there is a potential for wood, crops, and grasses to replace approximately one-third of the current U.S. petroleum consumption by 2030 (Hoekman 2009). Other estimates suggest that tropical areas of the world have the capacity to produce 20% of OECD liquid-fuel needs by 2020. For example, Brazil currently uses ethanol for approximately 30% of its transportation needs (Sagar and Kartha 2007). This would stabilize economic situations in these countries and could mitigate greenhouse emissions if deforestation is not embraced as a means of acquiring more land to produce biofuel feed stocks (Mathews 2008). However, using tropical land for biofuel raises significant ecological concerns (see below). Moreover, some reports suggest that energy from crops is unlikely to attain the projected output of 400 EJ/yr (EJ = 10^{18} J) required to avoid exceeding atmospheric CO_2 of 550 ppm. The more likely amount is 22 EJ from biofuels, and this could lower CO_2 release by 2070 Mt CO_2/yr compared to fossil fuel use, which is insufficient to limit atmospheric CO_2 to 550 ppm (Sims et al. 2006). With these CO_2 diminutions and likely near-term energy production from biofuels in mind, conversion of vehicles to electric power seems to be a major requirement in making transportation truly sustainable (Pro et al. 2005), but in the interim, biofuels are likely to assume a major portion of liquid fuel sources for vehicles.

3.4.2 Specific Fuel Types

3.4.2.1 Biodiesel

Biodiesel can use high-oil crop seeds or animal fat (Hoekman 2009). Conversion of waste oil from cooking processes into biodiesel is a well-known process across the globe (Tsai et al. 2007). Life cycles indicate that vegetable oils are ~2 times more sustainable than tallow as a biofuel feedstock (Niederl and Narodoslawsky 2006). Moreover, animal fat can carry many contaminants (Herrera et al. 1994; Hela et al. 2006; Jevsnik et al. 2004; Paton and Petterson 1997; Sallam and Morshedy 2008) that may be retained in biodiesel. Prions are one of the more exotic of these contaminants. Studies indicate that the hydrolysis and transesterification processes effectively destroy these molecules (10^7 reduction factor) (Mittelbach et al. 2007).

Biodiesel and hydrogen generation are the most prominent areas of biofuel research (Kajikawa and Takeda 2008). Jatropha oil (*Jatropha gossypiifolia* and *J. curcas)* and soapnut oil (*Sapindus mukorossi*, *S. emarginatus,* and *S. rarak*) may be used to replace normal food crops as biodiesel feedstocks (Chhetri et al. 2008). *J. gossypiifolia* and *J. curcas* have the potential to serve as good biodiesel sources in tropical areas such as India (Kumar and Sharma 2005) and Brazil (de Oliveira et al. 2009). However, oil-content cultivation and harvesting methods need to be improved for optimal crop utilization and minimization of ecological damage.

Although not widely publicized, the use of microalgae for biodiesel production provides much higher yields per acre than oil crop conversion and other biofuel production methods (Chisti 2008; (Patil et al. 2008). The concept of using algae as a means to sequester CO_2 is not a new idea (Gao and McKinley 1994). More recent efforts have expanded to include and improve the biofuel recovery process for alga. Biodiesel from microalgae has garnered a wide share of literature regarding biofuels (Kajikawa and Takeda 2008). This conversion technique has the potential to significantly relieve pressures to cultivate extensive tracts of land for biofuel feedstock generation.

3.4.2.2 Ethanol

Maize and sorghum are currently key crops for advancing our understanding and improvement of bioethanol production (Carpita and McCann 2008). However, using food crops for fuel production is widely viewed as a poor decision. Even so, obtaining sufficient biofuel feedstocks from non-food crops will require significant amounts of land and substantial natural resources to maintain them. These land use processes must be sustainable (Keyzer et al. 2008), especially if a goal of biofuel use is to mitigate any ecological impacts.

It should be noted that much of the inefficiency in producing fuel from ethanol accrues through transport costs of moving feedstocks to centralized conversion facilities. To efficiently move forward, biofuels must develop a consistent and reliable network of feedstock producers; efficient processes to harvest, separate, and transport feedstocks to production facilities; more efficient and widely dispersed production facilities; distribution capacity to deliver biofuels to blending and delivery facilities; and vehicles and other machinery that can effectively use these fuels (Hoekman 2009).

Lignocellulosic crops offer ecological improvements over most food crops, but may not offer habitat improvements over set-aside grasslands. Critical evaluations of costs and benefits in energy, habitat, and carbon cycling need to be evaluated for cellulosic grasses and short-term rotation coppice (Rowe et al. 2009). Switch grass was identified early in the search for biofuel feedstocks for cellulosic ethanol production (Parrish and Fike 2005) and has become an important target plant to be converted to ethanol. But other feedstocks are available, such as municipal wastes and several fast-growing shrub and tree varieties.

Miscanthus giganteus may serve as a viable alternative to or as an improvement over switchgrass. *M. giganteus* produced an average of 30 t/ha of harvestable biomass with a maximum of 61 t/ha. Switchgrass (Cave-in-Rock variety) achieved somewhat lower yields, averaging 10 t/ha (Heaton et al. 2008). However, *M. giganteus* is not native to most areas where it is proposed for use, and it is well documented that the introduction of invasive species alters ecosystems, most often in unwanted ways. The current means to control invasiveness is to plant sterile hybrids. It is important here to note that perennial crops are far more energy efficient than are annual crops, especially high-maintenance row crops. This is because there is no energy input needed to plant the crop after the first planting (Sagar and Kartha 2007). However, using annuals does provide more rapid spread of the crop, increasing the potential for annuals to become invasive. In the search for high biomass yield, feedstocks may become reliant on exotic plant species that may become invasive without careful measures. Such introductions of invasive species have already caused problems in Australia (Davis et al. 2008). Conversely, harvesting existing invasive species for biofuel production may be an effective strategy for reducing acreage impacted by the invasive species. A good example is described in a pilot study wherein kudzu (*Pueraria montana*) is shown to provide a feedstock capable of producing as much biofuel as that derived from corn (Sage et al. 2009).

Evaluations of *M. giganteus* production demonstrated that biomass yield did not change more than 5% over a 14-year study interval. In fact, 60 to 120 kg N/ha did not increase its biomass yield during any time interval of a 14-year trial. Pesticides, while not required every year, were required regularly for the first few years until the grass monoculture was established (Christian et al. 2008). The need for pesticides in biofuel feedstocks must be considered in the overall ecological considerations of conversion from petroleum to biofuels. Corn stover (~70% lignin) can serve as a useful feedstock for biofuels. Removal of the stover could adversely affect soil replenishment and erosion mechanisms (Johnson et al. 2004). Removal rates and frequencies of plant biomass from specific areas should be evaluated to minimize impacts on soil health, wildlife habitat, and contaminant introduction into areas generating feedstocks.

Application of residual materials from bioreactor processes to areas used for biomass harvest could mitigate soil replenishment and erosion concerns. However, the amount of residual bioreactor material needed to stabilize soil following row crop production is significantly higher than when stover is used. Therefore, the amount of stover to be used for fuel production must be carefully weighed against soil sustainability. Efforts to increase yields through crop rotation have met with some success (Smith et al. 2008), and many are likely to be ecologically beneficial (Sagar and Kartha 2007), normally as a result of nutrient replenishment and natural pest and disease control (see Chapter 4). However, specific factors controlling the effectiveness of rotations in controlling diseases are unclear (Liebig et al. 2007).

Production of ethanol derived from properly managed switchgrass can produce 94% less greenhouse gases than gasoline. This determination was based on managed switchgrass and an energy return of 540% biofuel from fossil fuel inputs (Schmer et al. 2008). It is possible that yields could go as high as 700%. From a nutrient and energy perspective, it is important to note that repeated switchgrass harvesting within a growing season increases yield more than does N fertilizer application (Thomason et al. 2004). Fall switchgrass harvest yields more biomass than does spring harvesting, but better wildlife management results from alternative harvesting times (Adler et al. 2006). August harvest of switchgrass provided quality cover for songbirds and altered species composition, compared with fields that were not harvested (Roth et al. 2005). We believe that a mosaic harvesting approach in switchgrass monocultures may provide continuous feedstock and reasonable habitat for a wide range of grassland bird and small mammal species.

Conversion of land use from arable crops to ligninocellulosic crops (LCs) has the potential to improve many ecologically important parameters, including energy cycles, greenhouse gas emissions, and biodiversity. However, LCs do not provide improvements when compared with permanent grasslands. Water use is of particular concern in all production of crops, including those to be used for fuel. Water use and reuse will be an important consideration for production as LCs become more widely used for biofuel production (Dominguez-faus et al. 2009; Rowe et al. 2009). Wheat straw and clover (*Trifolium* spp.) may be used to increase biomass production. When coupled with wet oxidation methods, this process more efficiently produces ethanol and can be decentralized to allow more efficient feedstock acquisition and fuel distribution (Thomsen and Hauggaard-Nielsen 2008).

Expanding the mixture of vegetation to include grasses and woody plants produces ecosystems that are preferred over grasslands alone. Mixed grasses are the next most beneficial, providing better habitat than short-rotation coppice, which itself is better than row crops. Some species will invariably benefit more from each management type (Bies 2006). Given these ecological parameters and the need to avoid expansive crop monocultures, biofuel production that can accommodate several different feedstocks is essential. Syngas production allows a broader range of feedstocks that could include materials such as household wastes (Sagar and Kartha 2007). The process can also produce a range of fuel products such as methanol, ethers, Fischer-Tropsch fuels, and hydrogen.

3.4.2.3 Forest Products

There is a large body of literature regarding wood and wood byproducts as fuel sources (Sathre and Gustavsson 2006; Scott and Tiarks 2008; Balat 2007; Champagne 2007; Rowe et al. 2009; Tomic et al. 2008; Wolf et al. 2006). As all natural resources are seen as fuel sources, care must be taken when considering forest products as a source of fuels. The world currently consumes 40% of sustainable forest biomass production (Parikka 2004). To utilize much more forest would require utilization of pristine forests, which is not an ecologically viable scenario. For example, an in-depth analysis of ways to balance carbon from fossil fuel with forest products considers all forest woods to be available for use and showed that all the most fuel-efficient scenarios involved harvesting trees for energy production. Of course this is not a viable approach. Transferring logged forests into palm oil stands may allow 150% more carbon fuel to be produced than is input (Wicke et al. 2008), but this is minimal compared to the carbon sequestration provided by the original forest that was logged. There is controversy regarding the effectiveness of measures being taken to improve palm plantation management and to protect forests in the regions conducive for their growth (Basiron 2007). Clearing of forests to plant palm trees (*Elaeis guineensis*) is rapidly destroying the remaining habitat for orangutans (*Pongo pygmaeus*) (Nantha and Tisdell 2009).

Many aspects of land conversion to biofuel feedstocks in tropical regions have been addressed (Malhi et al. 2008). Significant adverse impacts are likely to ensue if tropical forests are converted to pastures to provide biofuel feedstocks (Sawyer 2008). In temperate areas, short-term rotation of forests for biofuels has been suggested for cesium-contaminated lands in Belarus (Vandenhove et al. 2002). This would transfer the cesium from forest areas to processing areas, and residues from biofuel production would then be contaminated with cesium. Another example of impacts from forest use as biofuel involves the projected adverse impacts on European goose (*Branta* spp. and *Anser* spp.) populations (Wisz et al. 2008). Genetic modification to improve wood and biofuel production of tree species is also being considered (Groover 2007). However, large-scale introduction of genetic modification into ecosystems is controversial.

3.4.3 Diffuse Network of Biofuel Production

Incorporating small farms and farm cooperatives into biofuel processes can provide local economic improvements and establish a diffuse network of feedstocks (Hoekman 2009) and production facilities. The authors believe that a more complete life cycle assessment could identify areas in the production cycle to improve the efficiency of biodiesel production and distribution, thereby making decentralized processes more reliable and sustainable. Biodiesel and direct solar are renewable energy sources that can be implemented in a sustainable fashion at the community level, thereby eliminating much of the transport and distribution costs (Khan et al. 2007). These could also be implemented in tandem to minimize fossil fuel input into biodiesel production. By doing so, the amounts of resources expended producing biofuels would be minimized, thereby minimizing impacts on resources upon which wildlife depend.

Perhaps a significant improvement in the global environment could be made by implementing sustainable biofuel programs in developing countries, where approximately 2.5 billion people depend on biomass as a fuel for their primary energy source for cooking. The density of these biomass consumers is highest in Africa and Asia, where biomass is used by 85% and 75% of the populations, respectively (Ugarte and He 2007). It should be noted that many of the fuel sources used in these areas are listed among the most beneficial replacements for fossil fuels (Zah et al. 2007; Scharlemann and Laurance 2008). Within industrialized countries with smaller land mass, transportation and distribution costs may not greatly affect the efficiency of biodiesel production; however, application rates of nitrogen fertilizer require much energy, and moderation of that energy output can increase the energy efficiency of biodiesel production to 120–170% (Stephenson et al. 2008).

3.4.4 General Impacts

Adverse impacts of biofuel production include ecological damage from nutrient loading, soil erosion, and crop-protection products. Large-scale biofuel production may be unsustainable because of the accompanying degradation of soil conditions and the energy inputs needed to produce fertilizers that must be used to mitigate soil degradation (Ulgiati 2001). Lignocellulosic fuel production limits these concerns when compared with the use of row crops for bio-diesel or ethanol production (Hill 2007). As pressure grows to provide nutrients and organic matter to soils, biosolids of various types may be considered as potential supplements (Wang et al. 2008). It has also been suggested that to minimize soil nutrient depletion, all residual material from corn fermentation must be returned to the harvested fields. Toxic metal accumulation in feedstocks is a critical concern, especially since residual materials from fermentation and biosolids such as sewage sludge are considered fertilizers for these non-food crops (Reed et al. 1999). Another important factor is that returning residual materials from centralized fermentation facilities to the fields of origin requires energy, which has a negative impact on the energy derived from corn-based biofuels (Patzek 2004).

Water, nitrogen, and pesticide use in corn-based biofuels production is a major concern as we contemplate placing more acres into production to support biofuel production (Dominguez-faus et al. 2009; Hoekman 2009). Soil conservation and improvement must be addressed systemically to increase sustainable environments, not only in agricultural areas, but also in urban and industrial areas (Lal 2007). Crop rotations are well known to minimize plant diseases and crop-specific pest species.

Improvements in fermentation processes for cellulosic ethanol production are most likely to involve more efficient enzymatic catalysis. This may require engineering microbes with greater cellulosic degradation properties. We feel that use of these microbes must be carefully considered, since the sustainability of biofuel processes seems to include returning the unreacted biomass to the fields producing the feedstocks, thereby releasing the microbes into the environment where they could proceed to damage cellulosic feedstocks and other unintended plants (Sathre and Gustavsson 2006).

3.4.5 Contaminants and Byproducts

Normal production of biodiesel is producing a glut of glycerol, and uses for this byproduct are needed (Hoekman 2009). Catalytic hydroprocessing avoids the glycerin byproduct. Glycerin can be used to produce propylene glycol, which in turn can be a source of ethylene and propylene oxide (van Haveren et al. 2008). Efficient alcohol–water separation technologies are crucial for making bioethanol viable (Vane and Alvarez 2008). It is also possible that some of the known fermentation byproducts may also pose environmental risks (Table 3.4).

Biodiesels have lower amounts of heteroatomic contaminants, thereby lowering emissions (Bhatti et al. 2008), and blending diesel with ethanol can reduce particulate emissions and reduce other significant urban pollutants (Miraglia 2007). These are important considerations for contaminant release into ecosystems, specifically those

TABLE 3.4
Likelihood That Contaminants in Bio-Ethanol Will Cause Ecological Effects

Ecological Effects		
Unlikely	**Likely**	**Unknown**
Formic acid	Itaconic acid	*cis*-Aconitic acid
Malonic acid	Gallic acid	*trans*-Aconitic acid
Lactic acid	5-Hydroxymethylfurfural acid	Levulinic acid
Maleic acid	Furfural	2-Hydroxy-2-methylbutyric acid
Acetic acid	Adipic acid	2-Furoic acid
Methylmalonic acid	3,4-Dihydroxybenzoic acid	Vanillic acid
Succinic acid	3,5-Dihydroxybenzoic acid	Homovanillic acid
Fumaric acid	2,5-Dihydroxybenzoic acid	Syringaldehyde
Glutaric acid	3,4-Dihydroxybenzaldehyde	
Vanillin	Salicylic acid	
Benzoic acid	4-Hydroxybenzeldehyde	
ortho-Toluic acid	4-Hydroxyacetophenone	
para-Toluic acid	Caffeic acid	
	Syringic acid	
	4-Hydroxybenzoic acid	
	4-Hydroxycoumaric acid	
	4-Hydroxycoumarin	
	Ferulic acid	
	Sinapinic acid	
	3-Hydroxy-4-methoxycinnamic acid	

systems in or near urban or intensely managed agricultural settings. In addition, sulfated metal oxides provide promising catalysts for fatty acid esterification for biodiesel production. (Kiss et al. 2007) Using these materials to improve energy production should be carefully considered to ensure that energy savings are not outweighed by adverse impacts on ecosystem services. We believe that metals from catalysts could pose soil contamination issues if bioreactor sludges are reapplied as fertilizers, as has been suggested as a way to improve the sustainability of biofuel production. Such metal effects have been documented worldwide in cases with sewage sludges (Ahmad et al. 2005; Champagne 2007; Farrell and Jones 2009; Iwegbue et al. 2007; Parker and Laha 2005; Wong 2005), petroleum refining (Chilingar and Endres 2005; Parker and Laha 2005; Wong 2005), and other waste disposal processes.

Evaluations of biomass to fuel production in the United Kingdom suggest that methane production may be a more efficient method of fuel production and would produce far less hazardous combustion emissions than bioethanol or biodiesel (Patterson et al. 2008). Such evaluations are essential to adequately determine biofuel production practices that will minimize impacts on wildlife habitat. Too few such evaluations have been completed to date.

Wood and grass that are being stored for biofuel production have been shown to contain higher levels of fungal and bacterial "dusts" than fresh biomasses do. These natural microbes may already pose risks to workers (Sebastian et al. 2006) before even considering *engineered* microbes. No research has been conducted to determine the extent of microbial releases into the ambient environment, much less to evaluate the effects that could be manifested in wildlife or the ecosystems that support them. It is well known that microbes, especially fungi, present significant risks to wildlife (see Chapter 4).

3.5 SUMMARY AND RECOMMENDATIONS

Most experts would agree that habitat loss or alteration is by far the greatest risk to species diversity and abundance and is the most significant risk of driving a species to extinction. It seems logical, therefore, that we would seek crop production processes that require less land and fewer resources. This effort is particularly important in view of the pressure to produce food and renewable biofuels with limited arable land mass.

Until we develop better crop protection alternatives, a combination of pesticides and plant genetics that have been adequately tested for safety, assessed for non-target risks, and used according to product labels represents the best practice solution to the worldwide food and fiber requirements. We are rapidly adding fuel supply to this list of necessary crops. We do not contend that there is always low risk to wildlife associated with the use of pesticides and plant genetics. But industries continue to develop pesticides that are: efficacious at lower rates, less persistent, not bioaccumulative, and better formulated. Better application methods, more easily applied, and plant genetics that are of low risk to non-target organisms are also being developed to improve natural resource stewardship. As pesticides and practices with these improvements become more commonly used by consumers who become better at following label instructions, the risk of pesticides and plant genetics to wildlife should greatly improve.

REFERENCES

Adler, P. R., M. A. Sanderson, A. A. Boateng, P. I. Weimer, and H. J. G. Jung. 2006. Biomass yield and biofuel quality of switchgrass harvested in fall or spring. *Agronomy Journal* 98 (6): 1518–1525.

Ahmad, I., S. Hayat, and J. Pichtel. 2005. *Heavy Metal Contamination of Soil: Problems and Remedies.* Enfield, NH: Science Publishers, Inc.

Aspelin, A. L. 1997. *Pesticides Industry Sales and Usage: 1994 and 1995 Market Estimates.* Washington, DC: U.S. EPA.

Aspelin, A. L., and A. H. Grube. 1999. *Pesticides Industry Sales and Usage: 1996 and 1997 Market Estimates.* Washington, DC: U.S. EPA.

Babu, R. M., A. Sajeena, K. Seetharaman, and M. S. Reddy. 2003. Advances in genetically engineered (transgenic) plants in pest management: An overview. *Crop Protection* 22:1071–1086.

Balat, M. 2007. An overview of biofuels and policies in the European union. *Energy Sources Part B – Economics Planning and Policy* 2 (2): 167–181.

Basiron, Y. 2007. Palm oil production through sustainable plantations. *European Journal of Lipid Science and Technology* 109 (4): 289–295.

Bhatti, H. N., M. A. Hanif, M. Qasim, and A. U. Rehman. 2008. Biodiesel production from waste tallow. *Fuel* 87 (13–14): 2961–2966.

Bies, L. 2006. The biofuels explosion: Is green energy good for wildlife? *Wildlife Society Bulletin* 34 (4): 1203–1205.

Brookes, G., and P. Barfoot. 2005. GM crops: The global economic and environmental impact: The first nine years, 1996–2004. *AgBioForum* 8 (2–3): 187–196.

———. 2006. Global impact of biotech crops: Socio-economic and environmental effects in the first ten years of commercial use. *AgBioForum* 9 (3): 139–151.

Bullock, D., and E. Nitsi. 2001. GMO adoption and private cost savings: GR soybeans and Bt corn. In *Genetically Modified Organisms in Agriculture: Economics and Politics*, edited by G. D. Nelson. London: Academic Press.

Burnside, O. C. 1996. An agriculturalist's viewpoint of risk and benefits of herbicide-resistant cultivars. In *Herbicide-Resistant Crops: Agricultural, Environmental, Economic, Regulatory, and Technical Aspects*, edited by S. O. Duke. Boca Raton, FL: CRC Press.

CADPR. 2008. *Summary of Pesticide Use Report Data 2007.* Sacramento: California Department of Pesticide Regulation.

Carpenter, J. E., and L. P. Gianessi. 2001. Herbicide tolerant soybeans: Why growers are adopting Roundup Ready varieties. *AgBioForum* 2 (2).

Carpita, N. C., and M. C. McCann. 2008. Maize and sorghum: Genetic resources for bioenergy grasses. *Trends in Plant Science* 13 (8): 415–420.

CFR. 2008. Experimental Use Permit data requirements for terrestrial and aquatic non target organisms. Title 40, Vol. 23§158.243. US Government Printing Services.

Champagne, P. 2007. Feasibility of producing bio-ethanol from waste residues: A Canadian perspective feasibility of producing bio-ethanol from waste residues in Canada. *Resources Conservation and Recycling* 50 (3): 211–230.

Chhetri, A. B., M. S. Tango, S. M. Budge, K. C. Watts, and M. R. Islam. 2008. Non-edible plant oils as new sources for biodiesel production. *International Journal of Molecular Sciences* 9 (2): 169–180.

Chilingar, G. V., and B. Endres. 2005. Environmental hazards posed by the Los Angeles Basin urban oilfields: An historical perspective of lessons learned. *Environmental Geology* 47 (2): 302–317.

Chisti, Y. 2008. Biodiesel from microalgae beats bioethanol. *Trends in Biotechnology* 26 (3): 126–131.

Christian, D. G., A. B. Riche, and N. E. Yates. 2008. Growth, yield and mineral content of *Miscanthus x giganteus* grown as a biofuel for 14 successive harvests. *Industrial Crops and Products* 28 (3): 320–327.

Cobb, G. P., E. H. H. Hol, L. W. Brewer, and C. M. Bens. 2003. Diazinon in apple orchards: Dissipation from vegetation and exposure to non-target organisms. *Terrestrial Field Dissipation Studies: Purpose, Design and Interpretation* 842:170–188.

Costa, L. G. 2008. Toxic effects of pesticides. In *Casarett and Doull's Toxicology: The Basic Science of Poisons*, edited by C. D. Klaassen. New York: McGraw-Hill.

Craig, W., M. Tepfer, G. Degrassi, and D. Ripandelli. 2008. An overview of general features of risk assessments of genetically modified crops. *Euphytica* 164:853–880.

Culpepper, A. S., and A. C. York. 1998. Weed management in glyphosate-tolerant cotton. *Journal of Cotton Science* 2:174–185.

Davis, A. S., D. C. Brainard, and E. R. Gallandt. 2008. Introduction to the Invasive Plant Species and the New Bioeconomy Symposium. *Weed Science* 56 (6): 866–866.

Dikshith, T. S. S., ed. 1991. *Toxicology of Pesticides in Animals*. Boca Raton, FL: CRC Press.

Dominquez-faus R., S.E. Powers, J.G. Burken, P.J. Alvarez. 2009. The water footprint of biofuels: A drink or drive issue?, *Environmental Science & Technology*, 43:3005–3010.

Donaldson, D., T. Kiely, and A. Grube. 2002. *Pesticides Industry Sales and Usage: 1998 and 1999 Market Estimates*. Washington, DC: U.S. EPA.

Dong, X. B., S. Ulgiati, M. C. Yan, X. S. Zhang, and W. S. Gao. 2008. Energy and eMergy evaluation of bioethanol production from wheat in Henan Province, China. *Energy Policy* 36 (10): 3882–3892.

Dunwell, J. M. 2000. Transgenic approaches to crop improvement. *Journal of Experimental Botany* 51:487–486.

Ecobichon, D. J. 2001. Toxic effects of pesticides. In *Casarett and Doull's Toxicology: The Basic Science of Poisons*, edited by C. D. Klaassen. New York: McGraw-Hill.

Edmonds Institute. 1998. *Manual for Assessing Ecological and Human Health Effects of Genetically Engineered Organisms*. Scientists' Working Group on Biosafety. Edmonds, WA: Edmonds Institute.

Evaristo de Jesus, K.R., A. C. Lanna, F. D. Vieira, A. Luiz de Abreu, and D. Ubeda de Lima. 2006. A proposed risk assessment method for genetically modified plants. *Applied Biosafety* 11:127–137.

FAO. 2009. The lurking menace of weeds. Food and Agriculture Organization. http://www.fao.org/news/story/en/item/29402/icode/.

Farrell, M., and D. L. Jones. 2009. Critical evaluation of municipal solid waste composting and potential compost markets. *Bioresource Technology* 100 (19): 4301–4310.

Fawcett, R., and D. Towery. 2002. Conservation tillage and plant biotechnology: How new technologies can improve the environment by reducing the need to plow. *Conservation Technology Information Center (CTIC)* 1–24.

Gao, K., and K. R. McKinley. 1994. Use of macroalgae for marine biomass production and CO_2 remediation: A review. *Journal of Applied Phycology* 6 (1): 45–60.

Garcia-Alonzo, M., E. Jacobs, A. Raybould, T. E. Nickson, P. Sowig, H. Willekens, P. van der Kouwe, et al. 2006. A tiered system for assessing the risk of genetically modified plants to non-target organisms. *Environmental Biosafety Research* 5:57–65.

Gianessi, L. P., and C. S. Silvers. 2000. *Trends in Crop Pesticide Use: Comparing 1992 and 1997*. Washington, DC: National Center for Food and Agricultural Policy.

Giddings, J. M., K. R. Solomon, and S. J. Maund. 2001. Probabilistic risk assessment of cotton pyrethroids: II. Aquatic mesocosom and field studies. *Environmental Toxicology & Chemistry* 20 (3): 660–668.

Gilliom, R.J. 2007. Pesticides in U.S. streams and groundwater. *Environmental Science & Technology* 41: 3407–3413.

Groom, M. J., E. M. Gray, and P. A. Townsend. 2008. Biofuels and biodiversity: Principles for creating better policies for biofuel production. *Conservation Biology* 22 (3): 602–609.

Groover, A. T. 2007. Will genomics guide a greener forest biotech? *Trends in Plant Science* 12 (6): 234–238.

Hancock, J. F. 2003. A framework for assessing the risk of transgenic crops. *Bioscience* 53:512–519.

Hayes, T. B., A. Collins, M. Mendoza, N. Noriega, A. A. Stuart, and A. Vonk. 2002. Hermaphroditic demasculinized frog exposure to the herbicide atrazine at low ecologically relevant doses. Paper presented at National Academy of Science.

Hayes, T. B., K. Haston, M. Tsui, A. Hoang, C. Haeffele, and A. Vonk. 2003. Atrazine-induced hermaphroditism at 0.1 ppb in American leopard frogs (*Rana pipiens*): Laboratory and field evidence. *Environmental Health Perspective* 111:568–575.

Hayes, W. J., and E. R. Laws, eds. 1991. *Handbook of Pesticide Toxicology*. San Diego: Academic Press.

Heaton, E. A., F. G. Dohleman, and S. P. Long. 2008. Meeting US biofuel goals with less land: The potential of *Miscanthus*. *Global Change Biology* 14 (9): 2000–2014.

Hela, D. G., I. K. Konstantinou, T. M. Sakellarides, D. A. Lambropoulou, T. Akriotis, and T. A. Albanis. 2006. Persistent organochlorine contaminants in liver and fat of birds of prey from Greece. *Archives of Environmental Contamination and Toxicology* 50 (4): 603–613.

Hendley, P., C. Holmes, S. Kay, S. J. Maund, K. Z. Travis, and M. Zhang. 2001. Probabilistic risk assessment of cotton pyrethroids: III. A spatial analysis of the Mississippi, USA, cotton landscape. *Environmental Toxicology & Chemistry* 20 (3): 669–678.

Herrera, A., A. A. Arino, M. P. Conchello, R. Lazaro, S. Bayarri, and C. Perez. 1994. Organochlorine pesticide-residues in Spanish meat-products and meat of different species. *Journal of Food Protection* 57 (5): 441–444.

Hill, J. 2007. Environmental costs and benefits of transportation biofuel production from food- and lignocellulose-based energy crops: A review. *Agronomy for Sustainable Development* 27 (1): 1–12.

Hill, R. A., and C. Sendashonga. 2003. General principles for risk assessment of living modified organisms: Lessons from chemical risk assessment. *Environmental Biosafety Research* 2:81–88.

Hoekman, S. K. 2009. Biofuels in the US: Challenges and opportunities. *Renewable Energy* 34 (1): 14–22.

Hokanson, K., D. Heron, S. Gupta, S. Koehler, C. Roseland, S. Shantharam, J. Turner, et al. 2008. The concept of familiarity and pest resistant plants. In *Ecological Effects of Pest Resistance Genes in Managed Ecosystems*, edited by P.L. Trayner and J.H. Westwood. Organized by Information Systems for Biology.

Iwegbue, C. M. A., F. N. Emuh, N. O. Isirimah, and A. C. Egun. 2007. Fractionation, characterization and speciation of heavy metals in composts and compost-amended soils. *African Journal of Biotechnology* 6 (2): 67–78.

James, C. 2008. Global status of commercialized biotech/GM crops: 2008. ISAAA Briefs No. 39-2009. ISAAA, Ithaca, NY. Available at http://www.isaaa.org.

Jevsnik, M., V. C. Flajs, and D. Z. Doganoc. 2004. Evidence of organochlorine pesticide and polychlorinated biphenyl residues in Slovenian poultry tissues from 1997 to 1999. *Journal of Food Protection* 67 (10): 2326–2331.

Johnson, J. M. F., D. Reicosky, B. Sharratt, M. Lindstrom, W. Voorhees, and L. Carpenter-Boggs. 2004. Characterization of soil amended with the by-product of corn stover fermentation. *Soil Science Society of America Journal* 68 (1): 139–147.

Kajikawa, Y., and Y. Takeda. 2008. Structure of research on biomass and bio-fuels: A citation-based approach. *Technological Forecasting and Social Change* 75 (9): 1349–1359.

Kalaitzandonakes, N., and P. Suntornpithug. 2001. Why do farmers adopt biotech cotton? Paper presented at Beltwide Cotton Conference.

Kendall, R. J., and T. E. Lacher, Jr., eds. 1994. *Wildlife Toxicology and Population Modeling: Integrated Studies of Agroecosystems*. Chelsea, MI: CRC Press/Lewis Publishers.

Keyzer, M., M. Merbis, and R. Voortman. 2008. The biofuel controversy. *Economist–Netherlands* 156 (4): 507–527.

Khan, M. I., A. B. Chhetri, and M. R. Islam. 2007. Analyzing sustainability of community-based energy technologies. *Energy Sources Part B – Economics Planning and Policy* 2 (4): 403–419.

Kiely, T., D. Donaldson, and A. Grube. 2004. *Pesticides Industry Sales and Usage: 2000 and 2001 Market Estimates*. Washington, DC: U.S. EPA.

Kiss, A. A., G. Rothenberg, and A. C. Dimian. 2007. 'Green' catalysts for enhanced biodiesel technology. *Catalysis of Organic Reactions* 115:405–414.

Kjaer, C., C. Damgaard, G. Kjellsson, B. Strandberg, and M. Strandberg. 1999. *Ecological Risk Assessment of Genetically Modified Higher Plants (GMHP): Identification of Data Needs*. Copenhagen: DMU National Environmental Research Institute.

Kjellsson, G. 1997. Principles and procedures for ecological risk assessment of transgenic plants. In *Methods for Risk Assessment of Transgenic Plants II: Pollination Gene-Transfer and Population Impacts*, edited by G. Kjellsson, V. Simonsen, and K. Ammann. Basel: Birkhauser Verlag.

Knauft, D.A., and F.P. Gardner. 1998. Diversity and Genetics in the Development of Crop Species. In *Principles of Ecology in Plant Production*, edited by T. R. Sinclair and T. R. Gardner. New York: CAB International.

Krieger, R., ed. 2001. *Handbook of Pesticide Toxicology*. San Diego: Academic Press.

Kudsk, P., H. R. Andersen, N. Cedergreen, S. K. Mathiassen, F. Møhlenberg, J. C. Streibig, and A. M. Vinggaard. 2005. Combined effects of pesticides. Danish Environmental Protection Agency, Pesticide Research Report No. 98, 1–100.

Kumar, N., and P. B. Sharma. 2005. *Jatropha curcus*: A sustainable source for production of biodiesel. *Journal of Scientific & Industrial Research* 64 (11):883–889.

Lal, R. 2007. Soil science and the carbon civilization. *Soil Science Society of America Journal* 71 (5): 1425–1437.

Liebig, M. A., D. L. Tanaka, J. M. Krupinsky, S. D. Merrill, and J. D. Hanson. 2007. Dynamic cropping systems: Contributions to improve agroecosystem sustainability. *Agronomy Journal* 99 (4): 899–903.

Liska, A. J., and K. G. Cassman. 2008. Towards standardization of life-cycle metrics for biofuels: Greenhouse gas emissions mitigation and net energy yield. *Journal of Biobased Materials and Bioenergy* 2 (3): 187–203.

Malhi, Y., T. Roberts, and R. A. Betts. 2008. Climate change and the fate of the Amazon: Preface. *Philosophical Transactions of the Royal Society B – Biological Sciences* 363 (1498): 1727–1727.

Mann, R.M., R.V. Hyne, C.B. Choung, and S.P. Wilson. 2009. Amphibians and agricultural chemicals: Review of the risks in a complex environment. *Environmental Pollution* 157:2903–2927.

Margarit, E., M. I. Reggiard, and H. R. Permingeat. 2008. Bt protein rhizosecreted from transgenic maize does not accumulate in soil. *Electronic Journal of Biotechnology* 11 (2): 1–10.

Mathews, J. A. 2008. Biofuels, climate change and industrial development: Can the tropical South build 2000 biorefineries in the next decade? *Biofuels Bioproducts & Biorefining – Biofpr* 2 (2): 103–125.

Matsumura, F. 1985. *Toxicology of Insecticides*. New York: Plenum Press.

Maund, S. J., K. Z. Travis, P. Hendley, J. M. Giddings, and K. R. Solomon. 2001. Probabilistic risk assessment of cotton pyrethroids: V. Combining landscape-level exposures and ecotoxicological effects data to characterize risk. *Environmental Toxicology & Chemistry* 20 (3): 687–692.

McCloud, D. E. 1998. Development of agricultural ecosystems. In *Principles of Ecology in Plant Production*, edited by T. R. Sinclair and T. R. Gardner. New York: CAB International.

McKinley, T. L., R. K. Roberts, R. M. Hayes, and B. C. English. 1999. Economic comparison of herbicides for johnsongrass (*Sorghum halepense*) control in glyphosate-tolerant soybean (*Glycine max*). *Weed Technology* 13:30–36.

Miraglia, S. G. E. 2007. Health, environmental, and economic costs from the use of a stabilized diesel/ethanol mixture in the city of São Paulo, Brazil. *Cadernos De Saude Publica* 23:S559–S569.

Mittelbach, M., B. Pokits, H. Muller, M. Muller, and D. Riesner. 2007. Risk assessment for prion protein reduction under the conditions of the biodiesel production process. *European Journal of Lipid Science and Technology* 109 (1): 79–90.

Mortensen, S. R. 2006. Toxicity of organophosphorus and carbamate insecticides using birds as sentinels for terrestrial vertebrate wildlife. In *Toxicology of Organophosphate and Carbamate Compounds*, edited by R. C. Gupta. Amsterdam: Elsevier Academic Press.

Nantha, H. S., and C. Tisdell. 2009. The orangutan-oil palm conflict: Economic constraints and opportunities for conservation. *Biodiversity and Conservation* 18 (2): 487–502.

Nauen, R. 2007. Insecticide resistance in disease vectors of public health importance. *Pest Management Science* 63:628.

Nickson, T. E. 2005. Crop biotechnology: The state of play. In *Gene Flow from GM Plants*, edited by G. M. Poppy and M. J. Wilkinson. Oxford: Blackwell Publishing.

———. 2008. Planning environmental risk assessment for genetically modified crops: Problem formulation for stress-tolerant crops. *Plant Physiology* 147 (2): 494–502.

Nickson, T. E., and M. J. Horak. 2006. Assessing familiarity: The role of plant characterization. Paper presented at Ninth International Symposium on the Biosafety of Genetically Modified Organisms, Jeju Island, Korea.

Nickson, T. E., and M. J. McKee. 2002. Ecological assessment of crops derived through biotechnology. In *Biotechnology and Safety Assessment*, edited by J. A. Thomas and R. L. Fuchs. Amsterdam: Academic Press.

Niederl, A., and M. Narodoslawsky. 2006. Ecological evaluation of processes based on by-products or waste from agriculture: Life cycle assessment of biodiesel from tallow and used vegetable oil. *Feedstocks for the Future: Renewables for the Production of Chemicals and Materials* 921:239–252.

NRC. 1989. *Field Testing Genetically Modified Organism – Framework for Decisions*. Washington, DC: National Academy Press.

———. 2000. *The Future Role of Pesticides in US Agriculture*. Washington, DC: National Academy Press.

OECD. 1986. *Recombinant DNA Safety Considerations*. Paris: Organization for Economic Cooperation and Development.

———. 1993. *Safety Consideration for Biotechnology: Scale-up of Crop Plants*. Paris: Organization for Economic Cooperation and Development.

Parikka, M. 2004. Global biomass fuel resources. *Biomass & Bioenergy* 27 (6): 613–620.

Parker, W., and S. Laha. 2005. Biosolids and sludge management. *Water Environment Research* 77:1464–1534.

Parrish, D. J., and J. H. Fike. 2005. The biology and agronomy of switchgrass for biofuels. *Critical Reviews in Plant Sciences* 24 (5–6): 423–459.

Patil, V., K. Q. Tran, and H. R. Giselrod. 2008. Towards sustainable production of biofuels from microalgae. *International Journal of Molecular Sciences* 9 (7): 1188–1195.

Paton, M. W., and D. S. Petterson. 1997. Absorption by sheep of dieldrin from contaminated soil. *Australian Veterinary Journal* 75 (6): 441–445.

Patterson, T., R. Dinsdale, and S. Esteves. 2008. Review of energy balances and emissions associated with biomass-based transport fuels relevant to the United Kingdom context. *Energy & Fuels* 22 (5): 3506–3512.

Patzek, T. W. 2004. Thermodynamics of the corn–ethanol biofuel cycle. *Critical Reviews in Plant Sciences* 23 (6): 519–567.

Pro, B. H., R. Hammerschlag, and P. Mazza. 2005. Energy and land use impacts of sustainable transportation scenarios. *Journal of Cleaner Production* 13 (13–14): 1309–1319.

Qaim, M., and D. Zilberman. 2003. Yield effects of genetically modified crops in developing countries. *Science* 299:900–902.

Raybould, A. 2006. Problem formulation and hypothesis testing for environmental risk assessments of genetically modified crops. *Environmental Biosafety Research* 5:119–125.

Raybould, A. 2007. Ecological versus ecotoxicological methods for assessing the environmental risks of transgenic crops. *Plant Science* 173:589–602.

Reed, R. L., M. A. Sanderson, V. G. Allen, and A. G. Matches. 1999. Growth and cadmium accumulation in selected switchgrass cultivars. *Communications in Soil Science and Plant Analysis* 30 (19–20): 2655–2667.

Relyea, R.A. 2009. A cocktail of contaminants: how mixtures of pesticides at low concentrations affect aquatic communities. *Oecologia* 159:363–376.

Rissler, J., and M. Mellon. 1996. *The Ecological Risks of Engineered Crops*. Cambridge, MA: MIT Press.

Romeis, J., D. Bartsch, F. Bigler, M. P. Candolfi, M. Gielkens, S. E. Hartley, R. L. Hellmich et al. 2008. Assessment of risk of insect resistant crops to nontarget arthropods. *Nature Biotechnology* 26 (2): 203–208.

Romeis, J., M. Meissle, and F. Bigler. 2006. Transgenic crops expressing *Bacillus thuringiensis* toxins and biological control. *Nature Biotechnology* 24 (1): 63–71.

Roth, A. M., D. W. Sample, C. A. Ribic, L. Paine, D. J. Undersander, and G. A. Bartelt. 2005. Grassland bird response to harvesting switchgrass as a biomass energy crop. *Biomass & Bioenergy* 28 (5): 490–498.

Rowe, R. L., N. R. Street, and G. Taylor. 2009. Identifying potential environmental impacts of large-scale deployment of dedicated bioenergy crops in the UK. *Renewable & Sustainable Energy Reviews* 13 (1): 260–279.

Runge, C. F., and B. Ryan. 2003. The economic status and performance of plant biotechnology in 2003: Adoption, research and development in the United States. Council for Biotechnology Information (CBI).

Sagar, A. D., and S. Kartha. 2007. Bioenergy and sustainable development? *Annual Review of Environment and Resources* 32:131–167.

Sage, R. F., H. A. Coiner, D. A. Way, G. B. Runion, S. A. Prior, H. A. Torbert, R. Sicher, and L. Ziska. 2009. Kudzu [*Pueraria montana* (Lour.) Merr. Variety lobata]: A new source of carbohydrate for bioethanol production. *Biomass & Bioenergy* 33 (1): 57–61.

Sallam, K. I., and A. E. M. A. Morshedy. 2008. Organochlorine pesticide residues in camel, cattle and sheep carcasses slaughtered in Sharkia Province, Egypt. *Food Chemistry* 108 (1): 154–164.

Sathre, R., and L. Gustavsson. 2006. Energy and carbon balances of wood cascade chains. *Resources Conservation and Recycling* 47 (4): 332–355.

Sawyer, D. 2008. Climate change, biofuels and eco–social impacts in the Brazilian Amazon and Cerrado. *Philosophical Transactions of the Royal Society B – Biological Sciences* 363 (1498): 1747–1752.

Scharlemann, J. P. W., and W. F. Laurance. 2008. How green are biofuels? *Science* 319:43–44.

Schmer, M. R., K. P. Vogel, R. B. Mitchell, and R. K. Perrin. 2008. Net energy of cellulosic ethanol from switchgrass. *Proceedings of the National Academy of Sciences of the United States of America* 105 (2): 464–469.

Scott, D. A., and A. Tiarks. 2008. Dual-cropping loblolly pine for biomass energy and conventional wood products. *Southern Journal of Applied Forestry* 32 (1): 33–37.

Sebastian, A., A. M. Madsen, L. Martensson, D. Pomorska, and L. Larsson. 2006. Assessment of microbial exposure risks from handling of biofuel wood chips and straw: Effect of outdoor storage. *Annals of Agricultural and Environmental Medicine* 13 (1): 139–145.

Senseman, S. A., and K. Armbrust. 2007. *The Herbicide Handbook 2007*. Laurence, KS: Weed Science Society of America.

Shelton, A. M., J-Z. Zhou, and R. T. Roush. 2002. Economic, ecological, food safety and social consequences of the deployment of bt transgenic plants. *Annual Review of Entomology* 47:845–881.

Silvers, C. S., L. P. Gianessi, J. E. Carpenter, and S. Sankula. 2003. Current and potential role of transgenic crops in U.S. agriculture. *J. Crop Production* 9:501–530.

Sims, R. E. H., A. Hastings, B. Schlamadinger, G. Taylor, and P. Smith. 2006. Energy crops: Current status and future prospects. *Global Change Biology* 12 (11): 2054–2076.

Smith, R. G., K. L. Gross, and G. P. Robertson. 2008. Effects of crop diversity on agroecosystem function: Crop yield response. *Ecosystems* 11 (3): 355–366.

Solomon, K. R., J. M. Giddings, and S. J. Maund. 2001. Probabilistic risk assessment of cotton pyrethroids: I. Distributional analyses of laboratory aquatic toxicity data. *Environmental Toxicology & Chemistry* 20 (3): 652–659.

Solomon, K. R., J. A. Carr, L. H. Du Preez, J. P. Giesy, R. J. Kendall, E. E. Smith, and G. J. Van Der Kraak. 2008. Effects of atrazine on fish, amphibians, and aquatic reptiles: A critical review. *Critical Reviews in Toxicology* 38:721–772.

Stephenson, A. L., J. S. Dennis, and S. A. Scott. 2008. Improving the sustainability of the production of biodiesel from oilseed rape in the UK. *Process Safety and Environmental Protection* 86 (B6): 427–440.

Tank, S. L., L. W. Brewer, M. J. Hooper, G. P. Cobb, and R. J. Kendall. 1993. Survival and pesticide exposure of northern bobwhites (*Colinus virginianus*) and eastern cottontails (*Sylvilagus floridanus*) on agricultural fields treated with Counter (R) 15g. *Environmental Toxicology & Chemistry* 12 (11): 2113–2120.

Tapp, H., and G. Stotzky. 1998. Persistence of the insecticidal toxin from *Bacillus thuringiensis* subsp. *kurstaki* in soil. *Soil Biology and Biochemistry* 30 (4): 471–476.

Thomas, S., D. Burke, M. Gale, M. Lipton, and A. Weale. 2009. The use of genetically modified crops in developing countries: A follow-up discussion paper. Nuffield Council on Bioethics. 20032009. Available from http://www.nuffieldbioethics.org/home/.

Thomason, W. E., W. R. Raun, G. V. Johnson, C. M. Taliaferro, K. W. Freeman, K. J. Wynn, and R. W. Mullen. 2004. Switchgrass response to harvest frequency and time and rate of applied nitrogen. *Journal of Plant Nutrition* 27 (7): 1199–1226.

Thomsen, M. H., and H. Hauggaard-Nielsen. 2008. Sustainable bioethanol production combining biorefinery principles using combined raw materials from wheat undersown with clover-grass. *Journal of Industrial Microbiology & Biotechnology* 35 (5): 303–311.

Tiedje, J. M., P. C. Wong, K. A. R. Mitchell, W. Eberhardt, Z. G. Fu, and D. Sondericker. 1989. The planned introduction of genetically engineered organism: Ecological considerations and recommendations. *Ecology* 70:298–315.

Tomic, F., T. Kricka, and S. Matic. 2008. Available agricultural areas and the use of forests for biofuel production in Croatia. *Sumarski List* 132 (7–8): 323–330.

Travis, K. Z., and P. Hendley. 2001. Probabilistic risk assessment of cotton pyrethroids: IV. Landscape-level exposure characterization. *Environmental Toxicology & Chemistry* 20 (3): 679–686.

Tsai, W. T., C. C. Lin, and C. W. Yeh. 2007. An analysis of biodiesel fuel from waste edible oil in Taiwan. *Renewable & Sustainable Energy Reviews* 11 (5): 838–857.

Ugarte, D. D., and L. X. He. 2007. Is the expansion of biofuels at odds with the food security of developing countries? *Biofuels Bioproducts & Biorefining – Biofpr* 1 (2): 92–102.

Ulgiati, S. 2001. A comprehensive energy and economic assessment of biofuels: When 'green' is not enough. *Critical Reviews in Plant Sciences* 20 (1): 71–106.

UNEP. 1995. *UNEP International Technical Guidelines for Safety in Biotechnology*. Edited by U. N. E. Program.

USEPA. 1998. *Guidelines for Ecological Risk Assessment*. Washington, DC: U.S. Environmental Protection Agency.

———. 2003. *Refined (Level II) Terrestrial and Aquatic Models Probabilistic Ecological Assessments for Pesticides: Terrestrial*. Washington, DC: U.S. Environmental Protection Agency, FIFRA Science Advisory Panel.

———. 2006. *Event MIR604 Modified Cry3A Protein Bt Corn – Plant-Incorporated Protectant*. Washington, DC: U.S. Environmental Protection Agency, FIFRA Science Advisory Panel.

———. 2007. *White Paper on the Potential for Atrazine to Affect Amphibian Gonadal Development*. Washington, DC: U.S. Environmental Protection Agency, Office of Prevention, Pesticides, and Toxic Substances, Office of Pesticide Programs.

———. 2009. *What Is a Pesticide?* Washington, DC: U.S. Environmental Protection Agency.

USGS. 2008. *Wildlife Mortality Information: Wildlife Organophosphate/Carbamate Poisoning (OP/CARB)*. Washington, DC: U.S. Geological Survey.

van Haveren, J., E. L. Scott, and J. Sanders. 2008. Bulk chemicals from biomass. *Biofuels Bioproducts & Biorefining – Biofpr* 2 (1): 41–57.

Vandenhove, H., F. Goor, S. O'Brien, A. Grebenkov, and S. Timofeyev. 2002. Economic viability of short rotation coppice for energy production for reuse of caesium-contaminated land in Belarus. *Biomass & Bioenergy* 22 (6): 421–431.

Vane, L. M., and F. R. Alvarez. 2008. Membrane-assisted vapor stripping: Energy efficient hybrid distillation-vapor permeation process for alcohol-water separation. *Journal of Chemical Technology and Biotechnology* 83 (9): 1275–1287.

Walker, M. 2009. The world's new numbers. *Wilson Quarterly*.

Wang, H. L., S. L. Brown, G. N. Magesan, A. H. Slade, M. Quintern, P. W. Clinton, and T. W. Payn. 2008. Technological options for the management of biosolids. *Environmental Science and Pollution Research* 15 (4): 308–317.

Ware, G. W., and D. M. Whitacre. 2004. *The Pesticide Book*, 6th ed. Willoughby, OH: MeisterPro Information Resources.

Westin, D. P., R. W. Holmes, J. You, and M. J. Lydy. 2005. Aquatic toxicity due to residential use of pyrethroid insecticides. *Environmental Science Technology* 39 (24): 9778–9784.

Wicke, B., V. Dornburg, M. Junginger, and A. Faaij. 2008. Different palm oil production systems for energy purposes and their greenhouse gas implications. *Biomass & Bioenergy* 32 (12): 1322–1337.

Wilkinson, M. J., J. Sweet, and Poppy G. M. 2003. Risk assessment of GM plants: Avoiding gridlock? *Trends in Plant Science* 8:208–212.

Wisz, M., N. Dendoncker, J. Madsen, M. Rounsevell, M. Jespersen, E. Kuijken, W. Courtens, C. Verscheure, and F. Cottaar. 2008. Modelling pink-footed goose (*Anser brachyrhynchus*) wintering distributions for the year 2050: Potential effects of land-use change in Europe. *Diversity and Distributions* 14 (5): 721–731.

Wolf, A., A. Vidlund, and E. Andersson. 2006. Energy-efficient pellet production in the forest industry: A study of obstacles and success factors. *Biomass & Bioenergy* 30 (1): 38–45.

Wong, P. K. 2005. Practical issues of land application of biosolids. *Heavy Metal Contamination of Soil*. 1–23.

Yu, S. J. 2008. *The Toxicology and Biochemistry of Insecticides*. Boca Raton, FL: CRC Press, Taylor & Francis Group.

Zah, R., H. Böni, M. Gauch, R. Hischier, M. Lehmann, and P. Wäger. 2007. *Ökobilanz von Energieprodukten*. St. Gallen, Switzerland: Empa.

4 Influence of Pesticides and Environmental Contaminants on Emerging Diseases of Wildlife

Steven M. Presley, Galen P. Austin, and C. Brad Dabbert

CONTENTS

4.1 INTRODUCTION

As a result of increasing awareness and significant concern for public health, global efforts to investigate the rapidly growing numbers of emerging infectious diseases of humans have been ongoing for almost two decades. Equally concerning and oftentimes directly associated with the emergence of human pathogens are the rapid emergence and resurgence of diseases affecting wild and domestic species of animals. Many of these emerging diseases of animals are zoonoses (diseases normally existing in animal populations but capable of being transmitted to humans) and pose significant threats to human health, agricultural animal production, and wildlife ecology. Numerous factors are believed to be directly and indirectly contributing to the rapid increase in emerging diseases of wildlife, including ecosystem disruption, anthropogenic influences such as herbicide and pesticide usage, and other environmental contaminants.

Our use of the term *wildlife*, particularly for this chapter, refers primarily to vertebrate species, comprising amphibians, birds, fishes, mammals, and reptiles. Even though invertebrates such as arthropods are often involved to some extent in the maintenance and transmission dynamics of wildlife diseases, and may actually be the disease-causing organism in some cases, this chapter focuses on the vertebrate host species that may suffer the effects of the disease, as well as serve as an amplifying host or reservoir of pathogenic agents. Certainly wildlife populations serve as critical and essential components of any ecosystem, with each species occupying a certain niche and contributing to the overall integrity of the system. Similarly, diverse wildlife populations within an area or region may be critical and essential components for the maintenance and transmission dynamics of infectious pathogens that may occur, or are sustained, within an area or region.

Wildlife populations susceptible to or infected with infectious disease-causing pathogens are not limited to remote rural areas, pristine natural woodlots or forests, coastal marshlands, and so on, but occur in almost every landscape, including densely populated urban areas, suburban areas, and pastoral farm settings. Because many wildlife species are highly mobile, with some species having extensive territorial ranges, infected animals often serve as a means for the spread and introduction of the pathogen into naïve animal populations and geographic areas.

The exposure of wildlife to pesticides is not limited to direct contact through topical absorption, ingestion, or inhalation of toxicants applied to agricultural cropland, domiciles, golf courses, lawns, livestock, etc., but also through accidental or inadvertent introduction into the environment. Additionally, in many instances the benefits of pesticide use (such as improved public human health, increased food production, and avoidance of pestiferous arthropods) may or may not outweigh the costs resulting from pesticide use (such as toxicity, morbidity, and mortality in non-target organisms). An example of unintended consequences of pesticide usage to improve human health is the extensive distribution and use of permethrin (and other related compounds) to treat bednets for malaria (*Plasmodium* spp.) prevention and control throughout Africa, Asia, and South America. In many countries and regions where pesticide-treated bednets are distributed, poverty is rampant, and the bednets may be utilized for other purposes, such as seining fish. Also, people will launder their bednets in local streams, rivers, or other fresh-water sources. Permethrin and other pyrethroid insecticides are very toxic to fishes (NPIC 2009).

We discuss here the importance of healthy wildlife populations to many aspects of our lives, our society, and the environment. Specific areas addressed include human health, economic, recreational, and aesthetics, as well as the broader ecosystem function at large. We review the history of various diseases of wildlife, identify several highly relevant currently emerging and resurgent diseases of wildlife, and discuss how such diseases may affect ecosystems and wildlife population health. We review and characterize anthropogenic and naturally occurring factors that may be contributing to the emergence of new diseases in wildlife species and the expansion of such diseases into new geographic areas. Anthropogenic factors such as habitat disruption, modification, and elimination and the use of pesticides and other contaminants in the environment are evaluated as related to enhancing the susceptibility of wildlife populations to emerging and resurgent diseases. We discuss the various direct and indirect physiological and behavioral effects pesticides and other environmental contaminants may have on wildlife.

Additionally, the potential effects and influence of predicted global climate changes on the susceptibility of wildlife and transmission of wildlife diseases are presented. Finally, we discuss the potential long-term effects of and the inseparable relationships between emerging diseases of wildlife, domestic animals, and humans. Specific examples of complex relationships that exist between anthropogenic factors and emerging wildlife diseases discussed in this chapter include (i) wildlife and its role in transmission of human and domestic animal diseases (e.g., brucellosis and tuberculosis), (ii) indirect and collateral effects of insecticidal treatment of domiciles on vectors and reservoirs of arthropod-borne zoonoses, and (iii) increased potential for human and domestic animal exposure to emerging wildlife diseases as a result of natural wildlife movement as well as habitat encroachment.

4.1.1 Importance of Healthy Wildlife

Beyond the importance of healthy wildlife populations in maintaining ecosystem functions, such as decomposers, pollinators, predators, prey, scavengers, and soil aerators, wildlife directly and indirectly affect human health and domestic animal health, as well as the economic vitality of a locale, region, or country. Wildlife also greatly enhance the aesthetic beauty and enjoyment of the outdoor environment for many people. An unhealthy wildlife population, whether caused by malnutrition or by infectious diseases, causes a cascading negative effect on other wildlife species, as well as on the overall faunal and floral components of the ecosystem within which it functions.

Because many people enjoy observing, interacting with, hunting, and consuming wildlife, there is a significant potential that humans can be exposed to and infected with disease-causing zoonotic pathogens from infected animals. Additionally, domestic animals, including pets, service animals, sporting and working animals, and livestock species, are susceptible to many (if not most) of the diseases that infect wildlife species. These domestic species can become infected, causing significant economic and performance losses, and once infected can potentially transmit pathogens to humans. The potential for transmission of disease-causing pathogens from wildlife to people or domestic animals is not limited to direct animal-to-animal transmission, but also includes indirect transmission of the infectious agent from animal-to-intermediary then from intermediary-to-animal. The intermediary may be animate or inanimate (e. g., a surface or water). An example of this indirect transmission route can be related through a scenario in which great-tailed grackles (*Quiscalus mexicanus*), infected with a gastrointestinal bacterial pathogen (such as *Salmonella* spp.), feed and water from a pet dog's dishes or defecate on the dishes left outdoors while the dog is not present. When the dog returns to the dishes and eats and drinks, the pathogen-contaminated material is ingested and the dog becomes infected.

4.1.2 Human Health

An emerging disease is one that was previously unrecognized as causing disease in a particular species or geographic area, or is a previously recognized disease in new host species, or is a previously recognized disease occurring in a new geographic area. Zoonoses is diseases that typically occur in, are transmitted among, and are maintained in animal populations, but the causative pathogen may be transmitted to and cause disease in humans.

There is increasing interest in understanding the ecology of zoonotic disease transmission and how such transmission is influenced by anthropogenic modifications and influences on the environment (Grenfell and Harwood 1997). Human-induced changes in the environment, as well as the multitude of determinants directly and indirectly influenced by climate changes, alter disease dynamics by changing relationships between pathogens, hosts, and vectors, and by altering the spatial context in which these relationships occur. The accelerating pace of anthropogenic environmental change, coupled with the current threat of bioterrorism, has placed a premium on knowledge describing and predicting the dynamics and persistence of zoonotic pathogens, their vectors, and their reservoir/amplifying hosts as they relate to human risk.

More than 1,415 species of infectious agents are known to cause disease in humans, including 538 bacteria and rickettsia, 307 fungi, 287 helminthes, 66 protozoa, and 217 viruses and prions. More than 61% (868/1,415) of those infectious agents are zoonoses, and approximately 13% (175/1,415) are considered to be emerging pathogens. About 75% (132/175) of the recognized emerging pathogens are considered to be zoonoses, and 28% (37/132) of those emerging zoonotic pathogens are transmittable to humans by arthropod vectors (such as fleas, flies, mosquitoes, and ticks) (Taylor et al. 2001).

4.1.3 Agriculture and Domestic Animal Health

The Earth's surface area is 510,072 million square kilometers, of which only 29.1% is land; the other 70.9% is covered with water. Only 11.6% of Earth's land surface is usable for crop production, either in arable crop production (10.6%) or permanent crop production (1%) (CIA 2009). The other 88.4% of the Earth's land surface is either inhabited by humans or not suitable for crop production (such as forest, woodland, grassland, pastureland, arctic tundra, ice covered, swamp, and desert). However, some 26% of Earth's land surface that is not suitable for crop production, much of which is in semi-arid or arid regions, is used for grazing by livestock. Approximately 33% of the worldwide protein consumption is derived from animal agriculture (FAO 2006). Ruminant animals harvest forage, indigestible by humans, and convert it into a nutritious source of protein for human consumption on land otherwise not suitable for food crop production. While agriculture accounts for only 4% of the global gross domestic product, it employs 40% of the global labor force (CIA 2009). The animal

TABLE 4.1
Worldwide Livestock Production Estimates (2007)

Species	Head (× 1000)
Chickens	17,251,899
Ducks	1,069,108
Geese/Guinea fowls	334,851
Turkeys	470,077
Buffaloes	202,387
Camels	22,009
Cattle	1,389,590
Sheep	1,112,521
Goats	850,220
Pigs	989,884
Horses	58,409

Source: Adapted from FAO (2006).

agriculture industry employs approximately 1.3 billion people worldwide (FAO 2006). Table 4.1 provides recent estimates of worldwide livestock production.

The animal agriculture industry utilizes primarily two production methods: animals raised on pasture, and animals confined in pens or buildings for growing and finishing. Large-scale operations of the latter are referred to as confined or concentrated animal feeding operations, or CAFOs. While confinement in a CAFO affords some protection for animals from diseases, depending on the type of confinement implemented, they are still not completely protected from contact with wildlife or arthropod vectors of diseases. Animals reared on pasture or grasslands throughout the world cohabitate these areas with wildlife. Numerous studies have also documented inter-specific utilization of land by wildlife and domestic livestock species. Salter and Hudson (Salter 1980) reported observation of site use overlap by feral horses (*Equus ferus caballus*), cattle (*Bos primigenius tarus*), mule deer (*Odocoileus hemionus*), white-tailed deer (*O. virginianus*), elk (*Cervus canadensis*), and moose (*Alces alces*), and Barrett (1982) reported observation of site use overlap by cattle, deer, and feral swine (*Sus domesticus*). McInnis and Vavra (1987) reported observation of site-use overlap by feral horses, cattle, and pronghorn antelope (*Antilocapra americana*).

Menzano et al. (2007) reported inter-specific transmission of sarcoptic mange (*Sarcoptes scabiei*) from wild chamois (*Rupicapra rupicapra*) to a domestic goat (*Capra aegagrus hircus*) herd in the Italian Alps resulting from sympatric use of bedding areas and artificial salt licks placed by the goat farmer. The goat farmer and some members of his family, as well one of the authors and a Tarvisio Forest staff member, also contracted pseudoscabies from handling the goats. Gipson et al. (1999) reported feral swine fence-line contact with penned domestic swine (*Sus scrofa domesticus*) and incursion into pens containing domestic sows by feral swine boars.

Wyckoff et al. (2005) reported that GPS-collared feral swine occurred in close proximity to a domestic swine operation. In another study, Deck (2006) reported inter-specific use of pastures by cattle and feral swine by utilizing GPS collars and video cameras. These are just a few examples of the inevitability of direct or indirect contact between wildlife and domestic livestock and the opportunity for inter-specific disease transmission, which may occur through aerosols, insect vectors, or contact with or ingestion of contaminated feed, forage, water, or animal matter.

The enormous significance of the animal agriculture industry to the economic health and vitality of the global economy highlights the importance of prevention and early detection of diseases of both wildlife populations and agriculture-related animals. Additionally, foreign animal diseases (FAD) and other emerging and resurgent diseases of wildlife and livestock increasingly pose a real threat to the integrity and strength of the agriculture industry. Recent international events associated with unintentionally introduced livestock pathogens provide a realistic insight into the potential economic impact on agriculture globally. A 1997 outbreak of foot-and-mouth disease (FMD–*Aphthovirus*) in Taiwan spread to more than 6,000 farms in only 3 months, and resulted in the destruction of 3.8 million swine and a loss of approximately $5 billion US, as well as the total shutdown of swine exports from that country. Following the 2001 outbreak of FMD in Great Britain, the Purdue Agricultural Economics Report (5/01) stated that while "no one can come up with a specific dollar cost for the FMD outbreak … estimates range from $2 to 24 billion [USD]." More recently and within the United States (December 2003), the detection in Washington State of one Holstein cow initially imported from Canada with mad cow disease (bovine spongiform encephalopathy—BSE) caused massive disruption within the cattle industry, including a drop in live cattle prices of about 15% during the week immediately following detection. The resultant impact on the export market for U.S. beef and beef products was devastating.

The interactions between wildlife and domestic livestock is an obvious potential route of exposure and transmission of disease-causing pathogens between species, such as brucellosis-infected white-tailed deer grazing the same pasture as cattle or sheep. There are several less obvious or apparent pathogen transmission scenarios that have been brought to light in recent years. The *Escherichia coli* O157 outbreak in 2006 resulting from consumption of packaged spinach from California was initially traced back to cattle on an adjacent ranch, and subsequently *E. coli*-infected free-ranging feral swine were implicated in the outbreak. The cattle did not have access to the spinach fields, but feral swine did. Evidence (feces, rooting, and tracks) of their incursion into the spinach fields was found, and thus they were implicated as a possible source of contamination (Jay et al. 2007). Another scenario occurring in 2008 was the contamination of lettuce and other vegetables with *E. coli* by way of dairy cattle manure and contaminated facility water run-off.

Tuberculosis (TB) is a disease caused by *Mycobacterium* spp. bacteria; the primary causative pathogen in livestock and wildlife is *M. bovis*. Though eradication efforts have been ongoing in the United States since 1917, TB still persists. Even though most states in the United States are "free" of TB, routine surveillance occasionally discovers infected animals. Michigan is a state where eradication efforts have been hampered because white-tailed deer populations have a high prevalence of TB (Kaneene et al. 2002; Schmitt et al. 2002). Because of the difficulty of preventing contact between cattle and white-tailed deer, eradication of TB in Michigan has become a very daunting, if not impossible, task. Great Britain also serves as an example of how TB cycles between livestock and wildlife. Efforts to eradicate TB in cattle in Great Britain have been hampered because of the prevalence of TB in badgers (*Meles meles*). Numerous studies have shown that badgers frequent cattle feeding and watering points and feed storage areas (Garnett et al. 2002; Gormley and Costello 2003). A molecular epidemiological study of TB in several game reserves in South Africa suggests that TB in buffalo originated in cattle herds surrounding the reserves (Michel et al. 2009). As recently as June 2009, TB was reported in cattle in Nebraska and Texas (NDA 2009; TAHC 2009).

Brucellosis is a disease caused by *Brucella* spp. bacteria, with the most prevalent among livestock and wildlife in the United States being *B. abortus* and *B. suis*. In 1934 the U.S. Department of Agriculture began coordinated efforts to eradicate brucellosis in the United States, and as of 2008 most states were classified as "free" of brucellosis. While brucellosis has been eradicated in most U.S. states, it still persists in bison and elk herds in the Yellowstone National Park areas in the states of Montana and Wyoming, and continues to pose a threat to livestock, particularly cattle, pastured around the park (APHIS 2007).

Wildlife can and often do serve as reservoirs, amplifying hosts, and transmitters of infectious pathogens to domestic animals. The importance of animal agriculture to the health and nutrition of humans and the economy of nearly all nations is highly significant. There are 22 disease-causing organisms and microorganisms that are considered to pose the greatest threat to agriculture and the U.S. economy (see Table 4.2). Various wildlife species may play a significant role in the maintenance and transmission dynamics of these pathogens to domestic animals.

TABLE 4.2
Disease-Causing Organisms and Microorganisms Posing the Greatest Threat to U.S. Agriculture and the Economy (Listed in Order of Threat Status)

1. Foot-and-mouth disease	12. Vesicular stomatitis virus
2. Classical swine fever virus	13. Porcine enterovirus 1
3. African swine fever virus	14. Porcine enterovirus 9
4. Rinderpest virus	15. Lyssa viruses and rabies viruses
5. Rift Valley fever virus	16. Lumpy skin disease virus
6. Avian influenza virus	17. Porcine reproductive and respiratory syndrome virus
7. Newcastle disease virus	18. African horse sickness virus
8. Venezuelan equine encephalomyelitis virus	19. Anthrax
9. Bluetongue virus	20. *Chlamydia psittaci*
10. Sheep & goat pox virus	21. Heartwater, cowdriosis
11. Pseudorabies virus	22. Primary Screwworm

Source: Adapted from Wilson, T. M., et al. *Emerging Diseases of Animals*, edited by C. Brown and C. Bolin. Washington, DC: ASM Press.

4.2 RELEVANT EMERGING/RESURGENT WILDLIFE DISEASES AND THEIR SIGNIFICANCE

There are numerous factors that contribute to the phenomenal rate of increase in the number of emerging and resurgent diseases in both humans and animals. These factors are complex and integrated, but can basically be attributable to the ever-increasing size and "footprint" of the human population, particularly through urbanization of wildlife habitat and expansion of agricultural production areas.

The emergence and expansion of chronic wasting disease (CWD) of cervids is an example of how rapid and devastating an infectious disease of wildlife can be. Following initial recognition of CWD infections in mule deer in Colorado during the late 1960s, the disease has rapidly spread geographically throughout the central regions of the United States and Canada, and into black-tailed deer, white-tailed deer, and elk (Table 4.3).

Increasing attention and concern is being focused upon emerging and resurgent diseases globally. A significant number of emerging and resurgent diseases of wildlife are zoonoses, and parallel to emerging diseases only of humans, diseases infecting both domestic and wildlife species are emerging at an alarming rate. In particular, among all emerging infectious diseases, emerging and resurgent zoonotic pathogens have been determined to be the most abundant and of greatest concern (Epstein et al. 1998; Epstein 2001). Additionally, numerous zoonoses are considered to be potential bioterrorism agents, including plague and tularemia (Childs et al. 1998; Gubler 1998; O'Toole et al. 2002; Rotz et al. 2002).

The bacterium that causes plague, *Yersinia pestis*, is identified as a Category A biological threat agent by the U.S. Centers for Disease Control and Prevention (CDC) (Rotz et al. 2002). Human and animal exposure to *Y. pestis* typically involves the bite of an arthropod vector or direct contact with infective fluids or tissues from an infected host. The maintenance and transmission dynamics of sylvatic or enzootic plague typically involves rodent reservoir species and fleas that vector the bacteria between infective and uninfected hosts. Humans and other mammals are typically exposed to the pathogen through the bite of a flea. An estimated 1,700 cases of human plague occur annually worldwide, according to 50 years of data (Perry and Fetherston 1997). Plague infection in humans presents as bubonic, septicemic, or pneumonic, with bubonic progressing to septicemic and ultimately to pneumonic infection if untreated. Environmental and ecological determinants, particularly seasonal temperatures, precipitation, vegetation, predator species, and anthropogenic factors, may significantly influence the susceptibility and potential for epizootic events (die-offs).

Rodent-consuming predators also play a significant role in the transmission and spread of zoonotic pathogens. This transmission mechanism for plague was reported by Gage and Montenieri (1994), although secondary hosts including predators were found to spread *Y. pestis* only in the presence of a primary host such as prairie dogs (*Cynomys* spp.) (Biggins and Kosoy 2001). Predator-prey interactions through scavenging or hunting within infected colonies, whether active or abandoned, can result in transmission by way of infected fleas or inhalation of infected respiratory secretions from infected animals.

TABLE 4.3
Chronology of Significant Events in the History of Chronic Wasting Disease

Year	Event
Late 1960s	First recognition of a clinical syndrome termed "chronic wasting disease" in captive mule deer in Colorado
1977/1978[a]	Diagnosis of CWD in mule deer as a spongiform encephalopathy
1978/1979[a]	Diagnosis of CWD in captive mule deer and black-tailed deer in Wyoming
1979	Diagnosis of CWD in captive Rocky Mountain elk
late 1970s	Diagnosis of spongiform encephalopathy in captive mule deer in a zoo in Ontario (CWD did not persist in this location)
1980	First published report of CWD in captive mule deer
1981	Diagnosis of CWD in free-ranging Rocky Mountain elk in Colorado
1982	First published report of CWD in Rocky Mountain elk
1983	Start of hunter-harvest surveillance for CWD
1985	Diagnosis of CWD in free-ranging mule deer
1990	Diagnosis of CWD in free-ranging white-tailed deer
1992	First published report of CWD in free-ranging cervids
1996	Diagnosis of CWD in game farm elk in Saskatchewan
1997	Diagnosis of CWD in game farm elk in South Dakota
2000/2001[a]	Diagnosis of CWD in free-ranging mule deer in Saskatchewan, possibly associated with CWD-affected elk farm
2000/2001[a]	Diagnosis of CWD in free-ranging mule deer in Nebraska contiguous with the CWD endemic area of Colorado and Wyoming
2001	Extensive depopulation of game-farm elk in Saskatchewan due to CWD
2001	Diagnosis of CWD in an elk imported from Canada to Korea in 1997
2001	Declaration of a USDA animal emergency because of CWD in game farm elk
2002	Diagnosis of CWD in free-ranging deer associated with an affected game farm in Nebraska
2002	Diagnosis of CWD in free-ranging white-tailed deer in Wisconsin
2002	Diagnosis of CWD in a free-ranging mule deer in New Mexico

[a] Year samples collected/year diagnosis confirmed.

Source: Adapted from Williams, E. S., et al. *Journal of Wildlife Management* 66, 2002.

Plague-infected fleas have been recovered from prairie dog burrows 3 months to 1 year after the disappearance of prairie dogs (Lechleitner 1968; Cully et al. 1997). Among predators, seroconversion occurs in dogs (*Canis lupus familiaris*), foxes (*Vulpes vulpes*), coyotes (*C. latrans*), skunks (*Mephitis mephitis*), and raccoons (*Procyon lotor*) after exposure to *Y. pestis*, but mortality and morbidity are rare among these species (Barnes 1982). Conversely, domestic and feral cats (*Felis catus*) are unique among carnivores in that they are highly susceptible to plague, with published mortality rates of approximately 33% (Rust 1971; Eidson et al. 1991; Gasper et al. 1993). Additionally, humans can be exposed to *Y. pestis* via pneumonic transmission from cats. During the period 1947–2000 there were 16 cases of human plague resulting from exposure to infected cats, 4 of those developing as primary pneumonic plague (Werner et al. 1984; Eidson et al. 1988; Gasper et al. 1993; Gage et al. 1995; CDC 1997; Gage et al. 2000). In a case of fatal human pneumonic plague acquired from a domestic cat reported by Doll et al. (1994), no *Y. pestis*–infected fleas or rodents were found within the immediate proximity of the victim's dwelling. However, infected fleas and a ground squirrel carcass were recovered approximately 1 kilometer from the dwelling. Additionally, the victim's dog was found to have a positive titer (1:64) for *Y. pestis* antibodies, indicating that domestic predators, particularly in this case, a cat, are a critical link between plague epizootics and humans.

Table 4.4 provides a partial listing of several other globally emerging zoonoses in which wildlife may play a role in the natural enzootic transmission dynamics and lists factors that may be contributing to their respective emerging status.

TABLE 4.4
Global Emerging and Resurgent Zoonotic Diseases Involving Wildlife, and Factors Associated with Emergence and Resurgence

Disease	Pathogenic Organism	Principal Wildlife Species Involved in Transmission (vector, if any)	Geographic Distribution, and Regions of Emergence	Factors Associated with Emergence and Spread
		Bacterial Diseases		
Anthrax	*Bacillus anthracis*	Wild herbivores	Panglobal; Africa, Asia, South America, eastern Europe	Habitat disruption and wild host dispersion
Borreliosis (including Lyme disease)	*Borrelia* spp.	Wild rodents, deer (ticks and mites)	Panglobal	Increased populations of vectors associated with El Niño Southern Oscillation-related weather patterns
Brucellosis	*Brucella abortus, B. melitensis, B. canis*	Bison, elk, caribou, goats and sheep, canids (filth flies mechanically)	Panglobal	Habitat disruption and wild host dispersion, increased animal interactions (Merck 2005)
Plague	*Yersinia pestis*	Wild rodents and wide range of other wild mammals (fleas)	Panglobal, particularly India and SW U.S.A.	Enzootic foci disruption (Cully et al. 1997)

TABLE 4.4 (CONTINUED)
Global Emerging and Resurgent Zoonotic Diseases Involving Wildlife, and Factors Associated with Emergence and Resurgence

Disease	Pathogenic Organism	Principal Wildlife Species Involved in Transmission (vector, if any)	Geographic Distribution, and Regions of Emergence	Factors Associated with Emergence and Spread
Tuberculosis	*Mycobacterium tuberculosis*	Monkeys and other non-human primates	Panglobal, particularly Africa and SE Asia	Encroachment and disruption of habitat
	Mycobacterium bovis	Cattle and other ungulates	Panglobal	Livestock production and transportation practices
Tularemia	*Francisella tularensis*	Wild rabbits, rodents, cats (ticks, mosquitoes, and other hematophagous arthropods)	Americas, Europe, Asia	Habitat disruption and dispersion of wild hosts and vector species
		Protozoan Diseases		
African sleeping sickness (African trypanosomiasis)	*Trypanosoma brucei*	Wild dogs, ruminants, hyenas, carnivores (tsetse flies, *Glossina* spp.)	Africa	Disruption of habitat and dispersal of hosts and vectors
Chagas' disease (American trypanosomiasis)	*Trypanosoma cruzi*	Armadillos, bats, rodents, wild dogs and cats (triatomid bugs)	Western hemisphere, Americas, particularly SW U.S.A., Mexico	El Nino Southern Oscillation-related weather patterns
Cryptosporidiosis	*Cryptosporidium parvum*	Wild rodents and other mammals	Europe and North America	Farming practices, and cross-species transfer (Griffiths 1998)
Pulmonary protozoan infection	*Pneumocystis carinii*	Voles and shrews	Panglobal	Seasonal changes in immunity or density

TABLE 4.4 (CONTINUED)
Global Emerging and Resurgent Zoonotic Diseases Involving Wildlife, and Factors Associated with Emergence and Resurgence

Disease	Pathogenic Organism	Principal Wildlife Species Involved in Transmission (vector, if any)	Geographic Distribution, and Regions of Emergence	Factors Associated with Emergence and Spread
		Other Parasitic Diseases		
Intestinal nematode	*Trichostrongylus* spp.	Wild ruminants and rabbits	Panglobal	Seasonal changes in host species immunity or density (Cattadori et al. 2005)
Sarcoptic mange (acariasis)	*Sarcoptes scabiei*	Wild canids	Australia, U.K., Sweden	Dispersal of infected animals; interactions between domestic and wild canids (Lindstrom et al. 1994)
		Viral Diseases		
Ebola and Marburg hemorrhagic fever viruses	*Filoviruses*	Chimpanzees, gorillas, insectivorous and fruit bats suspected (unknown)	Sub-Saharan Africa, Indonesia, Philippines	Increased aggregation of insects and mammals; animal dislocation (Pinzon et al. 2004)
Hantavirus pulmonary syndrome	*Bunyaviruses*	*Peromyscus* spp. and other wild rodents	Americas	El Niño Southern Oscillation-related weather patterns (Trevejo et al. 1998)
Rabies	*Lyssaviruses*	Skunks	Panglobal; particularly prevalent in Africa, Asia, eastern Europe, South America	Dispersal and seasonal weather patterns (Guerra et al. 2003)

TABLE 4.4 (CONTINUED)
Global Emerging and Resurgent Zoonotic Diseases Involving Wildlife, and Factors Associated with Emergence and Resurgence

Disease	Pathogenic Organism	Principal Wildlife Species Involved in Transmission (vector, if any)	Geographic Distribution, and Regions of Emergence	Factors Associated with Emergence and Spread
Tick-borne encephalitis virus (Central European)	*Flavivirus*	Rodents, hedgehogs, birds (ixodid ticks)	Europe	Temperature and precipitation (Randolph et al. 2000)
West Nile virus	*Flavivirus*	Wild birds (mosquitoes)	Panglobal	Temperature and precipitation (Campbell et al. 2002)

Source: Compiled from Daszak et al. *Emerging Infectious Diseases* 5:735–748, 1999; Daszak, P., A.A. Cunningham. *Parasitology Today* 16:404–405, 2000; Daszak et al. *Science* 287:443–449, 2000; *The Merck Veterinary Manual*. Whitehouse Station, NJ: Merck & Co., Inc.; Altizer et al. *Ecology Letters* 9:467–84, 2006.

Climate change, particularly with regard to expansion of the geospatial range of a zoonosis, its reservoir or amplifying host, and its arthropod vectors, can greatly increase the exposure potential for humans and domestic and wildlife animal species (Jetten and Focks 1997; Epstein 2001). Introduction of an epizootic into various domestic and peridomestic animal species associated with urban environments increases the likelihood of disease crossover into human populations. Black-tailed prairie dogs (*Cynomys ludovicianus*) are increasingly associated with urban and suburban residential areas in the southwestern United States and are particularly vulnerable to *Y. pestis*–caused epizootics, while the canine and feline predators and other animals and arthropods associated with their colonies may be infected and serve to spread the pathogen.

Epizootics in predators are generally considered to occur most frequently in areas containing multiple populations of host species at high densities in diverse and patchy habitats (Gage et al. 1995) because predators maximize hunting efforts in areas of abundant prey. Rural prairie dog colonies have been found to support 3 to 4 times higher densities of deer mice (*Peromyscus maniculatus*) and grasshopper mice (*Onychomys leucogaster*) than non-colonized areas (Agnew et al. 1986; Anderson and Williams 1997). Cully et al. (1997) followed a plague epizootic of Gunnison's prairie dogs (*Cynomys gunnisoni*) in New Mexico and reported collecting *Y. pestis*-infected fleas from prairie dogs, deer mice, and thirteen-lined ground squirrels (*Spermophilus tridecemlineatus*). Urban colonies may be composed of hundreds of prairie dogs, and their burrows provide habitat for numerous commensal mammalian species (Lomolino and Smith 2001), including ground squirrels, deer mice, and voles (*Microtus* spp.). These species can harbor infected fleas and serve as enzootic hosts for *Y. pestis,* and can initiate outbreaks of plague or other zoonoses in prairie dogs (Cully and Williams 2001).

People in the United States at greatest risk of exposure to *Y. pestis* live among or near enzootic rodent populations that are common in several western states, including New Mexico, Arizona, Colorado (Gage et al. 1995), and Texas (Miles et al. 1952). Coyotes, foxes, raccoons, and other carnivores often maintain very large home ranges in order to accommodate nutritional requirements. A number of urban prairie dog colonies could realistically be included within overlapping home ranges of several predators, thus such colonies may be linked by their predators as well as by dispersal of rodents (Cully and Williams 2001). Although domestic predators (dogs and cats) likely do not maintain large foraging areas, they provide a direct linkage between zoonoses-infected prairie dog colonies and humans, as evidenced by the cat-related plague incidents previously described. Domestic predators directly interact with infected colonies, dispersing rodents and wild predators. Dramatic declines in prairie dog populations associated with plague outbreaks may force predators to seek food in residential or urban areas inhabited by humans and their pets, including cats. The potential for plague infection in human populations increases as wild and domestic predators come into contact with encroaching urban/residential prairie dog colonies.

The rapid transition from enzootic sylvatic plague to an epizootic outbreak and then to an explosive human epidemic of pneumonic plague was demonstrated in India during late 1994. Bubonic plague transmitted to humans from rodents by way of fleas first appeared in a known plague-enzootic area during August 1994, in the Beed District of western India. By late September 1994 an explosive epidemic of primary pneumonic plague was occurring in neighboring states. More than 6,300 unconfirmed but suspected cases were reported in early October, with 693 of those cases laboratory-confirmed by serologic evidence by late October 1994 (CDC 1994a, b; WHO 1994).

Tularemia, also known as rabbit fever, a zoonosis caused by the bacterium *Francisella tularensis,* is also a Category A biological threat agent (Rotz et al. 2002). Tularemia is typically found in wild animals, especially rodents, rabbits, and hares. In North America, tularemia is usually a rural disease caused by *F. tularensis* biovar *tularensis* (Type A). Humans may become infected through the bite of infected arthropods (most commonly ticks, deerflies, and mosquitoes), by handling infected sick or dead animals, by eating or drinking contaminated food or water, or by inhaling airborne bacteria. *Francisella tularensis* is highly infectious, with a very small number of bacteria (10–50 organisms) capable of establishing disease. Tularemia can be fatal if the person is not treated with appropriate antibiotics. Tularemia may present in various forms depending on the route of initial infection, including oropharyngeal (via ingestion), ulceroglandular and glandular (via direct contact or arthropod vectors), pleuropneumonic (via inhalation), and systemic (Reintjes et al. 2002).

4.3 VULNERABILITY AND EFFECTS OF DISEASES ON WILDLIFE CLASSES

Wildlife ecologists have traditionally considered disease processes to be associated with individual animals and not necessarily to be processes that can significantly influence population growth. This paradigm occurs because disease processes in wild animals are poorly understood compared with other processes such as food limitation or competition. Several factors contribute to this lack of understanding. First, unless a disease affects the animal's normal behavioral function, most symptomatic-diseased wild animals seek dense cover for shelter (epizootic events in waterfowl are an exception). Even those animals that die in the open are readily consumed by opportunistic predators or scavengers, generally within a matter of 48 hours. This reality makes it very difficult to assign ultimate causes of mortality when conducting cause-specific mortality studies. Such mortalities are frequently recorded as a depredation. Similarly, an animal can be affected by a disease organism that alters its swiftness or alertness and leads to the animal being more easily depredated than if it had been healthy. Again, though the proximate cause of death was depredation, ultimately the disease organism caused the mortality to occur. Disease organisms can also weaken animals, making them much less able to successfully compete for food or other resources. Though we generally consider disease to be a nonfactor in population growth because we generally see predominantly healthy animals, it is likely that disease organisms and their effects on wildlife play a significantly larger role than we understand.

If disease processes are not recognized as a significant factor in wildlife population growth today, marked changes in factors such as habitat availability, increasing availability and concentration of contaminants, and more probable and rapid introduction of diseases to naïve hosts could cause disease processes to be significant in the future. One reason many biologists believe disease processes are not significant in determining population dynamics is that animals have evolved with their pathogens, and so they, especially the adults, should be immune to common pathogens they encounter. Only when unusual circumstances compromise an animal's immune system should disease processes result in influences at the population level. Even if these concepts are currently correct, a plethora of changes are occurring from the local to the global level that could influence animal immunocompetence and exposure to novel pathogens. As human populations expand, wildlife habitats are being constricted and fragmented. This habitat reduction results in concentration of wildlife in the remaining habitat, which increases the probability of animal exposure to pathogens. Concentration can also lead to increased stress, which can result in decreased immunocompetence. This process is already believed to be occurring at small levels in waterfowl.

Most avian/fowl cholera (*Pasteurella multocida*) outbreaks occur when waterfowl are concentrated in wetlands during winter. At such locations many birds come together having experienced the rigors of migration, which is believed to be a stressful and potentially immune-system-compromising effect. Significant epizootics can occur when the pathogen is present and a multitude of susceptible birds are infected. As wetland habitats are lost, waterfowl will be further concentrated on the remaining acreage, and migration distances may become longer if traditional stopover sites have been eliminated, forcing birds to travel farther between stops to reach suitable areas. Avian cholera is not generally zoonotic and thus not a significant threat to human health, but the same principles can hold true for reduced immunocompetence and other pathogens of zoonotic concern such as the H5N1 strain of avian influenza. Unfortunately, one can derive a number of examples similar to the preceding one, in which environmental changes can lead to the stress of wildlife and result in decreased immunocompetence. How much these factors can affect immunocompetence compared with other ecological processes we understand to be stressful, such as migration, competition, and food limitation, is not known.

Whether disease interactions are seen as natural processes or as problematic intrusions into ecological systems that require human intervention, it is clear that diseases and ecosystem health are strongly related (Macdonald and Laurenson 2006). Environmental perturbations (climate change, habitat loss) can have cascading effects that lead to the emergence of disease events, both zoonoses and diseases only of wildlife. And conversely, the influence of disease on population processes can readily influence ecosystems. For instance, competition for resources could affect animal immunocompetence and the occurrence and outcome of infection. Smith (2007) reviewed a number of animal-pathogen models in the literature and found they support the hypothesis that nutrient limitation by way of competition for resources is a way in which density-dependent factors could influence population processes through disease.

Density-independent factors, such as weather patterns or climate change, which affect plant productivity and thus food availability, can affect animal immunocompetence in the same way. For instance, Pinzon et al. (2004) suggested that the trigger event for the transmission of Ebola hemorragic fever to humans was an acute dry period at the end of the normal rainy season. Plowright et al. (2008) reported that nutritional stress from habitat loss and climate change increased infection and transmission rates of Hendra virus (*Henipavirus*) among little red flying foxes (*Pteropus scapulatus*) in Australia. Schauber et al. (2005) reported that acorn production significantly influenced white-footed mouse (*Peromyscus leucopus*) population size and thus the potential for the transmission of Lyme disease (*Borrelia burgdorferi*) in the area. Snäll et al. (2008) developed a model predicting that plague epizootics in prairie dog colonies were related to increasing precipitation and were slowed by hotter temperatures. Habitat disturbance can have similar consequences. Gillespie et al. (2005) showed that selective logging in forests of Uganda is associated with higher parasite loads and infection rates in redtail guenons (*Cercopithecus ascanius*) when compared with non-logged areas. Similarly, Allan et al. (2003) reported that a reduction in small mammal diversity and an increase in the density of white-footed mice were results of forest fragmentation in the United States. This ecological shift resulted in an increased risk of Lyme disease transmission to humans because of a greater prevalence of infected tick nymphs in fragmented forest tracts.

Studies in which biodiversity was experimentally reduced resulted in increases in the prevalence of infection in wild, small mammal hosts of hantaviruses, as well as in the density of the reservoir population itself (Suzan et al. 2008). Ostfeld (2009) suggested that the reduction in biodiversity allowed arthropod vectors of disease-causing organisms to concentrate their feeding on hosts that were most susceptible or best suited for transmission of the pathogen. In diverse animal communities, this type of increased feeding activity may involve potential arthropod vector species that do not normally readily transmit the disease agent, as well as potential host species that are not typically highly susceptible to infection.

Additionally, human actions and efforts to influence disease transmission dynamics can cause unintended ecological disruptions that lead to other potential problems. Recent information from an experiment conducted in England best illustrates this point. Populations of Eurasian badgers were experimentally controlled to determine whether this would be an effective means of reducing the transmission of bovine tuberculosis from badgers to cattle (Donnelly et al. 2006). While badger culling in England as a means of controlling bovine tuberculosis has reduced the badger population, Woodroffe et al. (2006a, 2006b) reported that the practice has had the unintended consequence of increasing the incidence of *M. bovis* in badgers. Trewby et al. (2008) reported that the experimental reduction in Eurasian badgers led to an increase in red fox population densities. This increase in red fox density had the potential to increase the probability of the occurrence and transmission of other pathogens, including zoonoses such as rabies (*Lyssavirus*) and fox tapeworm (*Echinococcus multilocularis*) (Letkova et al. 2006). Population densities of red foxes have also increased in urban environments and as a result have established an urban wildlife cycle of *E. multilocularis* (Deplazes et al. 2004).

A relatively wide variety of wildlife species are moving into urban environments because of decreased preferred non-urban habitat availability. With this close contact between wildlife and humans, it is going to be challenging to prevent the spread of zoonoses to human populations. A recent report from Germany serves as an excellent example. It was found that 18% of Eurasian wild boars (*Sus scrofa scrofa*) collected within the city of Berlin (Germany) were infected with leptospirosis (*Leptospira interrogans* (sensu lato) as evidenced by leptospires present in their kidney tissue, thereby presenting significant potential as a source for human infection (Jansen et al. 2007). These are just a few examples of how environmental and anthropogenic disturbances can lead to a shift in ecosystem characteristics that lead to an increased probability of the maintenance and transmission of zoonotic organisms.

As people pursue recreation they may inadvertently come into contact with animals and their pathogens. For instance, Meites et al. (2004) reported that 85% of Californians infected with leptospirosis had contracted it through recreational pursuits. Subsistence hunters in Central Africa are believed to have contracted simian retrovirus from hunting and butchering non-human primates (Wolfe et al. 2004). Renter et al. (2006) showed that hunting was a potential source of salmonellosis infection for humans, when they detected four serovars of *Salmonella enterica* in white-tailed deer in southeastern Nebraska (U.S.A.). In the Czech Republic and throughout much of Europe, the European brown hare (*Lepus europaeus*) is harvested in great numbers. A survey by Treml et al. (2007) revealed that hunters taking and handling European brown hare carcasses may be exposed to the pathogenic microorganisms that cause tularemia, brucellosis, and leptospirosis.

The relative distance, speed of dispersal, and migration abilities of most birds make them unique among animals in their ability to geospatially spread disease-causing organisms across the landscape. Reed et al. (2003) point out that birds can harbor West Nile virus, *Ixodes* ticks harboring *Borrelia burgdorferi* (Lyme disease), avian influenza viruses, *Salmonella* spp. (salmonellosis), and *Campylobacter* spp., which can all cause disease in humans. They suggest that migratory birds are analogous to infected humans traveling by air, quickly spreading disease-causing organisms across the globe. Many bird species use habitats that are intended for disposal or treatment of contaminated substances, such as garbage landfills and sewage treatment plants, and may acquire pathogenic organisms from such sites. An example of the potential for transcontinental movement of disease-causing pathogens is the migration patterns of the Canada warbler (*Wilsonia canadensis*). It breeds in northern Canadian forests but winters in Latin America. Though no specific pathogens have been reported from this little-known species, it provides an example of the potential pathogen movement that could occur with hundreds of bird species.

The mourning dove (*Zenaida macroura*) has a more probable potential for transmitting disease organisms to humans. This species is ubiquitous, migratory, and adaptable to rural, suburban, and urban human environments (Otis 2008). They commonly use birdbaths and bird feeders, which can readily bring them into contact with humans. Mourning doves can harbor West Nile virus (Gerhold et al. 2007), and there is concern that additional information is needed regarding their potential to be reservoirs of other significant zoonotic pathogens (Otis 2008). Other behavioral factors such as flocking and their tendency to congregate in large roosting sites could facilitate the intra- and inter-flock transmission of pathogens. Approximately 20 million mourning doves are harvested by hunters in the United States annually, and an unknown number of these birds are harvested in Mexico and Central America annually; thus there is significant human contact with mourning dove tissues each year (Otis 2008).

4.4 EFFECTS OF PESTICIDES AND OTHER ENVIRONMENTAL CONTAMINANTS ON SUSCEPTIBILITY TO DISEASE

The depth and breadth of this section's topic is vast, with an extensive amount of relatively recent scientific literature available specifically addressing the principal factors relative to the wide range of influences and effects of pesticides and other environmental contaminants on wildlife susceptibility to disease. We limit this section to a broader and more general discussion of how pesticides and other environmental contaminants may directly and indirectly influence the susceptibility of wildlife to infectious and parasitic diseases, particularly emerging and resurgent pathogens.

Pesticides and other contaminants may directly and indirectly influence the ability of wildlife species to survive and thrive, whether in the aquatic, marine, or terrestrial environments, and whether the exposure to the contaminant or toxicant is acute or chronic. Directly, the effects of pesticides and other environmental contaminants on wildlife are manifest at all physiological levels (molecular, cellular, tissue, and organ) and may influence the overall fitness of the animal through parameters such as growth, locomotion, neurosensory, reproduction, and survival (Christian et al. 1997). The ability of wildlife to thrive, or compete and function at an optimum potential, and ultimately to survive may be indirectly affected by exposures to pesticides and other environmental contaminants that significantly modify animals' behavior and susceptibility to parasites and pathogens. Relative to both the direct and indirect effects of pesticides and other environmental contaminants on wildlife, unimpeded and unaltered neurochemical transmission within both the cholinergic and dopaminergic systems are critical to an animal's abilities for cognition, somatosensory, and motor function. Research studies have reported that prolonged and repeated disruptions of neurotransmission may ultimately affect the natural behavior of wildlife (Carlson et al. 1998; Basu et al. 2005).

A thorough review of recent literature relative to the direct and indirect effects of pesticides and other environmental contaminants on both the physiological and behavioral abilities of wildlife populations to survive and thrive revealed four key points:

- Acute and chronic exposures of wildlife to environmental contaminants may cause direct physiological damage to tissues and organs and impede major system function and processes (e.g., circulation, digestion, and respiration), as well as energy budget balance (acquisition and expenditure of energy).
- Prolonged inhibition of neurochemicals and disruption of neurotransmission may affect an animal's abilities to receive, process, and store information, and thereby influence animal behavior.
- Maternal exposure to contaminants during gestation may cause neurological developmental anomalies in offspring, and may cause neonatal exposure later through lactation.
- Indirectly, wildlife may become more susceptible to tumors and cancerous diseases, and to parasitic and infectious diseases, as a result of being compromised physiologically and immunologically.

4.4.1 Nutrition and Systems Function

Some pesticides and other environmental contaminants, especially many of the toxins that wildlife encounter through inhalation or ingestion, may disrupt or inhibit system function through physical and chemical modes of action. An example of acute exposure of an animal to a toxicant that prevents or severely limits its ability to feed or drink is the blistering and irritation of the oral cavity and digestive tract caused by the ingestion of ethylene-glycol-based antifreeze, oilfield chemicals, and various caustic pesticides. Additionally, other physiochemical and neurochemical products and functions within the exposed animal may cause it not to consume food, and may inhibit digestion and absorption of nutrients—thus disrupting the animal's energy budget (its ability to acquire sufficient energy to meet expenditures of energy for metabolism, growth, locomotion, and other requirements).

Wildlife exposed to various toxins and other chemicals, including heavy metals, may develop anemia as a result of an array of pathogenic mechanisms directly related to a specific toxicant. Table 4.5 provides a listing of the toxic causes of anemia. Anemia, an absolute decrease in the red blood cell mass, can develop from the loss, destruction, or lack of production of red blood cells. Acute anemia in an animal may result in reduced energy, weakness, and anorexia, or can result in shock and death in severe, prolonged disease (Merck 2005).

Bain et al. (2004) reported that Australian central bearded dragons (*Pogona vitticeps*) exposed to a single sub-lethal dose of organophosphates suffered substantially depressed total plasma cholinesterase and acetylcholinesterase activity for up to 21 days, but did not show any indications of significant effects on diurnal body temperature, standard metabolic rate, or feeding rate.

TABLE 4.5
Toxicants Associated with Anemia in Animals

Pathogenic Mechanism	Toxins or Chemicals	Heavy Metals
Oxidation	Crude oil, naphthalene	Copper (Cu), zinc (Zn)
Blood loss	Dicoumarol	
Immune-mediated hemolysis	Pirimicarb	
Hemolysis	Indole	Lead (Pb), selenium (Se)
Decreased marrow production	Benzene, trichloroethylene	Lead (Pb)

Source: Adapted from *The Merck Veterinary Manual*. Whitehouse Station, NJ: Merck & Co., Inc., 2005, p. 9.

Adult fitness of leopard frogs (*Rana pipiens*) is directly correlated to their growth rate as a tadpole, which is directly related to the food-seeking, feeding, and swimming performance of the tadpole. Leopard frog tadpoles exposed to carbaryl suffered devastating diminishment (90%) of activity within 24 hours of exposure, with only slight recovery after 48 hours (Bridges 1997). Western fence lizards (*Sceloporus occidentalis*) exposed to various doses of carbaryl were observed for effects on their metabolism over a 48-hour period, and their food consumption was observed for 96 hours. No total energy expenditure differences were determined; however, lizards receiving the lowest dose exhibited a 16–30% increase in standard metabolic rate, which was offset by a 45–58% decrease in additional energy expenditures. Lizards in the highest dose group exhibited a 30–34% decrease in energy acquisition compared to controls. The net result was a 1.83 kJ decrease in energy assimilation, equivalent to 5 times their daily standard metabolic rate requirements (DuRant et al. 2007).

French et al. (2001) examined whole animal energetic effects, among other effects, on white-footed mice exposed to polychlorinated biphenyls (PCBs). No net effects were reported for energy metabolism or for the growth of mice exposed to PCBs by way of dietary concentrations up to 25 ppm. Voltura and French (2000) measured food intake, digestive efficiency, and body mass in white-footed mice exposed to PCBs, and reported that after short-term exposure there were no significant differences between control and exposed animals. However after chronic (1-year) exposure, high doses of PCBs resulted in significantly higher food intake and oxygen consumption, indicating a chronic effect on energy allocation. In another study by Holliday et al. (2009) on the effects of diamondback terrapins (*Malaclemys terrapin*) exposed to PCBs in saline solutions, growth and metabolic rates were significantly reduced and altered respiratory patterns were observed.

Banded water snakes (*Nerodia fasciata*) collected from sites contaminated with high concentrations of trace minerals exhibited mean standard metabolic rates that were 32% higher than snakes from the reference site. As a result, snakes from the contaminated site would appear to have an elevated allocation of energy required for maintenance (Hopkins et al. 1999).

4.4.2 Neurotransmission and Behavior

An animal's behavior results from an integration of biochemical, physiological, and morphological processes and attributes linking cellular and molecular processes to ecological consequences (Carlson et al. 1998). There have been a large number of reports published from research on the effects of pesticides that inhibit neurotransmission, particularly regarding organophosphates and the effects on acetylcholine and cholinesterase. The effective and accurate transmission of signals between an animal's external environment and its nervous system is necessary for its ultimate survival. Studies of fishes have shown that exposure to pesticides and heavy metals altered components of the cholinergic and dopaminergic system and were directly correlated to behavioral changes. It is well established that mercury (Hg) can alter neurotransmission and ultimately influence behavior in wildlife, but little is known about the mechanisms that mediate these physiological changes (Basu et al. 2005). Mercury is a nonspecific cytotoxic compound, and studies of rodents have shown that organic and inorganic Hg can impair various aspects of neurotransmission and may cause behavioral changes (e.g., impaired ability to hunt, breed, or migrate) (Basu et al. 2005). Collectively, these findings suggest that prolonged disruptions to neurotransmission may ultimately influence animal behavior.

Cholinesterase is an important enzyme essential to the proper functioning of the nervous system of humans, other vertebrates, and insects. Organophosphate (OP) and carbamate (CM) classes of pesticides interrupt neurotransmission by inhibiting the production of, or by binding with, cholinesterase in many animal species. Cholinesterase-inhibiting pesticides (OPs and CMs) are among the most widely used insecticidal pest control chemicals in North America (applied at rates of 200 million acre-treatments per year in the United States) and throughout the world. Despite their rapid degradation and minimal potential for bioaccumulation, cholinesterase inhibitors are acutely toxic to a variety of non-target invertebrates as well as to wildlife (Hopkins et al. 2005).

Malathion is a commonly used OP in the control of insects such as the boll weevil (*Anthonomus grandis grandis*), Mediterranean fruit fly (*Ceratitis capitata*), and mosquitoes (Diptera: Culicidae). Holem et al. (2008) studied the effects of malathion on the locomotor performance of the western fence lizard (*Sceloporus occidentalis*). Results of their study showed that 85% of the lizards exhibited clinical symptoms of OP poisoning (e.g., body/limb tremors, twitching) within 4 hours of exposure, with symptoms generally subsiding within 24 hours. Additionally, they reported that lizard growth rate, food consumption, body condition index, and terrestrial locomotor performance were not significantly affected by malathion.

Hopkins et al. (2005) examined the effects on locomotor performance of a CM pesticide in two species of semi-aquatic snakes, the diamondback water snake (*Nerodia rhombifer*) and the black swamp snake (*Seminatrix pygaea*). They reported that *S. pygaea* was much more sensitive to the effects of carbaryl on swimming performance than was *N. rhombifer*. Although performance in carbaryl-exposed snakes was reduced by as much as 51% compared to unexposed controls, both species regained their pre-treatment level of swimming performance within 4 days of exposure. These findings are in general agreement with other studies of wildlife behavior and performance, in that animals typically recover within days of acute exposure to CM and OP (Hill and Camardese 1984; Dell'Omo and Shore 1996a, b).

There is a strong and growing body of evidence linking exposure to OP pesticides during gestation or the early postnatal period and neurodevelopmental effects in animals (Eskenazi et al. 1999). Neurobehavioral tests conducted post-natal found that animals exposed in utero demonstrated decreased balance (Muto et al. 1992), increased righting reflex time, and poorer cliff avoidance behaviors (Chanda et al. 1995).

Montie et al. (2009) screened opportunistically sampled short-beaked common dolphins (*Delphinus delphis*), Atlantic white-sided dolphins (*Lagenorhynchus acutus*), and gray seals (*Halichoerus grypus*) for the presence of brominated and chlorinated contaminants and metabolites in their cerebrospinal fluid (CSF). They reported detection of organochlorine (OC) pesticides, including *p,p'*-DDE (1,1-dichloro-2,2-bis(*p*-chlorophenyl)ethylene), hexachlorobenzene, *cis*-chlordane, *trans*-nonacholor, and *p,p'*-DDD (1,1-dichloro-2,2-bis(*p*-chlorophenyl). They also reported that PCBs are clearly transported into the cerebellum of white-sided dolphins despite the blood-brain barrier (BBB), and such exposure to PCBs also poses a potential risk. They concluded that marine mammals are not just exposed to one or a few organohalogen contaminants, but rather to a "cocktail" of metabolites, degradation products, and other developmental neurotoxicants (such as lead, methylmercury, perchlorate, and dioxins), as well as natural marine neurotoxins (associated with algal blooms).

DDT (1,1,1-trichloro-2,2-bis(*p*-chlorophenyl)ethane) is a persistent, widespread environmental contaminant found in most regions of the world. Chronic and acute exposure to DDT adversely affects the development, physiology, morphology, and behavior of both animals and humans (Iwaniuk et al. 2006). Eggs of most birds are highly sensitive to changes in maternal hormone balance and are extremely susceptible to endocrine disruption during early development. These effects have led to many behavioral abnormalities in birds exposed to DDT (Iwaniuk et al. 2006). Because the brain is the major site of hormone action for hormone-sensitive behaviors, these abnormal behaviors likely result from changes in the underlying neural structures.

PCBs, hydroxylated PCB, and polybrominated diphenyl ethers (PBDE) are considered to be developmental neurotoxicants. In a study by Schantz et al. (2003) there was strong evidence that PCB exposure is associated with negative effects in cognitive development in humans. This association was supported by the results of experiments on laboratory rats in which developmental exposure to PCBs caused hearing loss, locomotor deficits, and cognitive disorders related to learning and memory (Montie et al. 2009).

4.4.3 Susceptibility to Disease Due to Suppressed Immunity

The primary function and role of any organism's immune system is to identify, isolate, inactivate, and eliminate invading microorganisms, any byproducts they may have produced, and any other substances that are foreign to it through adaptive and innate methods. In-depth explanations of immune function and comparative immunity in various species are provided in the works of Flajnik (1998), Sharon (1998), and Turner (1994). For the purposes of this chapter, only a cursory overview of the basic structures and functions involved in competent immune system response is provided, and readers are encouraged to seek further information through listed and available literature.

Researchers have reported strong correlations between animal exposure to pesticides or other environmental toxicants and the suppression of an animal's immune response to specific pathogenic challenge through anticholinergic and non-cholinergic pathways, and responses such as hypersensitivity and autoimmunity. Histopathological changes to immune system tissues and organs, pathology at the cellular level, and alteration of lymphocyte population growth and function are attributable to exposure to pesticides (Barnett 1994; Vial et al. 1996; Voccia et al. 1999). These effects caused by pesticides and other environmental contaminants on the components and functions of the immune system have been linked to compromise and reduction of disease resistance in exposed animals. Below is a brief overview of recent research findings directly related to immunotoxic effects of pesticides in wildlife.

Galloway and Handy (2003) provide an excellent review of the scientific literature reporting the toxic effects on immune system tissues and organs and function of OP pesticides. They summarize that low level environmental exposures to OPs appear to stimulate an animal's immune system, while depressed immune function results from higher exposures. Relative to immunosuppressive effects of these pesticides on wildlife, they reported significant alterations of non-specific immunity, decreased host resistance, and hypersensitivity to pathogenic organisms.

High blood concentrations of PCBs and OC pesticides in polar bears (*Ursus maritimus*) were reported to be strongly associated with both impaired humoral and cell-mediated immune responses, implying that the animals may have increased susceptibility to infection and risks for disease (Lie et al. 2004; Fisk et al. 2005; Lie et al. 2005). From investigation of factors contributing to increased susceptibility and extensive mortality of dolphins and seals infected with *Morbillivirus* (Paramyxoviridae), it was determined that harbor seals (*Phoca vitulina*) exposed through consumption of fish to ambient environmental concentrations of bioaccumulated persistent lipophilic contaminants, including OCs and PCBs, suffered impaired natural-killer (NK) and specific T-cell immune response capacity (deSwart et al. 1996). This was one of the first reports of immunosuppression resulting from chronic exposure to environmental contaminants at ambient concentrations.

Albert et al. (2007) compared high- and low-dose laboratory exposures of DDT and dieldrin in northern leopard frogs (*Rana pipiens*), and reported that there were significant differences in the frogs' immune system responses between the high and low doses. Linzey et al. (2003) reported decreased B cells, a decrease in white pulp (spleen biomarker of exposure to OCs), and anatomical deformities in toads and frogs exposed to DDT/DDE and PCB in Bermuda. Additionally, Gilbertson et al. (2003) reported that DDT, malathion, and dieldrin in combination caused antibody suppression, enhanced delayed-type hypersensitivity reaction, and lowered respiratory bursts in northern leopard frogs in both controlled laboratory and field studies.

Decreases in T-lymphocytes in young Caspian terns (*Sterna caspia*) from the Great Lakes region (U.S.A.) were attributed to exposure to high concentrations of PCBs as reported by Grasman and Fox (2001). Grasman et al. (1996) reported organochlorine-associated immunosuppression in pre-fledgling Caspian terns and herring gulls (*Larus argentatus*) from the Great Lakes. Mayne et al. (2004) reported that fledgling tree swallows (*Tachycineta bicolor*) and eastern bluebirds (*Sialia sialis*) collected from apple orchards sprayed with non-persistent pesticides and DDE had significantly greater thymic lymphocyte densities and cortical/medulla ratios, and significant splenic B-cell hyperplasia.

These immune system compromises suggest songbird survival may be significantly reduced immediately after fledgling or during migration because of their resultant susceptibility to disease. Bustnes et al. (2004) reported that there was a significant or near significant positive relationship between persistent OCs and the number of heterophils and lymphocytes in the blood of gulls that had been naturally exposed to OCs. These findings suggest that the alterations to the immune function of these gulls were a result of OC exposure.

4.5 WILDLIFE DISEASE TRANSMISSION DYNAMICS, GEOGRAPHIC SPREAD, AND GLOBAL DISTRIBUTION

The specific dynamics associated with disease transmission and maintenance, particularly for zoonotic pathogens also associated with domestic animals that are vectored by arthropods, may be very complex and involve multiple life stages of the pathogen, multiple life stages of the arthropod vector, and even multiple growth stages of the host animal. Utilizing the relatively complex and multiple means for the transmission of plague, examples of potential routes or mechanisms of transmission of infectious diseases between wildlife and domestic animal species, as well as to humans, are illustrated in Figures 4.1 through 4.3.

The typical enzootic or sylvatic maintenance and transmission cycle of the bacterium that causes plague involves a primary reservoir host of wild mammals such as rodents, but may also involve canine and feline predators. Certain flea (Siphonaptera: Pulicidae, primarily) species that are associated with or are specific ectoparasites of the reservoir animals acquire *Y. pestis* through blood-feeding on infected hosts, accumulating an infective dose of the bacterium through repeated blood meals from the host and then transmitting the pathogen to a new host animal. Additionally, other wildlife such as canine and feline predators may become infected with *Y. pestis* through ingestion of infective tissues from plague-infected rodents or other animals (Figure 4.1).

Figure 4.2 illustrates the transmission routes that may occur through which other animals and humans may be exposed to *Y. pestis* infection, thereby expanding the disease beyond the enzootic or sylvatic cycle. Domestic pets as well as other wild or feral animals that are not typically involved in the maintenance of the enzootic cycle may become infected with *Y. pestis* through the ingestion of living (or very recently expired) rodents or other species that serve as primary reservoirs of the bacterium. Additionally, infective fleas may be transported out of the enzootic foci on animals (pets or other domestic animals transiting the area) and directly pass the bacterium to humans or other domestic animals while blood-feeding, thereby causing infection. Humans are also susceptible to plague infection through the inhalation

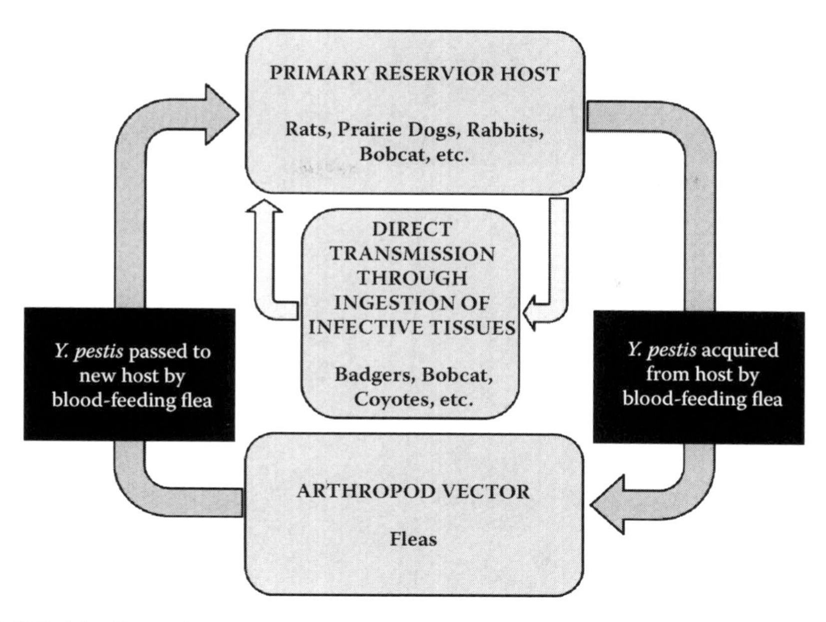

FIGURE 4.1 Enzootic or sylvatic transmission and maintenance cycle of plague (*Yersinia pestis*) involving mammalian primary hosts and flea vectors.

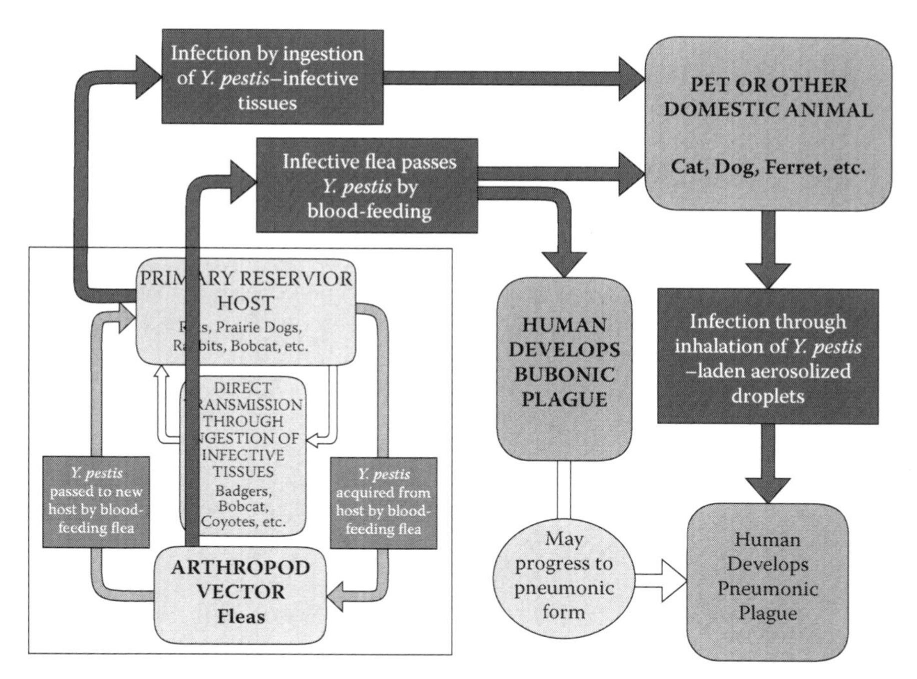

FIGURE 4.2 Potential secondary/indirect routes of transmission of *Yersinia pestis* from the enzootic or sylvatic cycle to other animal species and humans.

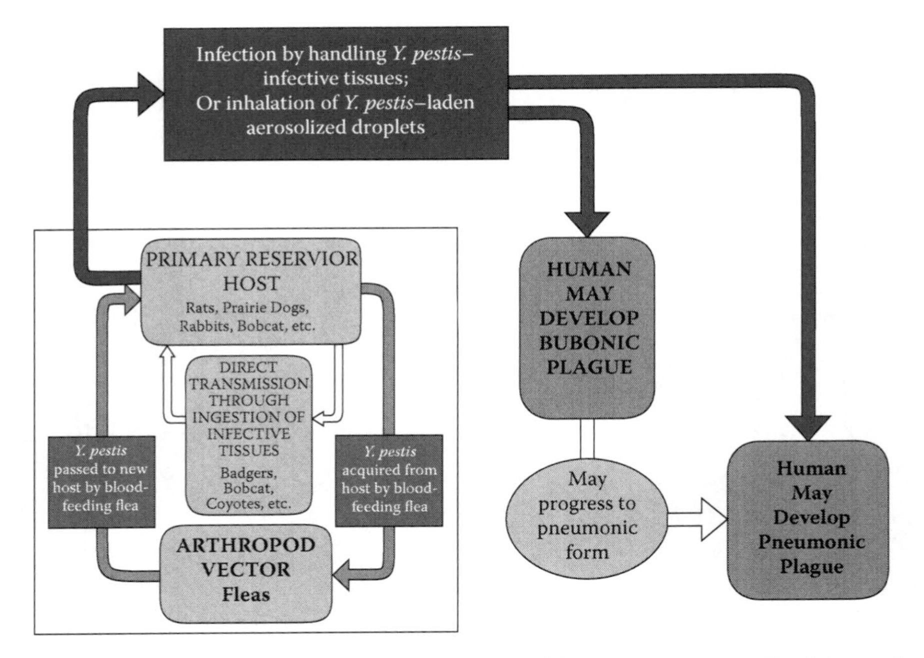

FIGURE 4.3 Potential accidental indirect routes of human exposure to *Yersinia pestis* infection.

of aerosolized droplets containing *Y. pestis* that are exhaled or through coughs or sneezes from other humans, cats, and black-footed ferrets suffering from the pneumonic infection of plague.

A third route of human exposure and infection is through the exposure to aerosolized blood or other bodily fluids from *Y. pestis*–infected animals. This form of exposure occurs typically during the field or laboratory handling of infected animals and tissues, and depending on the route of exposure (inhalation or entry through abrasion/laceration) may develop as systemic or pneumonic plague (Figure 4.3).

These various routes and mechanisms of transmission of infectious agents between wildlife and domestic animals and humans play an essential role in the emergence or resurgence of diseases through geographic spread of a pathogen into naïve populations and locations. Because of the potential for multiple species being involved, there may be enhanced or perpetuated genetic or proteomic changes in the virulence and other characteristics of the pathogen. An excellent example of this is the ongoing and continuous antigenic "shift" and "drift" of the virulence and infectivity of various influenza viruses, particularly H5N1, the highly pathogenic avian influenza virus.

4.5.1 Anthropogenic Factors — Movement of People and Animals

Emerging diseases of wildlife can be introduced into a particular population of animals in a multitude of ways, including through the transient movement through or permanent introduction of other infected wildlife species or arthropod vectors into the area, or by the movement or translocation of the previously naïve population into new geographic areas where the disease already exists. The movement/translocation, whether transient or permanent, of infected animals or the naïve population can be caused by a wide range of reasons, including natural disasters, a lack of adequate nutritional sources in an area, lack of adequate water, and human-induced factors. These human-induced factors may include changes in land use, such as cultivation of new areas for agricultural production, harvesting of woodlots, urbanization, and increased usage of an area for recreational purposes. There are numerous examples of how human factors and actions have caused disruption of the ecosystem function that ultimately led to the emigration of wildlife out of an area, as well as to the adaptation of wildlife to the new "pressures" (e.g., the movement of white-tailed deer, coyotes, and feral swine into suburban and urban areas).

The potential for the introduction of emerging disease pathogens into a new geographic area and previously naïve wildlife populations is significantly increased by the commercial movement of humans, domestic animals, and wildlife. There is growing concern regarding this increased potential for the introduction of disease agents, particularly for transborder movement of livestock and other animals as a result of open-market access through free-trade agreements between nation states (e.g., the North American Free Trade Agreement (NAFTA), and agreements negotiated by the World Trade Organization (WTO)). Additionally, international passenger travel into the United States is steadily increasing; in 1980 approximately 20 million passengers arrived in the United States, and in 1995 approximately 47 million passengers arrived there from international origins (an increase of 131%) (Murphy and Nathanson 1994). There is the real potential that international travelers may unknowingly transport contaminated animal products or other pathogen-harboring substances (such as FMD) from one country to other countries. Worldwide there is an average of approximately 2 billion people traveling by air annually; at any given moment there are approximately 500,000 people in transit on airliners (UK 2000). These numbers are anticipated to continue to increase as a result of the increasing convenience and speed at which we can travel throughout the globe. The speed of global travel in relation to world population growth is illustrated in Figure 4.4.

It is not only people transiting through or permanently coming to a country and inadvertently introducing a disease; the importation of domestic and wild animals may also serve as the vehicle for introduction. For example, within the United States, the U.S. Fish and Wildlife Service data on live animal importation in 2005 included 5.1 million amphibians, 259,000 birds, 203 million fish, 1.3 million reptiles, and 87,991 mammals (USFWS 2007). These animals are the known and documented live animals brought into the country: there are likely thousands or perhaps millions more that are not legally brought into the country and do not undergo the rigorous inspection and quarantine requirements that are in place to prevent the introduction of disease.

4.5.2 Climatic Influences

Projected global climate change, whether a result of carbon dioxide accumulation, ozone layer depletion due to aerosols and atmospheric composition, Earth orbital variations, land use and habitat alteration practices, solar activity, volcanic

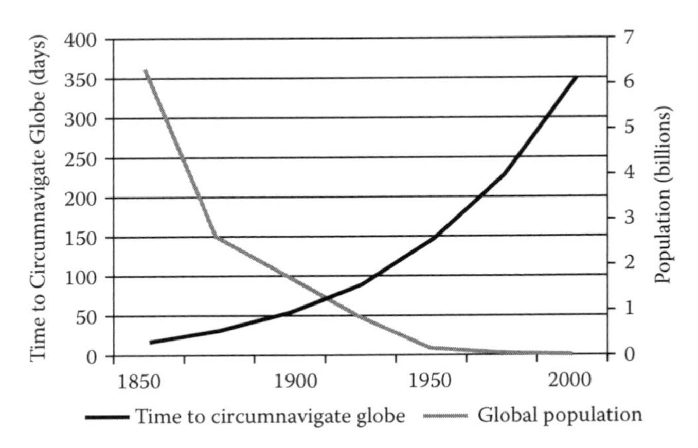

FIGURE 4.4 The inverse relationship between time required to circumnavigate the globe and human population growth. (Adapted from Murphy and Nathanson, *Seminars in Virology* 5:87–102, 1994).

activity, or unrecognized other factors, will likely have a significant influence on a wide array of ecological processes and relationships (Stenseth et al. 2002; Stenseth and Hurrell 2005; Stenseth et al. 2006; IPCC 2007; Christensen et al. 2007). The maintenance and transmission dynamics of infectious diseases, particularly zoonotic pathogens, may be significantly influenced and altered by ecological factors resulting from climate change. Particularly for three-agent zoonoses that involve an arthropod vector in the enzootic maintenance of the pathogen (such as fleas in the sylvatic plague model), the complexity involved in predicting the influence and extent of ecosystem disruption resulting from climate change becomes more severe.

Snäll et al. (2009) utilized dynamic ecological models to develop projections of the regional influence of climate change on the enzootic maintenance and transmission of plague within and between black-tailed prairie dog colonies in the prairies of the northern United States. Basing their model on the assumption that plague transmission dynamics are driven by precipitation and temperature (Parmenter et al. 1999; Enscore et al. 2002; Collinge et al. 2005; Snäll et al. 2008), they report that the likelihood of *Y. pestis* transmission from infectious to susceptible colonies increases with increases in the spring and summer precipitation that occurred in the preceding year, and that the rate of enzootic infection through secondary reservoir hosts (e.g., other small rodents and coyotes) decreases relative to increases in the number of hot days (>35°C) during the focal year. They reported that model scenarios of the most extensive increases in global warming predicted fewer plague-infected prairie dog colonies and smaller geographic areas in the region covered by colonies as a result of decreased precipitation and an increase in the number of hot days. Less intense global warming model projections predicted higher probabilities of increased numbers of infected colonies and expansion of the spatial distribution of colonies. They point out that their findings are geographically limited to and applicable only to regions of black-tailed prairie dog and plague distribution north of New Mexico's northern border influenced by seasonal monsoon rains and the resultant lesser understood climate and plague-transmission dynamics. Other researchers have reported predicted increases in human plague occurrence due to climate change projections for the same geographic region (Nakazawa et al. 2007), as well as in other regions such as Central Asia (Stenseth et al. 2006; Kausrud et al. 2007).

4.6 CONCLUSIONS

Wildlife and other animals naturally exposed to high concentrations of environmental toxicants, including pesticides and other contaminants such as toxic metals and various mycotoxins, may have an increased susceptibility to infection by disease-causing pathogens and parasitic organisms. Increased susceptibility to disease may result from direct and indirect toxic effects of pesticides and other environmental contaminants on both the physiological and behavioral abilities of wildlife populations to survive and thrive. Acute and chronic exposures of wildlife to a wide range of environmental toxicants, including pesticides, may cause direct physiological damage to tissues and organs and impede major system functions and processes (e.g., circulation, digestion, and respiration), as well as disrupt energy budget balance (acquisition and expenditure of energy). As a result of acute and chronic exposures, animals may suffer prolonged inhibition of neurochemicals and disruption of neurotransmission that may affect an animal's abilities to receive, process, and store information, and thereby influence animal behavior. Maternal exposure to contaminants during gestation may cause neurological developmental anomalies in an animal's offspring, and there may be neonatal exposure through lactation. Indirectly, wildlife may become more susceptible to tumors and cancerous diseases, as well as parasitic and infectious diseases, as a result of being compromised physiologically and immunologically.

Although there have been significant and laudable reductions in the misuse of pesticides, including insecticides, herbicides, algacides, and rodenticides, there remains a threat to wildlife resulting from indirect factors associated with pesticide use. It is not only pesticides that may directly and indirectly negatively affect wildlife, but also personal care products and pharmaceuticals that end up in the surface waters and feedstuffs they may consume. There is a rapidly expanding literature base on the immediate and potential long-term negative effects of these environmental contaminants on aquatic, marine, and terrestrial wildlife. As global human populations continue to increase and expand into previously wild environs, there will continue to be increasing risks posed to wildlife from pesticides and other environmental contaminants, both directly through toxicity and indirectly through immunosuppression and other more subtle compromises to an animal's ability to avoid infection by disease—and ultimately to thrive and survive.

ACKNOWLEDGMENTS

The authors express their sincere appreciation to Ms. Nina Dacko, Ms. Anna Hoffarth, and Mrs. Christena Stephens for their very valuable assistance in tirelessly collecting and assimilating pertinent literature and reference materials.

REFERENCES

[APHIS]. 2007. Animal and Plant Health Inspection Service. United States Department of Agriculture.

[CDC]. 1994a. U.S. Centers for Disease Control and Prevention. Human plague – India, 1994. *MMWR Morb Mortal Wkly Rep* 43:689–91.

[CDC]. 1994b. U.S. Centers for Disease Control and Prevention. Update: human plague – India, 1994. *MMWR Morb Mortal Wkly Rep* 43:761–2.

[CDC]. 1997. U.S. Centers for Disease Control and Prevention. Fatal human plague – Arizona and Colorado, 1996. *MMWR Morb Mortal Wkly Rep* 46: 617–20.

[CIA]. 2009. U.S. Central Intelligence Agency. *The World Fact Book.*

[FAO]. 2006. Food and Agriculture Organization. *Livestock's Long Shadow: Environmental Issues and Options.* Food and Agriculture Organization of the United Nations.

[IPCC]. 2007. Intergovernmental Panel on Climate Change. *Climate Change 2007: Impacts, Adaptations and Vulnerabilities.* Cambridge, UK: Cambridge University Press.

[NDA]. 2009. Nebraska Department of Agriculture. NDA announces additional quarantines in bovine tuberculosis case. Nebraska Department of Agriculture News Release.

[NPIC]. 2009. National Pesticide Information Center. Permethrin technical fact sheet. Oregon State University and U.S. Environmental Protection Agency.

[TAHC]. 2009. Texas Animal Health Commission. Cattle tuberculosis confirmed in Texas: Check with states of destination before shipping cattle. Texas Department of Agriculture News Release.

[UK]. 2000. United Kingdom Parliament Select Committee on Science and Technology, Fifth Report. United Kingdom Parliament.

[USFWS]. 2007. U.S. Fish and Wildlife Service. International wildlife trade. U.S. Department of Interior. Washington, DC.

[WHO]. 1994. World Health Organization. Plague in India. Regional Office for Europe, Communicable Disease Report (CD News) 5.

Agnew, W., D. W. Uresk, and R. M. Hansen. 1986. Flora and fauna associated with prairie dog colonies and adjacent ungrazed mixed-grass prairie in western South Dakota. *Journal of Range Management* 39:135–139.

Albert, A., K. Drouillard, G. D. Haffner, and B. Dixon. 2007. Dietary exposure to low pesticide doses causes long-term immunosuppression in the leopard frog (*Rana pipiens*). *Environmental Toxicology & Chemistry* 26:1179–1185.

Allan, B. F., F. Keesing, and R. S. Ostfeld. 2003. Effect of forest fragmentation on Lyme disease risk. *Conservation Biology* 17:267–272.

Altizer, S., A. Dobson, P. Hosseini, P. Hudson, M. Pascual, and P. Rohani. 2006. Seasonality and the dynamics of infectious diseases. *Ecology Letters* 9:467–484.

Anderson, S. H., and E. S. Williams. 1997. Plague in a complex of white-tailed prairie dogs and associated small mammals in Wyoming. *Journal of Wildlife Diseases* 33:720–732.

Bain, D., W. A. Buttemer, L. Astheimer, K. Fildes, and M. J. Hooper. 2004. Effects of sublethal fenitrothion ingestion on cholinesterase inhibition, standard metabolism, thermal preference, and prey-capture ability in the Australian central bearded dragon (*Pogona vitticeps*, Agamidae). *Environmental Toxicology & Chemistry* 23:109–116.

Barnes, A. M. 1982. Surveillance and control of the bubonic plague in the United States. *Symposium of the Zoological Society of London* 50:237–270.

Barnett, J. B., K.E. Rodgers. 1994. *Pesticides.* New York: Raven Press.

Barrett, R. H. 1982. Habitat preferences of feral hogs, deer, and cattle on a Sierra foothill range. *Journal of Range Management* 35:342–346.

Basu, N., K. Klenavic, M. Gamberg, M. O'Brien, D. Evans, A. M. Scheuhammer, and H. M. Chan. 2005. Effects of mercury on neurochemical receptor-binding characteristics in wild mink. *Environmental Toxicology & Chemistry* 24:1444–1450.

Biggins, D. E., and M. Y. Kosoy. 2001. Influences of introduced plague on North American mammals: Implications from ecology of plague in Asia. *Journal of Mammalogy* 82:906–916.

Bridges, C. M. 1997. Tadpole swimming performance and activity affected by acute exposure to sublethal levels of carbaryl. *Environmental Toxicology & Chemistry* 16:1935–1939.

Bustnes, J. O., S. A. Hanssen, I. Folstad, K. E. Erikstad, D. Hasselquist, and J. U. Skaare. 2004. Immune function and organochlorine pollutants in arctic breeding glaucous gulls. *Archives of Environmental Contamination and Toxicology* 47:530–541.

Campbell, G. L., A. A. Marfin, R. S. Lanciotti, and D. J. Gubler. 2002. West Nile virus. *Lancet Infectious Disease* 2:519–529.

Carlson, R. W., S. P. Bradbury, R. A. Drummond, and D. E. Hammermeister. 1998. Neurological effects on startle response and escape from predation by medaka exposed to organic chemicals. *Aquatic Toxicology* 43:51–68.

Cattadori, I. M., B. Boag, O. N. Bjornstad, S. J. Cornell, and P. J. Hudson. 2005. Peak shift and epidemiology in a seasonal host-nematode system. *Proceedings Biological Sciences* 272:1163–1169.

Chanda, S. M., P. Harp, J. Liu, and C. N. Pope. 1995. Comparative developmental and maternal neurotoxicity following acute gestational exposure to chlorpyrifos in rats. *Journal of Toxicology and Environmental Health* 44:189–202.

Childs, J., R. E. Shope, D. Fish, F. X. Meslin, C. J. Peters, K. Johnson, E. Debess, D. Dennis, and S. Jenkins. 1998. Emerging zoonoses. *Emerging Infectious Diseases* 4:453–454.

Christensen, J. H., T. R. Carter, M. Rummukainen, and G. Amanatidis. 2007. Evaluating the performance and utility of regional climate models: The PRUDENCE project. *Climatic Change* 81:1–6.

Christian, K. A., R. V. Baudinette, and Y. Pamula. 1997. Energetic costs of activity by lizards in the field. *Functional Ecology* 11:392–397.

Collinge, S. K., W. C. Johnson, C. Ray, R. Matchett, J. Grensten, J. F. Cully, Jr., K. L. Gage, M. Y. Kosoy, J. E. Loye, and A. P. Martin. 2005. Testing the generality of a trophic-cascade model for plague. *EcoHealth* 2: ISSN 1612-9202(print); 1612-9210(electronic).

Cully, J. F., A. M. Barnes, T. J. Quan, and G. Maupin. 1997. Dynamics of plague in a Gunnison's prairie dog colony complex from New Mexico. *Journal of Wildlife Diseases* 33:706–719.

Cully, J. F., and E. S. Williams. 2001. Interspecific comparisons of sylvatic plague in prairie dogs. *Journal of Mammalogy* 82:894–905.

Daszak, P., and A. A. Cunningham. 2000. More on the ecological impact of fungal infections on wildlife populations. *Parasitology Today* 16:404–405.

Daszak, P., A. A. Cunningham, and A. D. Hyatt. 2000. Wildlife ecology: Emerging infectious diseases of wildlife – Threats to biodiversity and human health. *Science* 287:443–449.

Daszak, P., L. Berger, A. A. Cunningham, A. D. Hyatt, D. E. Green, and R. Speare. 1999. Emerging infectious diseases and amphibian population declines. *Emerging Infectious Diseases* 5:735–748.

Deck, A. L. 2006. Spatio-temporal relationships between feral hogs and cattle with implications for disease transmission. M.S. thesis, Texas A&M University, College Station, TX.

Dell'Omo, G., and R. F. Shore. 1996a. Behavioral and physiological effects of acute sublethal exposure to dimethoate on wood mice, *Apodemus sylvaticus*. *Archives of Environmental Contaminant Toxicology* 31:91–97.

Dell'Omo, G., and R. F. Shore. 1996b. Behavioral effects of acute sublethal exposure to dimethoate on wood mice, *Apodemus sylvaticus*: II – Field studies on radio-tagged mice in a cereal ecosystem. *Archives of Environmental Contaminant Toxicology* 31:538–542.

Deplazes, P., D. Hegglin, S. Gloor, and T. Romig. 2004. Wilderness in the city: The urbanization of *Echinococcus multilocularis*. *Trends in Parasitology* 20:77–84.

deSwart, R. L., P. S. Ross, J. G. Voss, and A. Osterhaus. 1996. Impaired immunity in harbour seals (*Phoca vitulina*) fed environmentally contaminated herring. *Veterinary Quarterly* 18:S127–S128.

Doll, J. M., P. S. Zeitz, P. Ettestad, A. L. Bucholtz, T. Davis, and K. Gage. 1994. Cat-transmitted fatal pneumonic plague in a person who traveled from Colorado to Arizona. *American Journal of Tropical Medicine and Hygiene* 51:109–114.

Donnelly, C. A., R. Woodroffe, D. R. Cox, F. J. Bourne, C. L. Cheeseman, R. S. Clifton-Hadley, G. Wei, et al. 2006. Positive and negative effects of widespread badger culling on tuberculosis in cattle. *Nature* 439:843–846.

DuRant, S. E., W. A. Hopkins, and L. G. Talent. 2007. Impaired terrestrial and arboreal locomotor performance in the western fence lizard (*Sceloporus occidentalis*) after exposure to an AChE-inhibiting pesticide. *Environmental Pollution* 149:18–24.

Eidson, M., L. A. Tierney, O. J. Rollag, T. Becker, T. Brown, and H. F. Hull. 1988. Feline plague in New Mexico: Risk factors and transmission to humans. *American Journal of Public Health* 78:1333–1335.

Eidson, M., J. P. Thilsted, and O. J. Rollag. 1991. Clinical, clinicopathological, and pathological features of plague in cats: 119 cases (1977–1988). *Journal of the American Veterinary Medical Association* 199:1191–1197.

Enscore, R. E., B. J. Biggerstaff, T. L. Brown, R. F. Fulgham, P. J. Reynolds, D. M. Engelthaler, C. E. Levy, et al. 2002. Modeling relationships between climate and the frequency of human plague cases in the southwestern United States, 1960–1997. *American Journal of Tropical Medicine and Hygiene* 66:186–196.

Epstein, P. R. 2001. Climate change and emerging infectious diseases. *Microbes and Infection* 3:747–754.

Epstein, P. R., H. F. Diaz, S. Elias, G. Grabherr, N. E. Graham, W. J. M. Martens, E. Mosley-Thompson, and J. Susskind. 1998. Biological and physical signs of climate change: Focus on mosquito-borne diseases. *Bulletin of the American Meteorological Society* 79:409–417.

Eskenazi, B., A. Bradman, and R. Castorina. 1999. Exposures of children to organophosphate pesticides and their potential adverse health effects. *Environmental Health Perspectives* 107:409–419.

Fisk, A. T., C. A. de Wit, M. Wayland, Z. Z. Kuzyk, N. Burgess, R. Letcher, B. Braune, et al. 2005. An assessment of the toxicological significance of anthropogenic contaminants in Canadian arctic wildlife. *Science Total Environment* 351–352:57–93.

Flajnik, M. F. 1998. Churchill and the immune system of ectothermic vertebrates. *Immunological Reviews* 166:5–14.

French, J. B., M. B. Voltura, and T. E. Tomasi. 2001. Effects of pre- and postnatal polychlorinated biphenyl exposure on metabolic rate and thyroid hormones of white-footed mice. *Environmental Toxicology & Chemistry* 20:1704–1708.

Gage, K. L., R. S. Ostfeld, and J. G. Olson. 1995. Nonviral vector-borne zoonoses associated with mammals in the United States. *Journal of Mammalogy* 76:695–715.

Gage, K. L., D. T. Dennis, K. A. Orloski, P. Ettestad, T. L. Brown, P. J. Reynolds, W. J. Pape, C. L. Fritz, L. G. Carter, and J. D. Stein. 2000. Cases of cat-associated human plague in the western U.S., 1977–1998. *Clinical Infectious Diseases* 30:893–900.

Gage, K. L., J.A. Montenieri. 1994. The role of predators in the ecology, epidemiology, and surveillance of plague in the United States. *Vertebrate Pest Conference* 16:200–206.

Galloway, T., and R. Handy. 2003. Immunotoxicity of organophosphorous pesticides. *Ecotoxicology* 12:345–363.

Garnett, B. T., R. J. Delahay, and T. J. Roper. 2002. Use of cattle farm resources by badgers (*Meles meles*) and risk of bovine tuberculosis (*Mycobacterium bovis*) transmission to cattle. *Proceedings of the Royal Society of London Series B – Biological Sciences* 269:1487–1491.

Gasper, P. W., A. M. Barnes, T. J. Quan, J. P. Benziger, L. G. Carter, M. L. Beard, and G. O. Maupin. 1993. Plague (*Yersinia pestis*) in cats: Description of experimentally induced disease. *Journal of Medical Entomology* 30:20–26.

Gerhold, R. W., C. M. Tate, S. E. Gibbs, D. G. Mead, A. B. Allison, and J. R. Fischer. 2007. Necropsy findings and arbovirus surveillance in mourning doves from the southeastern United States. *Journal of Wildlife Diseases* 43:129–135.

Gilbertson, M. K., G. D. Haffner, K. G. Drouillard, A. Albert, and B. Dixon. 2003. Immunosuppression in the northern leopard frog (*Rana pipiens*) induced by pesticide exposure. *Environmental Toxicology & Chemistry* 22:101–110.

Gillespie, T. R., C. A. Chapman, and E. C. Greiner. 2005. Effects of logging on gastrointestinal parasite infections and infection risk in African primates. *Journal of Applied Ecology* 42:699–707.

Gipson, P. S., J. K. Veatch, R. S. Matlack, and D. P. Jones. 1999. Health status of a recently discovered population of feral swine in Kansas. *Journal of Wildlife Diseases* 35:624–627.

Gormley, E., and E. Costello. 2003. Tuberculosis and badgers: New approaches to diagnosis and control. *Journal of Applied Microbiology* 94:80S–86S.

Grasman, K. A., and G. A. Fox. 2001. Associations between altered immune function and organochlorine contamination in young Caspian terns (*Sterna caspia*) from Lake Huron, 1997–1999. *Ecotoxicology* 10:101–114.

Grasman, K. A., G. A. Fox, P. F. Scanlon, and J. P. Ludwig. 1996. Organochlorine-associated immunosuppression in prefledgling Caspian terns and herring gulls from the Great Lakes: An ecoepidemiological study. *Environmental Health Perspectives* 104:829–842.

Grenfell, B., and J. Harwood. 1997. (Meta)population dynamics of infectious diseases. *Trends in Ecology & Evolution* 12:395–399.

Griffiths, J. K. 1998. Human cryptosporidiosis: Epidemiology, transmission, clinical disease, treatment, and diagnosis. *Advances in Parasitology – Opportunistic Protozoa in Humans* 40:37–85.

Gubler, D. J. 1998. Resurgent vector-borne diseases as a global health problem. *Emerging Infectious Diseases* 4:442–450.

Guerra, M. A., A. T. Curns, C. E. Rupprecht, C. A. Hanlon, J. W. Krebs, and J. E. Childs. 2003. Skunk and raccoon rabies in the eastern United States: Temporal and spatial analysis. *Emerging Infectious Diseases* 9:1143–1150.

Hill, E. F., and M. B. Camardese. 1984. Toxicity of anticholinesterase insecticides to birds: Technical grade versus granular formulations. *Ecotoxicology and Environmental Safety* 8:551–563.

Holem, R. R., W. A. Hopkins, and L. G. Talent. 2008. Effects of repeated exposure to malathion on growth, food consumption, and locomotor performance of the western fence lizard (*Sceloporus occidentalis*). *Environmental Pollution* 152:92–98.

Holliday, D. K., A. A. Elskus, and W. M. Roosenburg. 2009. Impacts of multiple stressors on growth and metabolic rate of *Malaclemys* terrapin. *Environmental Toxicology & Chemistry* 28:338–345.

Hopkins, W. A., C. L. Rowe, and J. D. Congdon. 1999. Elevated trace element concentrations and standard metabolic rate in banded water snakes (*Nerodia fasciata*) exposed to coal combustion wastes. *Environmental Toxicology & Chemistry* 18:1258–1263.

Hopkins, W. A., C. T. Winne, and S. E. DuRant. 2005. Differential swimming performance of two natricine snakes exposed to a cholinesterase-inhibiting pesticide. *Environmental Pollution* 133:531–540.

Iwaniuk, A. N., D. T. Koperski, K. M. Cheng, J. E. Elliott, L. K. Smith, L. K. Wilson, and D. R. W. Wylie. 2006. The effects of environmental exposure to DDT on the brain of a songbird: Changes in structures associated with mating and song. *Behavioural Brain Research* 173:1–10.

Jansen, A., E. Luge, B. Guerra, P. Wittschen, A. D. Gruber, C. Loddenkemper, T. Schneider, et al. 2007. Leptospirosis in urban wild boars, Berlin, Germany. *Emerging Infectious Diseases* 13:739–742.

Jay, M. T., M. Cooley, D. Carychao, G. W. Wiscomb, R. A. Sweitzer, L. Crawford-Miksza, J. A. Farrar, et al. 2007. *Escherichia coli* O157:H7 in feral swine near spinach fields and cattle, central California coast. *Emerging Infectious Diseases* 13:1908–1911.

Jetten, T. H., and D. A. Focks. 1997. Potential changes in the distribution of dengue transmission under climate warming. *American Journal of Tropical Medicine and Hygiene* 57:285–297.

Kaneene, J. B., M. VanderKlok, C. S. Bruning-Fann, M. V. Palmer, D. L. Whipple, S. M. Schmitt, and R. Miller. 2002. Prevalence of *Mycobacterium bovis* infection in cervids on privately owned ranches. *Journal of the American Veterinary Medical Association* 220:656–659.

Kausrud, K. L., H. Viljugrein, A. Frigessi, M. Begon, S. Davis, H. Leirs, V. Dubyanskiy, and N. C. Stenseth. 2007. Climatically driven synchrony of gerbil populations allows large-scale plague outbreaks. *Proceedings of the Royal Society B – Biological Sciences* 274:1963–1969.

Lechleitner, R. R., L. Kartman, M.I. Goldenberg, B.W. Hudson. 1968. An epizootic of plague in Gunnison's prairie dogs (*Cynomys gunnisoni*) in south-central Colorado. *Ecology* 49:734–743.

Letkova, V., P. Lazar, J. Curlik, M. Goldova, A. Kocisova, L. Kosuthova, and J. Mojzisova. 2006. The red fox (*Vulpes vulpes* L.) as a source of zoonoses. *Veterinarski Arhiv* 76:S73–S81.

Lie, E., H. J. S. Larsen, S. Larsen, G. M. Johnsen, A. E. Derocher, N. J. Lunn, R. J. Norstrom, O. Wiig, and J. U. Skaare. 2004. Does high organochlorine (OC) exposure impair the resistance to infection in polar bears (*Ursus maritimus*)? Part 1: Effect of OCs on the humoral immunity. *Journal of Toxicology and Environmental Health, Part A: Current Issues* 67:555–582.

Lie, E., H. J. S. Larsen, S. Larsen, G. M. Johansen, A. E. Derocher, N. J. Lunn, R. J. Norstrom, O. Wiig, and J. U. Skaare. 2005. Does high organochlorine (OC) exposure impair the resistance to infection in polar bears (*Ursus maritimus*)? Part II: Possible effect of OCs on mitogen- and antigen-induced lymphocyte proliferation. *Journal of Toxicology and Environmental Health, Part A: Current Issues* 68:457–484.

Lindstrom, E. R., H. Andren, P. Angelstam, G. Cederlund, B. Hornfeldt, L. Jaderberg, P. A. Lemnell, B. Martinsson, K. Skold, and J. E. Swenson. 1994. Disease reveals the predator: Sarcoptic mange, red fox predation, and prey populations. *Ecology* 75:1042–1049.

Linzey, D. W., J. Burroughs, L. Hudson, M. Marini, J. Robertson, J. P. Bacon, M. Nagarkatti, and P. S. Nagarkatti. 2003. Role of environmental pollutants on immune functions, parasitic infections and limb malformations in marine toads and whistling frogs from Bermuda. *International Journal of Environmental Health Research* 13:125–148.

Lomolino, M. V., and G. A. Smith. 2001. Dynamic biogeography of prairie dog (*Cynomys ludovicianus*) towns near the edge of their range. *Journal of Mammalogy* 82:937–945.

Macdonald, D. W., and M. K. Laurenson. 2006. Infectious disease: Inextricable linkages between human and ecosystem health. *Biological Conservation* 131:143–150.

Mayne, G. J., P. A. Martin, C. A. Bishop, and H. J. Boermans. 2004. Stress and immune responses of nestling tree swallows (*Tachycineta bicolor*) and eastern bluebirds (*Sialia sialis*) exposed to nonpersistent pesticides and p,p′,-dichlorodiphenyldichloroethylene in apple orchards of southern Ontario, Canada. *Environmental Toxicology & Chemistry* 23:2930–2940.

McInnis, M. L., and M. Vavra. 1987. Dietary relationships among feral horses, cattle, and pronghorn in southeastern Oregon. *Journal of Range Management* 40:60–66.

Meites, E., M. T. Jay, S. Deresinski, W. J. Shieh, S. R. Zaki, L. Tompkins, and D. S. Smith. 2004. Reemerging leptospirosis, California. *Emerging Infectious Diseases* 10:406–412.

Menzano, A., L. Rambozzi, and L. Rossi. 2007. A severe episode of wildlife-derived scabies in domestic goats in Italy. *Small Ruminant Research* 70:154–158.

Merck. 2005. *The Merck Veterinary Manual*. Whitehouse Station, NJ: Merck & Co., Inc.

Michel, A. L., M. L. Coetzee, D. F. Keet, L. Mare, R. Warren, D. Cooper, R. G. Bengis, K. Kremer, and P. van Helden. 2009. Molecular epidemiology of *Mycobacterium bovis* isolates from free-ranging wildlife in South African game reserves. *Veterinary Microbiology* 133:335–343.

Miles, V. I., M. J. Wilcomb, Jr., and J. V. Irons. 1952. Rodent plague in the Texas South Plains, 1947–49; with ecological considerations. *Public Health Monograph* 6:39–53.

Montie, E. W., C. M. Reddy, W. A. Gebbink, K. E. Touhey, M. E. Hahn, and R. J. Letcher. 2009. Organohalogen contaminants and metabolites in cerebrospinal fluid and cerebellum gray matter in short-beaked common dolphins and Atlantic white-sided dolphins from the western North Atlantic. *Environmental Pollution* 157:2345–2358.

Murphy, F. A., and N. Nathanson. 1994. The emergence of new virus diseases: An overview. *Seminars in Virology* 5:87–102.

Muto, M. A., F. Lobelle, J. H. Bidanset, and J. N. D. Wurpel. 1992. Embryotoxicity and neurotoxicity in rats associated with prenatal exposure to dursban. *Veterinary and Human Toxicology* 34:498–501.

Nakazawa, Y., R. Williams, A. T. Peterson, P. Mead, E. Staples, and K. L. Gage. 2007. Climate change effects on plague and tularemia in the United States. *Vector-Borne and Zoonotic Diseases* 7:529–540.

O'Toole, T., M. Mair, and T. V. Inglesby. 2002. Shining light on 'Dark Winter.' *Clinical Infectious Diseases* 34:972–83.

Ostfeld, R. S. 2009. Biodiversity loss and the rise of zoonotic pathogens. *Clinical Microbiology and Infection* 15:40–43.

Otis, D. L., J.H. Schulz, D. Miller, R.E. Mirarchi, T.S. Baskett. 2008. Mourning dove (*Zenaida macroura*). *The Birds of North America Online*. Cornell Lab of Ornithology.

Parmenter, R. R., E. P. Yadav, C. A. Parmenter, P. Ettestad, and K. L. Gage. 1999. Incidence of plague associated with increased winter-spring precipitation in New Mexico. *American Journal of Tropical Medicine and Hygiene* 61:814–821.

Perry, R. D., and J. D. Fetherston. 1997. *Yersinia pestis*: Etiologic agent of plague. Clinical *Microbiology Reviews* 10:35–66.

Pinzon, J. E., J. M. Wilson, C. J. Tucker, R. Arthur, P. B. Jahrling, and P. Formenty. 2004. Trigger events: Enviroclimatic coupling of Ebola hemorrhagic fever outbreaks. *American Journal of Tropical Medicine and Hygiene* 71:664–74.

Plowright, R. K., H. E. Field, C. Smith, A. Divljan, C. Palmer, G. Tabor, P. Daszak, and J. E. Foley. 2008. Reproduction and nutritional stress are risk factors for Hendra virus infection in little red flying foxes (*Pteropus scapulatus*). *Proceedings of the Royal Society B – Biological Sciences* 275:861–869.

Randolph, S. E., R. M. Green, M. F. Peacey, and D. J. Rogers. 2000. Seasonal synchrony: The key to tick-borne encephalitis foci identified by satellite data. *Parasitology* 121:15–23.

Reed, K. D., J. K. Meece, J. S. Henkel, and S. K. Shukla. 2003. Birds, migration and emerging zoonoses: West Nile virus, Lyme disease, influenza A and enteropathogens. *Clinical Medicine Research* 1:5–12.

Reintjes, R., I. Dedushaj, A. Gjini, T. R. Jorgensen, B. Cotter, A. Lieftucht, F. D'Ancona, et al. 2002. Tularemia outbreak investigation in Kosovo: Case control and environmental studies. *Emerging Infectious Diseases* 8:69–73.

Renter, D. G., D. P. Gnad, J. M. Sargeant, and S. E. Hygnstrom. 2006. Prevalence and serovars of Salmonella in the feces of free-ranging white-tailed deer (*Odocoileus virginianus*) in Nebraska. *Journal of Wildlife Diseases* 42:699–703.

Rotz, L. D., A. S. Khan, S. R. Lillibridge, S. M. Ostroff, and J. M. Hughes. 2002. Public health assessment of potential biological terrorism agents. *Emerging Infectious Diseases* 8:225–230.

Rust, J. H., Jr., D.C. Cavanaugh, R. O'Shita, J.D. Marshall, Jr. 1971. The role of domestic animals in the epidemiology of plague: I. Experimental infection of dogs and cats. *Journal of Infectious Diseases* 124:522–526.

Salter, R. E., and R.J. Hudson. 1980. Range relationships of feral horses with wild ungulates and cattle in western Alberta. *Journal of Range Management* 33:266–271.

Schantz, S. L., J. J. Widholm, and D. C. Rice. 2003. Effects of PCB exposure on neuropsychological function in children. *Environmental Health Perspectives* 111:357–376.

Schauber, E. M., R. S. Ostfeld, and A. S. Evans. 2005. What is the best predictor of annual Lyme disease incidence: Weather, mice, or acorns? *Ecological Applications* 15:575–586.

Schmitt, S. M., D. J. O'Brien, C. S. Bruning-Fann, and S. D. Fitzgerald. 2002. Bovine tuberculosis in Michigan wildlife and livestock. *Domestic Animal/Wildlife Interface: Issue for Disease Control, Conservation, Sustainable Food Production, and Emerging Diseases* 969:262–268.

Sharon, J., ed. 1998. *Basic Immunology*. Baltimore, MD: Williams & Wilkeins.

Smith, V. 2007. Host resource supplies influence the dynamics and outcome of infectious disease. *Integrative and Comparative Biology* 47:310–316.

Snäll, T., R. E. Benestad, and N. C. Stenseth. 2009. Expected future plague levels in a wildlife host under different scenarios of climate change. *Global Change Biology* 15:500–507.

Snäll, T., R. B. O'Hara, C. Ray, and S. K. Collinge. 2008. Climate-driven spatial dynamics of plague among prairie dog colonies. *American Naturalist* 171:238–248.

Stenseth, N. C., and J. W. Hurrell. 2005. Global climate change: Building links between the climate and ecosystem impact research communities. *Climate Research* 29:181–182.

Stenseth, N. C., A. Mysterud, G. Ottersen, J. W. Hurrell, K. S. Chan, and M. Lima. 2002. Ecological effects of climate fluctuations. *Science* 297:1292–1296.

Stenseth, N. C., N. I. Samia, H. Viljugrein, K. L. Kausrud, M. Begon, S. Davis, H. Leirs, et al. 2006. Plague dynamics are driven by climate variation. *Proceedings of the National Academy of Sciences of the United States of America* 103:13110–13115.

Suzan, G., E. Marce, J. T. Giermakowski, B. Armien, J. Pascale, J. Mills, G. Ceballos, et al. 2008. The effect of habitat fragmentation and species diversity loss on hantavirus prevalence in Panama. *Animal Biodiversity and Emerging Diseases: Prediction and Prevention* 1149:80–83.

Taylor, L. H., S. M. Latham, and M. E. J. Woolhouse. 2001. Risk factors for human disease emergence. *Philosophical Transactions of the Royal Society of London, Series B – Biological Sciences* 356:983–989.

Treml, F., J. Pikula, H. Bandouchova, and J. Horakova. 2007. European brown hare as a potential source of zoonotic agents. *Veterinarni Medicina* 52:451–456.

Trevejo, R. T., J. G. Rigau-Perez, D. A. Ashford, E. M. McClure, C. Jarquin-Gonzalez, J. J. Amador, J. O. de los Reyes, et al. 1998. Epidemic leptospirosis associated with pulmonary hemorrhage — Nicaragua, 1995. *Journal of Infectious Diseases* 178:1457–1463.

Trewby, L. D., G. J. Wilson, R. J. Delahay, N. Walker, R. Young, J. Davison, C. Cheeseman, P. A. Robertson, M. L. Gorman, and R. A. McDonald. 2008. Experimental evidence of competitive release in sympatric carnivores. *Biology Letters* 4:170–172.

Turner, R. J. 1994. *Immunology: A Comparative Approach*. Chichester, UK: Wiley.

Vial, T., B. Nicolas, and J. Descotes. 1996. Clinical immunotoxicity of pesticides. *Journal of Toxicology and Environmental Health* 48:215–229.

Voccia, I., B. Blakley, P. Brousseau, and M. Fournier. 1999. Immunotoxicity of pesticides: A review. *Toxicology and Industrial Health* 15:119–132.

Voltura, M. B., and J. B. French. 2000. Effects of dietary polychlorinated biphenyl exposure on energetics of white-footed mouse, *Peromyscus leucopus*. *Environmental Toxicology & Chemistry* 19:2757–2761.

Werner, S. B., C. E. Weidmer, B. C. Nelson, G. S. Nygaard, R. M. Goethals, and J. D. Poland. 1984. Primary plague pneumonia contracted from a cat at South Lake Tahoe, Calif. *Journal of the American Medical Association* 251:929–931.

Williams, E. S., M. W. Miller, T. J. Kreeger, R. H. Kahn, and E. T. Thorne. 2002. Chronic wasting disease of deer and elk: A review with recommendations for management. *Journal of Wildlife Management* 66:551–563.

Wilson, T. M., L. Logan-Henfrey, R. Weller, and B. Kellman. 2000. Agroterrorism, biological crimes, and biological warfare targeting animal agriculture. In *Emerging Diseases of Animals,* edited by C. Brown and C. Bolin: 23–57. Washington, DC: ASM Press.

Wolfe, N. D., W. M. Switzer, J. K. Carr, V. B. Bhullar, V. Shanmugam, U. Tamoufe, A. T. Prosser, et al. 2004. Naturally acquired simian retrovirus infections in central African hunters. *Lancet* 363:932–937.

Woodroffe, R., C. A. Donnelly, D. R. Cox, F. J. Bourne, C. L. Cheeseman, R. J. Delahay, G. Gettinby, J. P. McInerney, and W. I. Morrison. 2006a. Effects of culling on badger (*Meles meles*) spatial organization: Implications for the control of bovine tuberculosis. *Journal of Applied Ecology* 43:1–10.

Woodroffe, R., C. A. Donnelly, H. E. Jenkins, W. T. Johnston, D. R. Cox, F. J. Bourne, C. L. Cheeseman, et al. 2006b. Culling and cattle controls influence tuberculosis risk for badgers. *Proceedings of the National Academy of Sciences of the United States of America* 103:14713–14717.

Wyckoff, A. C., S. E. Henke, T. Campbell, D. G. Hewitt, K. VerCauteren. 2005. Preliminary serologic survey of selected diseases and movements of feral swine in Texas. *Wildlife Damage Management Working Conference* 11:23–32.

5 Impacts of Contaminants and Pesticides on Biodiversity and Ecosystem Structure and Function

Thomas E. Lacher, Jr., John W. Bickham, Claude Gascon, Rhys E. Green, Robin D. Moore, and Miguel Mora

CONTENTS

5.1 INTRODUCTION

5.1.1 The Common History of Ecotoxicology and Biodiversity

It was 1962, and the term *biodiversity* did not yet exist, nor did the term *ecotoxicology*, yet the relationship between pesticides and biodiversity was clearly and eloquently framed with the publication of *Silent Spring* by Rachel Carson (1962). The book presented the possibility that large-scale declines in avian populations were due to the application of DDT in the environment, with subsequent cascading impacts on the larger ecosystem. Biodiversity was first mentioned in a publication by Raymond Dasmann in 1968 but did not come into wide usage until the 1980s, most notably with the publication of *Biodiversity* (Wilson and Peter 1988), and the word *ecotoxicology* was coined by Truhaut in 1969 (Truhaut 1977). Although *Silent Spring* raised awareness of the effects of toxic substances on wildlife, the focus of environmental toxicology for many years was on single species effects, in many cases targeting impacts on individuals and extrapolating those to estimate the risk to populations.

The early days of environmental toxicology were heavily focused on elucidating the chemical processes related to the transfer and transformation of compounds in the natural environment (Kendall et al. 1996) with a heavy emphasis on analytical chemistry. An approach much like the cancer-related research that focused on specific chemical impacts on metabolic process dominated this early, more mechanistic stage of research. These results were often tested on model organisms such as *Gambusia affinis* for aquatic systems and bobwhite quail (*Colinus virginianus*), mallard ducks (*Anas platyrhynchos*), and European starlings (*Sturnus vulgaris*) for terrestrial systems. Studies were heavily focused on dose-response relationships, frequently with mortality as an endpoint (Yu 2005). Because more subtle effects can occur that impact reproduction, survivorship, and behavior, research now focuses on the assessment of the impact of chemicals on specific biomarkers ("a xenobiotically induced alteration in cellular or biochemical components or processes, structures, or functions that is measurable in a biological system or sample," (Kendall et al. 1996)), which has allowed for the detection of more subtle effects, although the ecological significance is still uncertain (Forbes et al. 2006). There has also been an emphasis on ecological and population-level effects (Barnthouse and Sorenson 2007). The use of biomarkers other than mortality has resulted in more extensive field-testing of the impacts of chemicals on a wider range of species. This has led to the development of more sophisticated toxicity testing and an effective integration of controlled laboratory and well-designed field experimental studies. This integration is necessary to generate relevant data for the protection of wildlife and fish populations in the field (Kendall and Lacher 1991, 1994).

Data generated in these studies are often then used in the development of generalized probabilistic risk assessments. Risk assessments combine data on the fate of toxic substances and their route and rate of exposure with estimates of the hazard to wildlife (U.S. EPA 1992; Chapter 9, this volume). Probabilistic risk assessments rely heavily on modeling approaches and estimates of uncertainty to extrapolate risk estimates to the population and community level. Combining risk assessment approaches with basic information on ecological processes such as food webs and other interspecific interactions reveals that impacts can often affect multiple species in ways that impact ecosystem function and ecosystem services and result in the decline of biodiversity at regional and global scales. In many respects the study of ecotoxicology developed along a timeframe very similar to that of the growth in interest in biodiversity as an environmental service and as a target for conservation. Current concerns about the loss of critical components of biodiversity have in fact indicated that environmental contaminants play a key role in the declines of many species across taxa (Stuart et al. 2004; Carpenter et al. 2008; Schipper et al. 2008), and can argue for the alteration of the more traditional ecotoxicology research designs for evaluating the larger impacts of toxic substances on terrestrial and aquatic biodiversity. As a better understanding of the links between biodiversity and ecosystem function emerges, research on the impact of contaminants on biodiversity can assist in the preservation of valuable ecosystem services (Balmford et al. 2002) and will result in better integration of ecotoxicology and biodiversity conservation.

5.1.2 Definitions of Biodiversity

Biodiversity is a term that, in spite of wide usage, does not have a universally accepted quantifiable definition. The most widely used definition is in the foreword of the *Handbook of the Convention of Biological Diversity* (Secretariat of the Convention on Biological Diversity 2005):

> Biological diversity—or biodiversity—is the term given to the variety of life on Earth, including plants, animals, and microorganisms, as well as the ecosystems of which they are a part. Biodiversity includes genetic differences within species, the diversity of species and the variety of ecosystems.

This definition, or minor variations of it, is used in most of the literature of the Convention on Biological Diversity (CBD), including the *Global Biodiversity Outlook 2* (Secretariat of the Convention on Biological Diversity 2006), which defines targets for the conservation and management of biodiversity. Generally three levels of biodiversity are targeted in research and conservation activities: genetic diversity, species diversity, and ecosystem diversity (Alonso et al. 2001); but these targets present problems of measurement for the generation of basic data, as well as in obtaining metrics of conservation success, particularly at the genetic and ecosystem levels. All three components are listed as goals within the 2010 Biodiversity target of the CBD (www.biodiv.org/2010-target), and there are specific conservation objectives defined, though the metrics needed to measure progress toward these targets are less specific. In order for ecotoxicologists to integrate biodiversity as an endpoint in studies of impacts of contaminants and pesticides, these metrics need to be clearly defined. However, for the research to be relevant to the conservation of biodiversity on a global scale, the metrics need to provide useable input toward the achievement of the conservation goals outlined in the 2010 targets of the CBD. This presents significant challenges and trade-offs.

In addition to meeting the goals of the 2010 target, the relationship between loss of biodiversity and ecosystem function is relevant to the Millennium Ecosystem Assessment (www.millenniumassessment.org) since this assessment and subsequent recommendations attempt to clarify the relationship between biodiversity and ecosystem services that provide benefits to human populations (Alcamo et al. 2003). This in turn feeds into Goal 7: Environmental Sustainability, of the Millennium Development Goals, which coordinates resources and efforts of many United Nations and other international agencies and organizations (www.un.org/millenniumgoals). Clear definitions of the metrics of monitoring and assessment of biodiversity are clearly critical to the generation of research results that benefit major global initiatives.

5.1.3 Metrics of Biodiversity

To be relevant to ecotoxicology, there need to be measurable indicators of the impact of stressors on aspects of biodiversity. These can be generated at all three levels of biodiversity: genetic, species, and ecosystem; and the case studies in this chapter present examples at all levels. Genetic impacts can be measured at the level of chromosomal aberrations, site mutations, and loss of genetic diversity, with these studies falling under the field of evolutionary toxicology (Bickham et al. 2000). An example of the use of genetic markers to study the impact of stressors on natural populations is presented in Case Study 3 in this chapter. Measurements of impacts at the species level focus on mortality, reduced reproductive output, or altered behavior in populations when the focus is a single species. Alternatively, biodiversity-related impacts will generally look at impacts on assemblages of species. Early studies in aquatic ecology used changes in diversity measures to evaluate impacts on stream and lake ecosystems (Hoffman et al. 2002). Common indices include the Shannon-Weiner Diversity Index, Simpson's Index, and other derived variables. All of these indices generate a single number that involves a combination of a measure of the number of species in a system (richness) and the proportional distribution of individuals among species (evenness). As a consequence, changes in the magnitude of these measures are difficult to interpret (Hurlbert 1971) and also suffer from methodological difficulties, such as the assumption of equal survey effort across all groups used in the calculation of the index, an assumption rarely, if ever, met (Magurran 1988). As a result, the

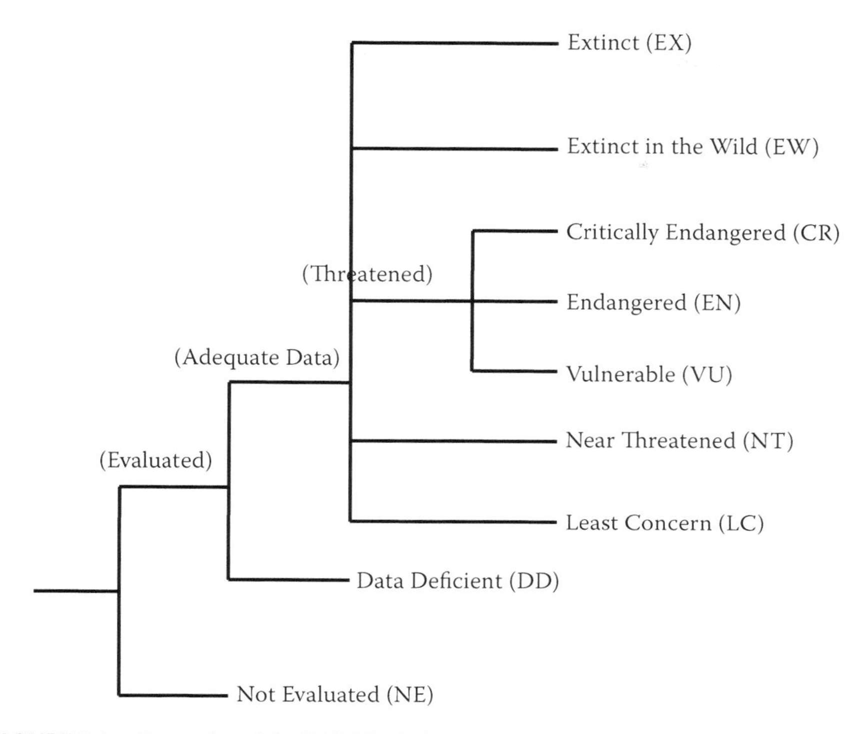

FIGURE 5.1 Categories of the IUCN Red List. For species with adequate data for classification, the relevant categories proceed from Least Concern through Extinct.

most widely used measure of the impact on biodiversity is a change in the number of species present, or richness. Richness is also not without fault, since it weights the importance of all species equally. Nevertheless, intensive sampling and the use of rarefaction approaches can generally generate reliable estimates of richness for more easily observed species (Gotelli and Colwell 2001).

Tracking the risk to biodiversity, even when the focus is on species, is a challenging process. There are far too many species, even of vertebrates, to gain a complete inventory of trends. In addition, it is important to use a consistent methodology when categorizing the conservation status of species and to monitor trends over time. The IUCN Red List (Baillie and Groombridge 1996; Lamoreux et al. 2003) is the global standard for evaluating the conservation status of species (Figure 5.1). The Red List process provides a rigorous, quantitative standard for categorization of the conservation status across taxa. The Red List has been used by IUCN since 1948, and originally the assessments were published in hard-copy books. More recently, the IUCN has embarked on an ambitious project of global assessments (Baillie et al. 2004) with the objective of assessing entire taxonomic groups and posting the results on line, available free to the public and reviewable by experts. The first group completed was the threatened and endangered birds (Birdlife International 2000), and recent assessments were completed for amphibians (Stuart et al. 2004), hard corals (Carpenter et al. 2008), and mammals (Schipper et el. 2008). All assessments undergo periodic complete revision and updates (the amphibian assessment has just undergone the first full update: www.iucnredlist.org/amphibians).

These assessments can contribute greatly to ecotoxicological studies, in particular those that assess the risk of toxic substances to populations of these species. The three most recent assessments in fact indicated significant risk to amphibians, marine mammals, and corals from pollution and contaminants. Although the assessments are snapshots in time, albeit with periodic revisions, a new tool has been proposed to track trends over time. The Red List Index (Butchart et al. 2004) develops annual trend assessments for selected species, showing trends in conservation status from year to year. The use of a Red List Index tool to track changes in status of those species under risk from toxic substance exposure will be a valuable tool in future ecotoxicological studies. However, the Red List index still presents problems when extrapolating abundance and rarity across spatial scales, although methods have been proposed that can address these concerns (Kunin 1998).

Another bias associated with the use of changes in richness as a metric is the fact that not all species have the same impact and function within an ecosystem. Often the focus is on surrogates, preferentially on indicator, umbrella, or economically important species (Fleishman et al. 2000; Rodrigues and Brooks 2007). In truth, all conservation planning is based in part on surrogates (since not all biodiversity can ever be assessed), even though the use of surrogates as indicators of other taxa or larger sets of species has mixed results (Rodrigues and Brooks 2007). Because much ecotoxicological work is focused on observable species that utilize altered or disturbed landscapes, the need to use surrogates is probably limited, and most affected species can be observed and used in estimating impacts on richness.

Perhaps the greatest challenge in using biodiversity as an endpoint in ecotoxicological research is in studies focused on the impact of stressors on ecosystem processes. This is particularly the case when the impacted component is biodiversity as an indicator of these processes. Ecosystem processes provide ecosystem services, and there is growing interest in the role of biodiversity in providing these services (Balmford et al. 2002) and in the economic value of these services (Costanza et al. 1997). The growing interest in the valuation of ecosystem services has shifted the emphasis of many international NGOs away from the conservation of species biodiversity toward biodiversity as a benefit to ecosystem services and human welfare (Turner et al. 2007; Wu 2009). The key issues concern selection of the species or suites of species that best reflect the ecosystems services of interest coupled with an understanding of the relationship of the measured impact on biodiversity and the concomitant response in the ecosystem service of interest. There is significant debate over whether suites of species or smaller subsets of species dominate ecosystem service performance and overall ecosystem function (Hooper et al. 2005; Gamfeldt et al. 2008). Thus, although there is great value in linking the impacts of stressors to ecosystem function through biodiversity, much research still needs to be done to define mechanisms of action and to elucidate causal or functional relationships between exposure and response.

5.2 CASE STUDIES: THE IMPACT OF CONTAMINANTS AND PESTICIDES ON BIODIVERSITY AND ECOSYSTEM PROCESSES

Effects of contaminants can occur on a number of levels: impacting guilds of organisms with similar feeding strategies or resource use; affecting entire taxonomic groups of organisms; impacting community function broadly defined by disturbing evolutionary process; and impacting communities and mosaics of ecosystems in complex ways. We will address these impacts by using four selected case studies.

The following case studies each address aspects of the impact of contaminants and pesticides on biodiversity. These impacts occur at all measurable levels of biodiversity, including genetic, species/population, and ecosystem. The particular compounds involved range from veterinary pharmaceuticals to petroleum residues to pesticides, and the modes of action vary from case study to case study. Although these case studies do not present an exhaustive treatment of impacts on biodiversity, they illustrate the complex pathways, unexpected consequences, and diversity of modes of action that must be considered when using biodiversity as an endpoint. They also present a foreboding picture of the consequences of the uncontrolled release of toxic substances into the environment.

5.3 CASE STUDY 1: IMPACTS ON AN IMPORTANT ECOLOGICAL GUILD

The decline in Asian vultures represents the loss of an ecologically important guild of birds: scavengers on large carcasses (Figure 5.2). In addition, this particular case study addresses a new chemical threat to wildlife: the consumption of, or exposure to, pharmaceuticals by non-target organisms. The release of estrogen or estrogenic compounds into the environment, with the possible disruption of endocrine function in non-target organisms (Colborn et al. 1996), is a similar phenomenon. All these

FIGURE 5.2 Case studies on the impacts of contaminants and pesticides, broadly defined, include the decline in *Gyps* vultures in India and global declines in amphibians—a *Hypsiboas fasciatus*, from southern Ecuador (photo by Robin D. Moore) (a); petroleum residues and the impact on freshwater vertebrates in Azerbaijan—a Soviet era oilfield at the edge of Baku Harbor (photo by John Bickham) (b); and the impacts of agriculture and agricultural pesticides on avian communities.

compounds represent a new class of anthropogenic contaminants that will require new methods of assessment and regulation.

5.3.1 Impacts of a Veterinary Pharmaceutical for Livestock on Biodiversity and Ecosystem Function at the Guild Level: Veterinary Diclofenac and the Decline of Asian Vultures

Rhys E. Green

Vultures have played an important role in Eurasian and African ecosystems for millions of years. The eight vulture species in the genus *Gyps* are widely distributed across Europe, Asia, and Africa, and all are obligate scavengers on the carcasses of large animals, especially ungulates. Formerly, they depended upon deaths in large herds of wild migratory ungulates, and this is still the case in restricted parts of Africa. However, the disappearance of these herds and their replacement by pastoralism has led vultures to depend to an increasing extent on the carcasses of domesticated ungulates. The Indian subcontinent holds hundreds of millions of domesticated ungulates, especially cattle and water buffalo, which are not killed for food in many areas because of religious traditions. Carcasses of dead animals are typically disposed of by leaving them to be scavenged where they died or by taking them to carcass dumps where skinners collect the hides before scavenging animals consume the flesh. Hence, there is a huge potential food supply for vultures. *Gyps* vultures were abundant in the region, probably numbering tens of millions, until the late 20th century. However, beginning in the 1990s, populations of all three *Gyps* vulture species endemic to the Indian subcontinent (*Gyps bengalensis*, *G. indicus*, and *G. tenuirostris*) collapsed and all became listed as Critically Endangered by the IUCN. Population surveys in India, Nepal, and Pakistan showed that the rate of decline was rapid, up to 50% per year for one species, and that it is still in progress (Baral et al. 2004; AVPP 2007; Prakash et al. 2007). The current population level of one of the Asian endemic species in India (*G. bengalensis*) is estimated at 0.1% of that in the early 1990s, with about 3% of the former populations of the other species remaining (Prakash et al. 2007). The decline has been so rapid that elevated mortality of adult birds must be the key underlying demographic mechanism. The normal life expectancy of adult vultures is high, so even drastic impairment of reproduction or the survival of immature birds would produce population declines at rates of only 5% to 10% per year, which is much slower than those observed (Pain et al. 2008).

Because of the importance of increased adult mortality, post-mortem studies of the causes of death of wild vultures have provided vital clues as to the cause of the population decline. From 1999 onward, accumulated post mortems performed in India, Pakistan, and Nepal consistently showed that the majority of dead *Gyps* vultures had extensive visceral gout, an accumulation of uric acid crystals in the tissues caused by kidney failure (Shultz et al. 2004). The proportion of birds with gout was highest for the oldest age classes (Gilbert et al. 2002), as would be expected if it resulted from a new cause of death with equal impact with respect to age and in addition to the mortality agents acting previously. Such a high incidence of gout is rarely seen in wild birds, so the puzzle then became to establish its etiology. Researchers looked hard for causes of renal damage. Novel vulture pathogens were identified (Oaks, Donahoe et al. 2004; Cardoso et al. 2005), but there was no indication that these were associated with gout. Tissues of dead vultures collected in Pakistan were analyzed for a wide range of toxic environmental pollutants, including heavy metals, organophosphates, organochlorines, and carbamates, but this failed to find significant numbers of birds contaminated at high levels or an association with gout (Oaks et al. 2001; Oaks, Gilbert et al. 2004). At the beginning of 2003, the most likely cause of the declines seemed, by a process of elimination, to be an unknown infectious disease (Pain et al. 2003; Cunningham et al. 2003). However, later that year attention was focused by studies in Pakistan on the possible effects of veterinary nonsteroidal anti-inflammatory drugs (NSAIDs). Analysis of tissues from the carcasses of *G. bengalensis* found dead in the wild showed that all individuals with extensive visceral gout also had detectable residues of the NSAID diclofenac in the kidneys (Oaks, Gilbert et al. 2004). None of the carcasses of vultures without visceral gout had detectable diclofenac. Later, the same association of diclofenac contamination and gout was found in India for both *G. bengalensis* and *G. indicus* (Shultz et al. 2004). Experimental dosing of captive *G. bengalensis* resulted in rapid death (Oaks, Gilbert et al. 2004). The experimentally treated vultures were affected by diclofenac in a dose-dependent way. Death occurred rapidly in all of the birds exposed to high doses and many of those given low doses. In all cases, the dead birds had visceral gout. Histological examination revealed kidney damage similar to that found in the carcasses of wild vultures with gout (Oaks, Gilbert et al. 2004; Meteyer et al. 2005). Later experimental studies showed that other *Gyps* species were similarly affected (Swan, Cuthbert et al. 2006).

Some NSAIDs were already known to be nephrotoxic to birds and mammals. Surveys of veterinarians and veterinary pharmaceutical retailers in Pakistan found that diclofenac, which is used to reduce pain and swelling in livestock, was widely available and cheap (Oaks, Gilbert et al. 2004). It had been available for veterinary use in Pakistan only since 1998. Later surveys in India indicated that veterinarians had been using the drug there since about 1994 (Pain et al. 2008). Diclofenac was then no longer covered by patent, and veterinary formulations were being manufactured by more than 50 companies in India. Across the subcontinent, it has been the welfare drug of choice for veterinarians treating livestock for a range of conditions, including inflammation of the limbs and mastitis. It is generally administered as an intramuscular injection, although an ingestible bolus form also exists. The hypothesis was that contamination with diclofenac of tissues of domesticated ungulates treated with the drug led to the death of vultures that scavenged the carcasses of treated animals. These deaths were frequent enough to account for the rapid population declines.

Despite the powerful evidence for the involvement of diclofenac, there remained good reasons to question whether it could be the sole cause of the vulture declines. NSAIDs tend to have short residence times in mammalian tissue, including ungulates. In cattle that receive a standard veterinary course of diclofenac, tissue contamination declines to undetectable levels within a week (Taggart et al. 2006; EMEA 2004). Hence, it seemed that an improbably large number of animals would have to be treated with diclofenac just before they died to pose a serious threat to vultures.

However, despite the rapid clearance of diclofenac, its average concentration in edible livestock tissues is still sufficient to kill more than 10% of vultures feeding from the carcass of an animal treated after a day or two of treatment (Green et al. 2006).

The proportion of livestock carcasses that would need to contain levels of diclofenac lethal to vultures to have caused the observed population crash was estimated with a simulation model of a vulture population (Green et al. 2004). Using a range of plausible assumptions about normal vulture mortality rates and intervals between meals, it was shown that less than 1% of livestock carcasses (0.1% to 0.7%, depending upon the vulture population and model parameter values), would have to carry lethal concentrations of diclofenac to have caused the observed rates of population decline in India and Pakistan between 2000 and 2003. The modeled proportions of dead adult and subadult vultures that would have visceral gout, the characteristic sign of diclofenac poisoning, if the observed declines were caused only by diclofenac, were found to be consistent with the observed proportions of dead vultures with gout in Pakistan and India (Green et al. 2004).

Although these analyses demonstrated that a sufficiently high proportion of dead vultures showed signs of diclofenac poisoning to account for the declines and that the proportion of contaminated ungulate carcasses need only be low, they did not show that sufficient ungulate carcasses really were contaminated with high enough diclofenac concentrations. The only way to do that convincingly was to collect tissue samples from a representative sample of dead domesticated ungulates from many sites across India. A survey of 1,848 liver samples collected from domesticated ungulates from carcass dumps in 12 states in India in 2004 and 2005 found diclofenac in 10% of carcasses. Diclofenac was found in cattle, water buffaloes, goats, and horses, but not in sheep (Taggart et al. 2007). These observed levels of contamination were then used, in combination with dose-response toxicity results (Oaks, Gilbert et al. 2004) and the vulture population model described above, to estimate the rate of population decline expected for a population of *G. bengalensis* with this exposure. The expected rate of decline was consistent with that observed in road transect surveys of vultures carried out a few years before the carcass surveys (Green et al. 2007). Hence, there was sufficient diclofenac in ungulate carcasses available to vultures in India to cause their populations to decline at the observed rate without the necessity to invoke any other causes.

The publication and dissemination of research evidence about the role of diclofenac by a group of conservation bodies from 2004 to 2006 led to recommendations that government authorities in all vulture-range states introduce legislation or regulations to prevent all veterinary uses of diclofenac that pose a risk to vultures, and that captive populations of all three affected *Gyps* species be established immediately in South Asia for the purposes of conservation breeding and subsequent reintroduction to a diclofenac-free environment (ISARPW 2004). After careful evaluation of these recommendations and the supporting evidence, the government of India withdrew the registration of diclofenac for veterinary use in May 2006, and banned its production for veterinary purposes. The government of Nepal took similar action in August 2006, followed shortly thereafter by the government of Pakistan. Captive breeding programs have been established for all three of the endangered vulture species in India, Pakistan, and Nepal to ensure that birds are available for restocking and population restoration when diclofenac has been removed from their food supply. The first successful captive breeding of *Gyps bengalensis* occurred in 2008. Attempts have been made in Pakistan and Nepal to prevent deaths caused by diclofenac by provisioning wild vultures with carcasses of ungulates known not to have been treated with the drug. These interventions probably reduce mortality rates, but satellite-tagged vultures in the areas covered by them have been shown to continue to range widely and potentially to feed on contaminated carcasses up to several hundred kilometers from the intervention sites (Gilbert et al. 2007). Hence, the effectiveness of such measures in arresting declines is uncertain.

Given the widespread veterinary use, low cost, and perceived efficacy of diclofenac, it is not surprising that the effect of the change in regulations has been disappointing so far. In India, the drug is still widely available to livestock keepers, and formulations intended for humans are being used to replace the banned veterinary formulations. A survey of domesticated ungulate carcasses in India in 2006, after the ban, found a proportion of diclofenac-contaminated carcasses very similar to that before the ban. A safety-testing program on vultures led to the identification in 2006 of an alternative NSAID, meloxicam, with similar efficacy in ungulates and no detectable adverse effects on vultures and other scavenging birds at doses that they are likely to encounter in the wild (Swan, Naidoo et al. 2006; Swarup et al. 2006). No other NSAID has yet been shown by a thorough testing program to be safe for vultures. There is evidence that some NSAIDs other than diclofenac, including carprofen and flunixin, can cause kidney damage and death in vultures and other birds (Cuthbert et al. 2006). Efforts to implement the ban are now focused both on discouraging the use of diclofenac and encouraging the use of meloxicam.

The impact of the collapse of vulture populations on human well-being is difficult to measure, but is likely to have been substantial, given the millions of tons of carrion per year that were formerly disposed of by vultures. The flesh from domesticated ungulate carcasses that is not now consumed by vultures supports a larger population of dogs than hitherto, and this may have increased costs to humans associated with disease, especially rabies (Markandya et al. 2008). There are also likely to have been substantial effects on populations of other pests, such as rats and insects, and on the nuisance and other health risks associated with decaying carrion in areas with high human populations. Quantitative studies of these effects have not been done, primarily because of the scarcity of relevant statistics on an appropriate spatial scale. However, it seems probable that vultures formerly performed a large-scale ecosystem service to humans in removing carrion. Although the birds also posed a nuisance, such as the risk of collision with aircraft, this was probably much less than the benefits they provided and also less than the nuisance and costs caused by their ecological replacements, especially feral dogs. The case of diclofenac and Asian vultures illustrates the potential for technological innovation to pose novel and unexpected major hazards to wildlife populations. This is the first case of a veterinary pharmaceutical being shown to have such an effect, and testing protocols and regulatory mechanisms to safeguard against similar cases remain inadequate to prevent repetition.

5.4 CASE STUDY 2: THE DECLINE OF A VERTEBRATE CLASS

The widespread and rapid decline in amphibian populations is one of the most pressing biodiversity crises of the century. In this case, an entire class is suffering worldwide declines. The recent Global Amphibian Assessment by the IUCN (www.iucnredlist.org/amphibians) revealed that nearly 1,200 of the 6,260 species in the assessment are threatened by some form of pollution. Contaminants cause direct mortality and might also weaken the immune system of amphibians, making them more susceptible to fungal diseases (Figure 5.2a).

5.4.1 Taxon-Wide Impacts — Pesticides and Contaminants, and Their Roles in the Decline of Amphibian Populations

Robin D. Moore and Claude Gascon

In 2004 the Global Amphibian Assessment revealed that an alarming 32.5% of some 6,000 described amphibian species are threatened with extinction, and some 43% are declining (Stuart et al. 2004). There are a number of factors that contribute to these declines, including possible complex interactions between climate change and the global spread of the chytrid fungus *Batrachochytrium dendrobatidis* (Alexander and Eischerd 2001; Pounds et al. 2006; Laurance 2008; Lips et al. 2008; Rohr et al. 2008). Chytrid alone has caused significant declines in the populations of species globally, including the extinction of at least 67 species of *Atelopus* since 1980 (Rohr et al. 2008) and the severe decline and possible extinction of 14 species of upland rainforest frogs in eastern Australia (Laurance 2008). While habitat loss and alteration and the spread of chytrid remain the largest threats to the survival of the majority of threatened species, there is a growing body of evidence that chemical contaminants are in some way responsible for many amphibian declines (Blaustein et al. 2003), possibly also contributing to susceptibility to diseases such as chytridiomycosis.

Amphibians are frequently cited as good model organisms for studying environmental changes because of characteristics that make them well suited for experimental manipulation (see Hopkins 2007). They have been touted as excellent model organisms for testing pesticide exposure (Storrs and Kiesecker 2004), and there is evidence that amphibians are prone to high rates of bioaccumulation of trace elements relative to other taxonomic groups of similar trophic position (Unrine et al. 2007). This tendency toward bioaccumulation is likely a result of their feeding ecology, and could have serious implications for their susceptibility to environmental contaminants. Additionally, standard toxicity tests have shown that for some contaminants, amphibians are more sensitive than fish (Birge et al. 2000) that are used to establish safe environmental levels.

The effect of toxins, such as pesticides, heavy metals, acidification, and nitrogen-based fertilizers, on amphibians may be lethal, sublethal, direct, or indirect. Historically, the direct effect of toxins on amphibians has been investigated by determining lethal concentrations of isolated compounds (reviewed in Harfenist et al. 1989). One of the most commonly used measures of toxicity is the LC_{50}; this is the toxicity of the surrounding medium that will kill half of the animals being tested. LC_{50} tests offer definitive test of cause and effect, and have been extremely useful for determining the lethality of different contaminants and the relative susceptibility of different organisms. While contaminant effects that result in direct mortality would most blatantly cause population declines, contaminants could also affect other responses that could lead to declines. LC_{50} tests reveal nothing about these non-lethal toxic effects, however. A chemical may have a large LC_{50}, yet may produce illness at very small exposure levels. In reality, most relevant concentrations in the environment are not directly lethal, and the effects may be more indirect than mortality detected by the LC_{50} tests. It is therefore important that LC_{50} tests be corroborated with experiments and observations that reflect environmentally relevant exposures. Many amphibian declines have occurred in remote areas where pesticide levels may be well below the known lethal levels based on lab studies (Davidson 2004), and there is mounting evidence that the role pesticides are playing in amphibian declines is probably via sublethal effects and long-term exposures (Gilbertson et al. 2003).

Conducting ecotoxicological research on pesticides and their impact on natural amphibian populations poses many challenges. One approach to testing the effects of ecotoxins under semi-natural conditions has been through the use of mesocosms (Rowe and Dunson 1994). Mesocosms essentially comprise replicated outdoor artificial systems that provide a halfway house between controlled laboratory experiments, which may not reflect realities in nature, and uncontrolled field observations. Importantly, they can provide replication and the potential to implement rigorous experimental design and statistical analysis.

Community-based mesocosm experiments have highlighted the importance of many indirect consequences of contamination, including impacts on food resources, predation, and competition (Boone and James 2005). Some studies have highlighted the importance of abiotic factors such as hydroperiod in influencing toxicity (Roe et al. 2006), while others have indicated that, while individuals may appear outwardly healthy when exposed to contaminants, subtle latent effects can emerge later in ontogeny (Rohr and Palmer 2005). What is clear from these experiments is that contamination can influence communities in complex ways. Often, as sensitive taxa decrease in abundance, insensitive taxa actually increase in abundance (Relyea and Hoverman 2006). Some contaminants may act to eliminate or reduce amphibian food resources, while others may increase resources (Boone and James 2003; Mills and Semlitsch 2004; Metts et al. 2005; Relyea 2005). Boone and Bridges (2003) found carbaryl to have an indirect positive effect on Woodhouse's toad (*Bufo woodhousii*) as a result of a trophic cascade instigated by the pesticide. Zooplankton, which feed on algae, are very sensitive to carbaryl, and their reduction or elimination can lead to algal blooms. Because Woodhouse's toad tadpoles feed on algae, these blooms have a positive effect on tadpole growth and mass at metamorphosis. Conversely, carbaryl is likely to have a negative indirect effect on salamanders by reducing their invertebrate food supply. Thus, the impacts in a community context are complex and dependent on which components of the food web are affected by the toxicant.

Exposure to contaminants during the early tadpole stages has been found to influence size at metamorphosis and, subsequently, juvenile and adult fitness. Bridges (1997 and 1999b) found that sublethal concentrations of carbaryl alter tadpole behavior, decreasing feeding rates and resulting in a smaller size at metamorphosis. Reduced mass at metamorphosis is negatively associated with future fitness in amphibians, and has been shown to influence time to first reproduction and reproductive potential (Smith 1987; Semlitsch et al. 1988; Berven 1990; Scott 1994). Negative effects on growth could theoretically lead to population declines in affected amphibians.

While many studies have assessed the influence of one contaminant on individuals, populations, and communities, in reality contaminants rarely act in isolation. Tens of thousands of contaminants are released into the environment (Kiely et al. 2004; Pimentel 2005), and amphibians are typically exposed to a cocktail of toxicants. Pesticide type used, application rate, breakdown products, and exposure levels all fluctuate spatially and temporally, making it problematic to decipher the relative contribution of different toxicants to observed responses in amphibian populations. To test the relative effects of multiple toxins, Relyea (2004) tested four pesticides (diazinon, carbaryl, malathion, and glyphosate) on five species of amphibian larvae and related responses to pesticides applied individually. The impact of combining the pesticides was found to be similar to that predicted by the total concentration of pesticides in the system, suggesting that synergistic interactions were not at play. However, further research is required to determine the effects of different combinations of contaminants that may be encountered in nature.

Sublethal effects of contaminants on the growth, development, and behavior of amphibian larvae (Bridges 1997, 2000; Goleman et al. 2002) may alter susceptibility to predation and competition (Bridges 1999a). The presence of contaminants may alter predator-prey dynamics by influencing activity levels of predators and/or prey, or by altering predator avoidance mechanisms. Some contaminants decrease activity in tadpoles (Freda and Taylor 1992; Bridges 1997), making them less susceptible to predation by visually oriented predators (Skelly 1994). Other contaminants have been found to increase activity in tadpoles and, consequently, predation rates (Cooke 1971). In some cases contaminants influence the behavior of both predator and prey in such a way that overall predation rates are not affected (Bridges 1999b). Relyea et al. (2005) found that the presence of low levels of pesticides can actually increase the survival of tadpoles in mesocosms by removing predatory insects, yet another way in which contaminants can influence predator-prey dynamics in complex and often unpredictable ways.

The susceptibility of tadpoles to toxicants may be influenced by interactions with other stressors. Relyea and Mills (2001), for instance, demonstrated that predator-induced stress makes the pesticide carbaryl more deadly to treefrog tadpoles (*Hyla versicolor*). When predator cues were present in the water, the pesticide became 2 to 4 times more lethal. This phenomenon was found to exist in six other species tested (Relyea 2003), suggesting that it may be widespread among amphibians. Evidence exists that contaminants may influence the susceptibility of amphibians to disease (Gilbertson et al. 2003), and it is likely that in many cases, multiple stressors are driving amphibian declines (e.g. Carey et al. 2001; Linder et al. 2003). Some evidence suggests the presence of multiple stressors does have a greater impact than the additive effects of each individual stress would predict (Little et al. 2000; Relyea and Mills 2001; Boone and James 2003). Contaminants therefore need to be considered in relation to other stressors impacting amphibian populations to assess their role in amphibian population declines and species extinctions.

The effects of pesticides on the growth and development of amphibian larvae may have particularly important implications for amphibians breeding in ephemeral pools. Contaminants that lengthen the larval period in the aquatic environment could increase the duration of time the larvae are vulnerable to both aquatic predators and desiccation in drying ponds—both of which may decrease juvenile recruitment and lead to population declines (Boone and Semlitsch 2002).

The presence of contaminants in a natural system can manifest itself in dramatic ways, as is the case with parasite-induced deformities in amphibians. Kiesecker (2002) used field experiments to demonstrate that exposure to trematode infection was required for the development of limb deformities in wood frogs (*Rana sylvatica*) and that deformities were more common at sites adjacent to agricultural runoff. Laboratory experiments corroborated the association between pesticide exposure and increased infection, with pesticide-mediated immunocompetency as the apparent mechanism. These results indicate that such mechanisms may explain pathogen-mediated amphibian declines in many regions.

Runoff from agricultural lands can expose pond-breeding amphibians to varying levels of pesticides over the course of time. The persistence and toxicity of contaminants in a natural system, and hence their potential impact on populations and communities, may vary according to numerous abiotic factors, such as environmental temperatures and pH and the amount and timing of rainfall. For example, atrazine, the most widely used pesticide in the United States, has been reported at levels from 0.1 to 6.7 ppb in amphibian breeding ponds in mid- to late July (Hayes et al. 2003), but during storm events agricultural runoff has been reported to be as high as 480 ppb (Huber 1993). Pulses of exposure may therefore be dependent not only on pesticide application regimes, but also on weather patterns. It is problematic, therefore, to draw clear links between any particular pesticide or any exposure scenario and amphibian declines because field conditions will almost certainly be different over the course of seasons and years.

Amphibian declines and extinctions typically happen beyond the scale of the breeding site, yet only a handful of studies have examined the relationships between amphibian declines and potential causes on regional scales. Davidson et al. (2001, 2002), however, demonstrated that population declines for numerous amphibian species in the Sierra Nevada mountains are significantly correlated with chemicals carried by winds from agricultural areas in California. Other predictions for declines, such as increased ultraviolet light or climate change, did not explain the observed patterns of declines, although habitat destruction did to some extent for some species (Davidson et al. 2002). Subsequent analyses have suggested that pesticides that inhibit acetyl-cholinesterase were most strongly correlated with declines (Davidson 2004). This regional study suggests that contaminants could be playing an important role in declines that occur miles away, providing the first conclusive link between population declines of many species and historical pesticide application.

A growing body of evidence, therefore, points to severe impacts of environmental contamination on amphibians at the individual, population, and community levels. There are still many pieces of the puzzle that need to be in place before we can fully assess the relative contribution of contaminants to global amphibian declines, however.

Investigating further negative effects of herbicides—the most commonly applied pesticides worldwide (Kiely et al. 2004)—on amphibian populations and communities should be a priority for future research. Research is additionally needed to examine patterns of contamination worldwide and how these relate to amphibian declines to fully assess the impact of contaminants acting in combination with other stressors. Studies that have demonstrated a link between contamination and amphibian declines in California (Davidson 2004) should be replicated in the tropics—in particular, in high-elevation "pristine" environments where the causes of population declines are mysterious. Finally, predictions need to be made, based on existing studies, for where agrochemical use will change and how this will affect regional amphibian populations.

Research into the effects of contaminants on amphibians has focused almost exclusively on the aquatic larval stage. While there is some reason to believe that the larval stage may be more sensitive than the terrestrial stage, contaminants could have effects that influence populations in the terrestrial environment (Boone et al. 2007). For instance, James et al. (2004) found that terrestrial exposure resulted in increased mortality in hibernating juvenile American toads. There is a paucity of published studies on the effects of contamination on the terrestrial stage or interactions between aquatic and terrestrial stages of amphibians, and further studies are encouraged to help make these links.

Studies focusing on the impact of contaminants over generations are needed to increase our understanding of long-term consequences of exposure to contamination, as are studies addressing physiological or genetic adaptation to long-term chemical exposure and how adaptation to a chemical stressor may influence population persistence. While most studies on the effects of contaminants of amphibians have focused on the effects of one or two contaminants, there are tens of thousands of contaminants to which amphibians are exposed in nature, and little to nothing is known about the effects of many of these contaminants on individuals, populations, or communities. Investigating the effects of chemicals from different contaminant classes to assess the potential impacts would augment the growing body of knowledge on the effects of contaminants on amphibians.

5.5 CASE STUDY 3: GENETIC IMPACTS WITH POPULATION AND EVOLUTIONARY CONSEQUENCES

Contaminants also have the potential to impact long-term population and community stability and function through broad-scale, multi-taxon genetic damage. In Azerbaijan, petroleum residues in aquatic systems have impacted freshwater communities of fish, reptiles, and amphibians, resulting in the depression of gene flow through impacts on reproduction and the generation of heteroplasmy via elevated mutation rates. Impacts therefore occur both at the cellular level and at the population level, with unforeseen community-level consequences (Figure 5.2c).

5.5.1 Evolutionary Toxicology–Biodiversity Impacts of Contaminants in Aquatic Ecosystems of a Biodiversity Hotspot

John W. Bickham

Azerbaijan has been the location for extensive studies on the effects of pollution on fish and wildlife populations (Bickham et al. 2003). Azerbaijan is a small country but is rich in ecosystem diversity, ranging from the Caspian Sea lowlands through arid and semiarid regions to montane forests and alpine tundra. The country sits amid the Caucasus biodiversity hotspot (Myers et al. 2000), recognized as one of the world's 25 richest and most threatened areas. As part of the former Soviet Union, Azerbaijan faces difficult economic challenges as well as social and political adjustments. The legacy of poor environmental safety practices during the Soviet era is severe in places such as the capital Baku, the industrial city of Sumgayit, and many of the oil fields around the country. In Sumgayit there is contamination of the environment emanating from various industrial sources and the industrial wastewater treatment plant. Contamination is extensive and threatens the health of local residents as well as refugees from the occupied areas of western Azerbaijan, many of whom live in the industrial region among the factories. Sumgayit is recognized by the Blacksmith Institute (2007) as one of the world's ten most contaminated sites.

The impacts of contaminants on organisms begin at the molecular level and proceed up through higher levels of biological organization, including cellular effects, tissue effects, health effects, effects on reproduction, population effects (including evolutionary responses), and ecosystem and community effects (Bickham et al. 2000). At lower levels, ranging from molecular through reproductive effects, impacts are largely predictable, based upon knowledge gained from traditional toxicological and ecotoxicological approaches. For example, one might be able to predict the impact on relative reproductive success of birds exposed to pesticides because individual response (the ability to reproduce) is likely to be a function of relatively few factors, including levels of exposure and the mechanisms of toxicity of the pollutant. However, the effects of contaminants on populations, ecosystems, and communities are less predictable because of the myriad factors that would affect the degree of impact and the complexity of these levels of organization. If a pesticide caused an expected 25% decrease in reproductive output due, say, to eggshell thinning, it could cause either a population decline or no observed effect, depending on whether or not the population has compensatory responses (e.g., density dependent survival). Whether or not the population has excess reproduction to begin with might be determined by absence of predators, productivity of the environment, diseases, parasites, or many other factors. So, populations under different environmental conditions are expected to respond differently to similar insults. Likewise, the losses of populations from simple ecosystems are likely to have more profound impacts than from complex ecosystems, as the ecosystem services such as pollination are more effectively buffered by high biodiversity in the latter than in the former.

The impacts of contaminants on evolutionary processes are also highly unpredictable. Evolutionary toxicology (Bickham and Smolen 1994; Bickham et al. 2000) is the study of the effects of pollutants on the genetics of natural populations. In effect, it is a study of the ways by which organisms adapt to uniquely stressful environments. Early studies of this nature involved the use of protein electrophoresis to detect effects of metal exposure on allelic diversity of protein coding loci (Nevo et al. 1986; Newman and Jagoe 1998), whereas most recent studies make use of molecular techniques such as DNA sequencing (Matson et al. 2006). Environmental pollution poses unique problems to organismal adaptation in that many of the chemicals used today in agricultural and industrial processes do not occur in nature. Thus, all organisms are naïve to these chemicals, and their capabilities to detoxify them depend on physiological systems designed for other, perhaps similar or dissimilar, ones.

Although biodiversity is frequently equated to taxonomic diversity, genetic variation within species is also a component, and the loss of genetic variation is of key concern in conservation biology. This is because genetic diversity buffers populations from environmental changes, and loss of genetic variation is correlated with increased probability of extinction. So, molecular population genetics methods can be used to measure diversity in populations exposed to pollution in order to detect the potential loss of a critical component of a species' ability to adapt to environmental change. Studies using this approach have demonstrated the degree to which a variety of species have been impacted by contamination at the level of population genetics.

In order to study the impact of contaminants on wildlife in and around Sumgayit, studies were conducted to determine the presence and levels of contaminants in environmental media, including sediment and tissues (Swartz et al. 2003). A total of 15 contaminants or classes of contaminants that exceeded one or more sediment-quality guidelines were identified from pond sediments taken adjacent to the industrial wastewater treatment plant. Those included the following: aldrin, biphenyl, chlordane, total DDT, β-endosulfan, fluorine, heptachlor, α-HCH, γ-HCH, Hg, naphthalene, phenanthrene, PCBs, total PAHs, and pyrene. A total of 75 chemicals, including congeners and breakdown products, were identified from tissues of two species of turtles—the European pond turtle (*Emys orbicularis*) and the Caspian turtle (*Mauremys caspica*)—taken from the same ponds as the sediments (Swartz et al. 2003). A comparison was made with European pond turtles taken from a relatively uncontaminated reference site, and the following chemicals were found to be significantly lower in tissues from the reference site compared to the wastewater treatment plant site: Hg, PCB, heptachlor, DDD, HCB, chlordane, pentachlorobenzene, total PAH, *trans*-nonachlor, α-HCH, and aldrin. These analyses show that there is a complex mixture of contaminants in the pond sediments, many of which have bio-accumulated in the tissues of resident turtles. Thus, the potential for impact was demonstrated for the ecosystem in general and the turtles in particular. The presence of such a complex mixture of contaminants in the environment and in animal tissues is cause for concern, but by itself it does not confirm damage to the wildlife populations.

Biomarker studies were conducted of turtles and marsh frogs (*Rana ridibunda*) from the contaminated wetlands in order to determine whether resident wildlife species were impacted. Swartz et al. (2003) and Matson et al. (2005b) used biomarkers of genetic damage, including the micronucleus assay and the flow cytometric method, to reveal genotoxic effects in blood from European pond turtles. Similarly, Matson et al. (2005a) documented chromosome damage in marsh frogs from Sumgayit using the same two biomarkers. The significance of these studies is that they confirm impact of contaminants on the somatic tissues of resident wildlife populations.

In order to investigate whether or not the impacts measured by biomarkers have resulted in population-level effects, population genetic analyses of marsh frogs based on nucleotide sequences of the mitochondrial DNA (mtDNA) control region were conducted (Matson et al. 2006). Frogs from three regions of Azerbaijan, Sumgayit, Ali Bairamly (to the south of Sumgayit), and Alti Agach (to the north of Sumgayit), were studied. The latter two sites served as reference or pristine sites since frogs from those areas did not show evidence of genotoxic damage using FCM or micronuclei (Matson et al. 2005a). The mtDNA is located in the mitochondria outside the cell nucleus. It exhibits strict maternal inheritance, and individuals typically possess only one form of mtDNA, inherited from the maternal parent. Based on this analysis, it was observed that frogs from Sumgayit have lower genetic diversity, including both haplotype and nucleotide diversity, compared with frogs from Ali Bairamly and Alti Agach. Moreover, the distributional patterns of the different forms (haplotypes) of mtDNA revealed that gene flow, which results from the migration of female frogs from one site to another and subsequent successful reproduction, was predominantly into the Sumgayit region from both the north and the south. Sumgayit, it seems, is an ecological sink and has likely experienced a long-term problem with successful reproduction compared with the pristine sites to the north and south. Excess reproductive output from surrounding areas has provided the source for migrants into Sumgayit, and the resident frogs likely have had relatively poor reproductive success or survival. In this case, genetics data provided insight into an important ecological effect that chemical exposure has likely caused.

An additional form of evidence of genetic impact was the presence of new mtDNA haplotypes that might indicate an increased mutation rate of the mtDNA. This was observed at the ponds adjacent to the wastewater treatment plant, ostensibly the most contaminated site in Sumgayit. Frogs from these ponds had the highest levels of haplotype and nucleotide diversity of the eight sites examined within Sumgayit. Here, a few individuals were observed that possessed two forms of mtDNA. Having more than a single form of mtDNA is called heteroplasmy, and it was concluded that these heteroplasmic frogs possess new mutations. It is assumed that within a few generations these new mutations will become fixed. That is, the heteroplasmy will disappear and the descendants of these heteroplasmic frogs will possess only a single form of mtDNA, the new mutation. Heteroplasmy is rare in nature (Pakendorf and Stoneking 2005). Matson et al. (2006) examined a total of 207 marsh frogs. Of that total, only 7 frogs were observed to be heteroplasmic. Of these, 6 were from the wastewater treatment plant ponds, and 1 was from an adjacent area. No other ponds in Sumgayit or from the reference areas showed evidence of heteroplasmy.

In studies of evolutionary toxicology it is necessary to distinguish naturally occurring from pollution-induced mutations. Mutations cannot be directly observed to result from a contaminant, but the association of mutations with contaminant distribution is strong evidence of cause and effect. Matson et al. (2006) observed heteroplasmy, unique rare alleles, and higher diversity estimates in marsh frogs from the most contaminated sites in Sumgayit. Thus the weight of evidence supports the conclusion that these are contaminant-induced mutations.

Ecotoxicological studies have shown that wildlife from the Sumgayit industrial zone have expressed genetic damage at the cellular level as well as at the population level. Both genotoxic and non-genotoxic chemicals were found in sediments and in tissue of frogs and turtles. Biomarkers of genetic damage confirmed the impact of contaminants at the cellular level, and correlation analyses implicated mercury and PAHs, both known genotoxins, as the root cause of the chromosomal damage. Effective female migration was estimated using population genetics methods. Gene flow was primarily into Sumgayit, both from the north and the south, which is indicative of reduced reproduction and evidence of an ecological sink at Sumgayit. At a regional scale, Sumgayit possessed reduced genetic variability compared with reference areas north and south. This diversity loss was interpreted to have been caused by a historical bottleneck in which genetic drift likely removed low-frequency haplotypes. This could have happened as a result of the establishment of the chemical plants beginning in the 1940s. In addition, on a finer geographic scale, heteroplasmic individuals and higher diversity estimates were observed at the wastewater treatment plant. This was interpreted as evidence of induced mutations and possibly indicative of an increased mutation rate. The ecotoxicology and evolutionary toxicology studies in Azerbaijan illustrate the progression of analyses from low to high scales of complexity that is necessary to confirm the emergent population effects of environmental contamination.

5.6 CASE STUDY 4: HABITAT ALTERATION AND AGRICULTURAL CHEMICALS IN THE DECLINE OF AVIAN POPULATIONS

There is actually a double impact on biodiversity related to the expansion of agricultural activities. Increasing human food demands lead to deforestation and habitat conversion, and attempts to increase productivity per unit area have resulted in the expansion of the use of agricultural chemicals. Both field and modeling studies (Kendall and Lacher 1994; Lacher 1994) reveal significant impacts of agricultural pesticides on avian populations that utilize agroecosystems (Figure 5.2). The impact has been especially severe in grassland birds adapted to use agroecosystems.

5.6.1 Community-level Effects of Pesticide Use — An Evaluation of the Impacts of Farmlands (Agricultural Intensification and Pesticide Use) on Bird Declines and Avian Biodiversity

Miguel Mora

Declines of individual species have been documented throughout the world. Effects of pesticides on individual species have been better documented than those on avian biodiversity, and noticeable episodes are known worldwide. For example, the decline of the grey partridge (*Perdix perdix*) in England was attributed to the decline of invertebrate prey as a consequence of herbicide use (Potts 1986). Other cases involve declines of red-winged blackbirds in Ohio (Blackwell and Dolbeer 2001), significant reductions in populations of Swainson's hawks wintering in Argentina due to the use of monocrotophos for insect control (Goldstein et al. 1999; Hooper et al. 1999), and the near demise of Asian vultures due to the use of the anti-inflammatory drug diclofenac in cattle (Green et al. 2006).

Birds are recognized as good indicators of overall farmland diversity (Donald et al. 2001). Birds respond in an integrated way to agricultural pressures (Firbank et al. 2008); therefore, it is more realistic to use more than one single taxon to indicate biodiversity changes responding to agricultural change. Chamberlain et al. (2000) considered over 30 agricultural variables in the analyses of annual changes of bird populations in England and Wales between 1962 and 1996. Five of the variables were fungicides, three herbicides, and insecticides.

It is broadly recognized that over the last 60 years, agricultural intensification has contributed to a decline in bird populations and biodiversity in Europe and North America (Benton et al. 2003; Chamberlain et al. 2000; Siriwardena et al. 2000). Both British and U.S. bird surveys indicate bird population declines in agricultural areas, likely exacerbated by pesticide use. Some farmland bird species in England suffered range contractions and population declines between 1960 and 1990 (Fuller et al. 1995; Siriwardena et al. 1998). The decline in farmland avian biodiversity is related to changing farming practices, including habitat loss, decreased food abundance, agrochemical use, agricultural type and activity (such as organic vs. non-organic farming, tillage vs. no-tillage, hedge rows vs. no hedge-rows), and direct mortality by farming operations. Habitat changes and intensified agricultural practices are probably the key factors in explaining the decline of many agricultural bird species (Chamberlain et al. 2000).

Land use is expected to be the primary driver of environmental change in the coming century (Sala et al. 2000). The main processes of agricultural intensification include the transformation between non-agricultural and agricultural habitats, the transformation of agricultural landscapes into new arrangements of crops, and the management of crops to increase productivity. Many variables related to agricultural farming influence bird diversity, and, as indicated earlier, most evaluations of the impacts of agricultural intensification and pesticide use on bird biodiversity have been conducted in Western Europe (Benton et al. 2003; Freemark and Boutin 1995).

Although a single indicator of the effects of agricultural intensification is not sufficient (Firbank et al. 2008), it is recognized that pesticide use is one of the three major variables that affect avian biodiversity (Boutin et al. 1999; Chamberlain et al. 2000; Benton et al. 2003; Newton 2004). Pesticides have affected birds directly through mortality and reproductive failures and indirectly through reduction of food supplies. Insecticides and herbicides reduce habitat and food available for many farmland birds. There is enough evidence to suggest that insecticides applied during the breeding season affect breeding performance of various bird species (Boatman et al. 2004). There is also strong evidence about the negative effects of herbicides on bird-species abundance. Pesticide use, particularly herbicides, has been one of the major drivers of bird population declines in Britain (Newton 2004). Groups of birds with ecological and phylogenetic similarities are more often affected in the same way by one or more variables associated with agricultural intensification (Siriwardena et al. 1998). For example, the abundance of insectivorous birds is affected by reduction of their insect food source sprayed by pesticides, whereas seed-eaters' abundance is more impacted by herbicides, which also reduce the birds' food source. Siriwardena et al. (1998) indicate that species specializing in farmland habitats have declined more than species that have more diverse habitat preferences. However, they concur in that a variety of factors, rather than only one, are responsible for the declines in abundance of farmland birds in Britain (Siriwardena et al. 1998; Benton et al. 2003).

Although pesticide use is a major factor contributing to decreased bird populations, the multivariate effects of agricultural practices interact strongly and should be considered collectively (Benton et al. 2003). However, Tilman et al. (2001) assert that pesticide use in agriculture will be one of the seven main variables that will significantly impact the environment over the next 20 to 50 years. Mineau and Whiteside

(2006) argue that the risk indices derived from all the insecticide treatments at the state level are better correlates of agricultural species declines than any other variable indicators of agricultural intensification.

The assessment of ecotoxicological effects of agricultural pesticides on bird biodiversity in North America was considered an emerging area of interest in the mid-1990s (Freemark and Boutin 1995), and not much progress has been achieved to date. However, the impacts of pesticides on bird abundance and biodiversity have been documented since the beginning of the use of pesticides in agriculture, and such impacts were thoroughly documented in Rachel Carlson's book *Silent Spring*, published in 1962.

In North America, declines of farmland bird species have been reported mostly based on the analysis of the Breeding Bird Survey data (Peterjohn and Sauer 1999; Sauer et al. 2000). La Sorte and Boecklen (2005) determined that from 1968 to 2003, bird-species richness increased at the local scale in non-urban areas in North America; however, they pointed out that biological diversity was more complex than just the number of species observed. They also determined that bird abundance declined significantly during the same period and that there was a more even distribution of individuals. However, the lack of data on avian use of farmland is one of the main impediments for assessing the impact of agricultural practices in reducing bird populations in North America (Boutin et al. 1999). An analysis of patterns of bird-species abundance in relation to insecticide use in the Canadian prairies determined that the abundances of nine species were negatively associated with a granular insecticide index derived from the amount of insecticide used and area of application (Mineau et al. 2005). They linked changes in bird-species abundance in the Canadian prairies to the use of toxic granular insecticides, particularly carbofuran and terbufos. Mineau et al. (1999) also pointed out that there was probably a significant loss of birds of prey to cholinesterase-inhibiting insecticides.

There are many studies that have examined whether there are differences in avian biodiversity between organic and conventional farms. The results have been contradictory, but for the most part species richness has been determined to be significantly higher on organic than on conventional farms (Belfrage et al. 2005; Bengtsson et al. 2005; Freemark and Kirk 2001; Genghini et al. 2006). Bengtsson et al. (2005) also showed through meta-analysis that species richness was higher on organic than on conventional farms. In a study conducted on farmlands in Ontario, of 25 species associated with farmland, 12 were at high or medium risk of exposure to pesticides, and 5 of them were showing population declines (Boutin et al. 1999). All the identified species at high or medium risk were ground foragers; of the 13 species at low risk, 3 were also declining significantly. Mineau et al. (2005) estimated that millions of birds were killed annually by carbofuran and carbamate compounds applied in the U.S. Midwest cornbelt region. Organophosphates, particularly monocrotophos, were associated with significant mortality of Swainson's hawks wintering in Argentina (Goldstein et al. 1999; Hooper et al. 1999). Clearly, OPs and carbamate pesticides have contributed significantly to a decrease in avian biodiversity in the United States.

The toxicity of pesticides to birds decreased in the United States from 1991 to 2003 (Mineau and Whiteside 2006), primarily because pesticides have become less long lived. In 1997, the 10 crops of most concern in terms of risk per hectare to birds were corn, cotton, alfalfa, wheat, potato, peanut, sugar beet, sorghum, tobacco, and citrus, with corn and cotton considered the two most important contributors to farmland bird mortality (Mineau and Whiteside 2006). The risk for grassland species, estimated as the ratio of bird kill per hectare to total agricultural area, was greater for some states in the southeastern United States, particularly in Louisiana, Mississippi, Georgia, South Carolina, and North Carolina (Mineau and Whiteside 2006).

Because of the lack of information regarding the potential impacts of pesticides on avian biodiversity in North America, a few models have been developed for predicting impacts of pesticides on many bird species, including game birds and threatened species in the state of Texas. Corson et al. (1998) observed that granivorous birds are more likely affected by organophosphorus pesticides than insectivorous or carnivorous species are, mostly because vegetation contains higher residues than insects or mammals, particularly right after spraying. This model predicted that smaller birds with greater food intake/body mass ratios were likely at greater risk of exposure to OPs and other pesticides because they ingest larger amounts of pesticide residue. More recently, Pisani et al. (2008) demonstrated that rainfall events after an insecticide application represented a great risk of exposure to pesticides by avian species drinking water along agricultural fields in the lower Rio Grande Valley, one of the main agricultural regions in Texas. The probability of exposure and risks increased with the intensity of irrigation and rainfall events (Pisani et al. 2008). In another model, we determined that species such as the burrowing owl (*Athene cunicularia*) wintering in south Texas were likely to be significantly affected by the OP insecticides chlorpyrifos, dicrotophos, and disulfoton, as well as other insecticides, herbicides, and defoliants applied to cotton and sorghum fields (Engelman 2008).

From the many studies reviewed here and many others, it is clear that pesticides in general are likely to affect aquatic and terrestrial biodiversity (Relyea 2005). Habitat heterogeneity is beneficial to biodiversity; it is suggested that reductions in agricultural pesticide use will significantly contribute to an increase in bird biodiversity in the agricultural landscape. Greater effort needs to be placed in documenting the impacts of agricultural pesticide use on avian communities so that a better assessment of their impacts on biodiversity can be obtained.

5.7 CONCLUSIONS: THE ROLE OF ECOTOXICOLOGY IN BIODIVERSITY CONSERVATION — THREATS, RISK, AND RESEARCH PRIORITIES

5.7.1 CONTAMINANTS, PESTICIDES, AND THREATS TO BIODIVERSITY

Any true assessment of the impact of contaminants and pesticides on the biota of the planet must be viewed in the context of biodiversity. Recent reports and assessments have revealed that these impacts are not trivial. The assessment of 845 species of reef-building corals revealed that 32.8% are in one of three categories of threat (vulnerable, endangered, and critically endangered) based upon the IUCN Red List classification (Carpenter et al. 2008). Major threats to these corals include climate change, elevated sea temperatures, ocean acidification, sewage discharges, and nutrient loading and eutrophication from agrochemicals (Carpenter et al. 2008), all related directly or indirectly to contaminants, pollutants, and pesticides.

Case Study 4 in this chapter presented the complex impact of agriculture and associated pesticide use on birds in agroecosystems. The conservation status of birds continues to decline worldwide, as indicated by trends in the Red List Index. In Europe, farmland birds have declined 44% in 26 years (BirdLife International 2008a). Agriculture is the major threat globally, and pollution is responsible for declines in nearly 200 species (BirdLife International 2008a). Impacts from agricultural and industrial chemicals have led to increased mortality in 6% of globally threatened birds, reduced reproductive success in 3%, and degradation of habitat in an additional 11% (BirdLife International 2008a). Of the 190 species of birds listed as critically endangered, nearly 30 are at risk from pollution, with an additional 36% threatened by small-scale agriculture and 26% by large-scale agroindustry (BirdLife International 2008b). Because declines attributable to agriculture often involve impacts from agricultural chemicals as well, the actual risk posed by pesticides is likely higher than what is reported in these assessments. As Case Study 1 indicated, however, there are other novel threats to avian populations, such as veterinary pharmaceuticals in the decline of *Gyps* vultures in India and the broader implications of a broad class on endocrine-disrupting compounds for vertebrates in general (Colburn et al. 1996).

Of the major taxa evaluated, mammals seem to experience less threat from contaminants and pesticides than other groups, with one major exception: marine mammals. An estimated 36% of marine mammals are listed as threatened (vulnerable, endangered, or critically endangered), and marine mammals are much more poorly studied than terrestrial mammals. As a result, the uncertainty of this estimate ranges from 23% to 61% (Schipper et al. 2008). Fully 60% of marine mammals are threatened by pollution, broadly defined (contaminants, marine debris, noise, climate change; Schipper et al. 2008). The oceans have become a complex soup of compounds and debris that pose significant threats to marine mammals and fisheries.

The Global Amphibian Assessment (Stuart et al. 2004) indicated that 32.5% of the approximately 6,000 species of amphibians were threatened and 45% were declining. Pollution was listed as the second most important threat behind habitat loss (Figure 5.3), affecting approximately 1,200 species (http://www.iucnredlist.org/initiatives/amphibians/analysis/major-threats). Case Study 2 in this chapter addresses these issues in great detail, and Case Study 3 presents examples of the impact of petroleum residues on freshwater amphibians, resulting in genetic damage to frogs at both the cellular and the population level.

It is clear the contaminants and pollution have a significant effect on causing declines in abundance and elevating the risk of extinction to many species across taxonomic groups. This impact is even greater if one includes the additional risk of anthropogenically produced CO_2 and the implications for extinction risk attributed to climate change (Thomas et al. 2004).

The impact of contaminants on biodiversity has a significant global component. A major concern in bird conservation is the impact of stressors on migratory birds, which in many cases utilize agricultural habitats in both their breeding and overwintering grounds. For example, many migrants forage in banana plantations

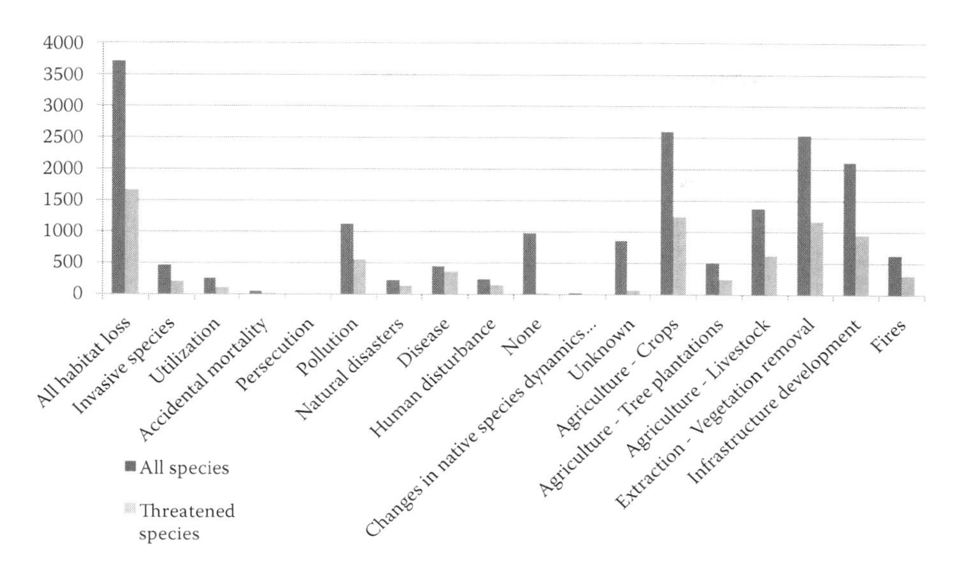

FIGURE 5.3 Threats contributing to the global decline in amphibian populations.

in Central America and are exposed there to a complex mix of herbicides, fungicides, and pesticides (Henriques et al. 1997; Lacher et al. 1997; Mortensen et al. 1998). A complete evaluation of the risk to biodiversity will require international collaboration and transboundary conservation initiatives (Mittermeier et al. 2005). These initiatives have proved to be challenging because of the complex political, economic, and social differences that often span international boundaries. In addition, the majority of the world's biodiversity resides in the tropics, often in developing countries with little capacity for the costly analytical work that ecotoxicology requires. The financing of international research capacity in toxicological research has not been part of most international development projects focused on biodiversity conservation, and as a consequence countries rely on international collaboration often complicated by difficult permitting processes that impede international work. Tropical ecotoxicology will require a new model of effective collaboration (Lacher and Goldstein 1997).

The time to implement new strategies for assessing threats to biodiversity is limited, as the developed world rushes to expand economies, which often means heavy exploitation of natural systems. A recent evaluation of large-scale development plans (IIRSA) for the Amazon Basin (Killeen 2007) presents a gloomy picture for the future of the regional biodiversity in the absence of quick action. Nine of the ten major corridors planned in IIRSA intersect high-biodiversity wilderness areas or biodiversity hotspots, much of them covered in tropical rainforest, Cerrado savannas, or Andean montane forests. Killeen (2007) projects near complete loss of Cerrado by 2030 without aggressive conservation action and argues that multilateral funding agencies and partner governments must proactively assess impacts and develop plans for mitigation, which should include new research and the development of scientific capacity in developing tropical countries.

5.7.2 The Application of Risk Assessment Procedures and Uncertainty to Biodiversity Conservation

Ecotoxicology has a great deal to contribute to biodiversity conservation as we move forward with new assessments, identify threats, and develop mitigations to reduce extinction risk. The development of the on-line global assessments as a part of the IUCN Red List process provides an opportunity for researchers from many disciplines to examine and evaluate assessments and make contributions to future assessments with new data on threats and risks. Ecotoxicology has a history of research targeted toward the goal of evaluating risk, and has protocols that address the ongoing assessment of risk and the incorporation of uncertainty in these evaluations (U.S. EPA 1992; Salice, Chapter 9, this volume). The company Applied Biomathematics (http://www.ramas.com/) has developed software for both ecotoxicology (RAMAS Ecotoxicology) and IUCN Red List procedures (RAMAS Red List Professional), which include tools for managing variability and uncertainty in developing estimates. Like the common history in the development of ecotoxicology and biodiversity conservation, there is a convergence in methodology, driven by the difficulties of sampling, incomplete data, and the need to reduce the uncertainty of risk estimates. Both disciplines must also deal with complex spatial dimensions. The cost of errors is extremely high for both disciplines as well. There are clearly areas of research, both theoretical and applied, that would benefit from the collaboration between environmental toxicologists and biodiversity scientists (Table 5.1).

TABLE 5.1
Areas of Research Collaboration in Ecotoxicology and Biodiversity Science

Areas of Research Collaboration
I. Field Survey and Design
Incorporation of sampling of tissues for toxicological work into biotic surveys
Use of rapid assessments across gradients of potential exposure to contaminants and pesticides
II. Assessment of Threats
Greater collaboration with environmental toxicologists in the development of IUCN threat assessments
Expansion of biotic surveys into toxicological "hot zones" for more effective assessment of threat
III. Collaborative International Research
Increased emphasis on the development of laboratories and research capacity in ecotoxicology in developing countries
Enhanced transboundary collaboration in biodiversity research and threat assessment
IV. Risk Assessment Approaches
Expanded collaboration on the development of spatial tools for the assessment of richness, risk, and exposure
Development of new tools for the assessment of uncertainty in processes in ecotoxicology and biodiversity science (risk, extinction, etc.)

As a beginning we need researchers with mutual concerns about threats to amphibians, marine mammals, corals, and other organisms to look for ways of collaborating to evaluate and mitigate these threats and seek common solutions for effective conservation action. If, in the common history of ecotoxicology and biodiversity conservation, we can find new methodologies to advance the common goals of protecting the Earth's biota, then both disciplines will have advanced.

REFERENCES

Alcamo, J. et al. 2003. *Ecosystems and Human Well-being: A Framework for Assessment*. Washington, DC: Island Press.

Alexander, M.A., and J.K. Eischerd. 2001. Climate variability in regions of amphibian declines. *Conservation Biology* 15:930–942.

Alonso, A., F. Dallmeier, E. Granek, and P. Raven, eds. 2001. *Biodiversity: Connecting with the Tapestry of Life*. Washington, DC: Smithsonian Institution Monitoring and Assessment of Biodiversity Program / President's Committee of Advisors on Science and Technology.

AVPP. 2007. Asian Vulture Population Project, February 2007. www.peregrinefund.org/vulture.

Baillie, J., and B. Groombridge. 1996. *1996 IUCN Red List of Threatened Animals*. Gland, Switzerland: IUCN.

Baillie, J.E.M., C. Hilton-Taylor, and S.N. Stuart. 2004. *A Global Species Assessment*. Gland, Switzerland: IUCN.

Balmford, A., A. Bruner, P. Cooper, R. Costanza, S. Farber, R.E. Green, M. Jenkins, et al. 2002. Economic reasons for conserving wild nature. *Science* 297:950–953.

Baral, H.S., J.B. Giri, and M.Z. Virani. 2004. On the decline of oriental white-backed vultures *Gyps bengalensis* in lowland Nepal. In *Raptors Worldwide,* Chancellor, R.D. and Meyburg, B.-U., eds. Berlin: World Working Group on Birds of Play and Owls and Budapest: MME Birdlife Hungary 215–219.

Barnthouse, L., and M. Sorenson. 2007. *Population-Level Ecological Risk Assessment*. Boca Raton, FL: CRC Press.

Belfrage, K., J. Bjorklund, and L. Salomonsson. 2005. The effects of farm size and organic farming on diversity of birds, pollinators, and plants in a Swedish landscape. *Ambio* 34:582–588.

Bengtsson, J., J. Ahnstrom, and A.C. Weibull. 2005. The effects of organic agriculture on biodiversity and abundance: A meta-analysis. *Journal of Applied Ecology* 42:261–269.

Benton, T.G., J.A. Vickery, and J.D. Wilson. 2003. Farmland biodiversity: Is habitat heterogeneity the key? *Trends in Ecology and Evolution* 18:182–188.

Berven, K.A. 1990. Factors affecting population fluctuations in larval and adult stages of the wood frog (*Rana sylvatica*). *Ecology* 71:1599–1608.

Bickham J.W., C.W. Matson, A. Islamzadey, G.T. Rowe, K.C. Donnelly, C.D. Swartz, W.J. Rogers, et al. 2003. Editorial: The unknown environmental tragedy in Sumgayit, Azerbaijan. *Ecotoxicology* 12:507–510.

Bickham, J.W., S.S. Sandhu, P.D.N. Hebert, L. Chikhi, and R. Anthwal. 2000. Effects of chemical contaminants on genetic diversity in natural populations: Implications for biomonitoring and ecotoxicology. *Mutation Research* 463:33–51.

Bickham, J.W., and M.J. Smolen. 1994. Somatic and heritable effects of environmental genotoxins and the emergence of evolutionary toxicology. *Environmental Health Perspectives* 102 (Suppl.) 12: 25–28.

BirdLife International. 2000. *Threatened Birds of the World*. Barcelona and Cambridge, UK: Lynx Edicions and Birdlife International.

BirdLife International. 2008a. *State of the World's Birds: Indicators for our Changing World*. Cambridge, UK: BirdLife International.

BirdLife International. 2008b. *Critically Endangered Birds: A Global Audit*. Cambridge, UK: BirdLife International.

Birge, W.J., A.G. Westerman, and J.A. Spromberg. 2000. Comparative toxicology and risk assessment of amphibians. In *Ecotoxicology of Amphibians and Reptiles*, Sparling, D.W., Linder, G. and Bishop, C.A., eds. Pensacola, FL: SETAC Press.

Blackwell, B.F., and R.A. Dolbeer. 2001. Decline of the red-winged blackbird population in Ohio correlated to changes in agriculture (1965–1996). *Journal of Wildlife Management* 65:661–667.

Blacksmith Institute. 2007. The world's most polluted places: The top ten of the dirty thirty. New York: Blacksmith Institute.

Blaustein, A.R., J.M. Romansic, J.M. Kiesecker, and A.C. Hatch. 2003. Ultraviolet radiation, toxic chemicals and amphibian population declines. *Diversity & Distributions* 9:123–140.

Boatman, N.D., N.W. Brickle, J.D. Hart, T.P. Milsom, A.J. Morris, A.W.A. Murray, K.A. Murray, and P.A. Robertson. 2004. Evidence for the indirect effects of pesticides on farmland birds. *Ibis* 146:131–143.

Boone, M.D., and R.D. Semlitsch. 2002. Interactions of an insecticide with competition and pond drying in amphibian communities. *Ecological Applications* 12:307–316.

Boone, M.D., and C.M. Bridges. 2003. Effects of pesticides on amphibian populations. In *Amphibian Conservation,* R.D. Semlitsch, ed.; 152–167. Washington, DC: Smithsonian Institution.

Boone, M.D., and S.M. James. 2003. Interactions of an insecticide, herbicide, and natural stressors in amphibian community mesocosms. *Ecological Applications* 13:829–841.

Boone, M.D., and S.M. James. 2005. Use of aquatic and terrestrial mesocosms in ecotoxicology. *Applied Herpetology* 2:231–257.

Boone, M.D., D. Cowman, C. Davidson, T. Hayes, W. Hopkins, R. Relyea, L. Schiesari, and R. Semlitsch. 2007. Evaluating the role of environmental contamination in amphibian population declines. In *Amphibian Conservation Action Plan*. Proceedings of IUCN/SSC Amphibian Conservation Summit 2005. C. Gascon, J.P. Collins, R.D. Moore, D.R. Church, J.E. McKay, and J.R. Mendelson III, eds.

Boutin, C., K.E. Freemark, and D A. Kirk. 1999. Farmland birds in southern Ontario: Field use, activity patterns and vulnerability to pesticide use. *Agriculture Ecosystems & Environment* 72:239–254.

Bridges, C.M. 1997. Tadpole swimming performance and activity affected by acute exposure to sublethal levels of carbaryl. *Environmental Toxicology & Chemistry* 16:1935–1939.

Bridges, C.M. 1999a. Predator–prey interactions between two amphibian species: Effects of insecticide exposure. *Aquatic Ecology* 33:205–211.

Bridges, C.M. 1999b. Effects of a pesticide on tadpole activity and predator avoidance behavior. *Journal of Herpetology* 33:303–306.

Bridges, C.M. 2000. Long-term effects of pesticide exposure at various life stages of the southern leopard frog (*Rana sphenocephala*). *Archives of Environmental Contamination and Toxicology* 39:91–96.

Butchart, S.H.M., A.J. Stattersfield, L.A. Bennun, S.M. Shutes, H.R. Akçakaya, J.E.M. Baillie, S.N. Stuart, C. Hilton-Taylor, and G.M. Mace. 2004. Measuring global trends in the status of biodiversity: Red List indices for birds. *PLoS Biology* 2:e383.

Cardoso M., A. Hyatt, P. Selleck, S. Lowther, V. Prakash, D. Pain, A.A. Cunningham, and D. Boyle. 2005. Phylogenetic analysis of the DNA polymerase gene of a novel alphaherpesvirus isolated from an Indian Gyps vulture. *Virus Genes*. 2005 May, 30 (3): 371–81.

Carpenter, K.E., M. Abrar, G. Aeby, R.B. Aronson, S. Banks, A. Bruckner, A. Chiriboga, et al. 2008. One-third of reef-building corals face elevated extinction risk from climate change and local impacts. *Science* 321:560–563.

Carson, R. 1962. *Silent Spring*. Boston: Houghton Mifflin.

Carey, C., W.R. Heyer, J. Wilkinson, R.A. Alford, J.W. Arntzen, T. Halliday, L. Hungerford, et al. 2001. Amphibian declines and environmental change: Use of remote-sensing data to identify environmental correlates. *Conservation Biology* 15:903–913.

Chamberlain, D.E., R.J. Fuller, R.G.H. Bunce, J.C. Duckworth, and M. Shrubb. 2000. Changes in the abundance of farmland birds in relation to the timing of agricultural intensification in England and Wales. *Journal of Applied Ecology* 37:771–788.

Colborn, T., D. Dumanoski, and J.P. Myers. 1996. *Our Stolen Future*. New York: Penguin Books.

Cooke, A.S. 1971. Selective predation by newts on frog tadpoles treated with DDT. *Nature* 229:275–276.

Corson, M.S., M.A. Mora, and W.E. Grant. 1998. Simulating cholinesterase inhibition in birds caused by dietary insecticide exposure. *Ecological Modelling* 105:299–323.

Costanza, R., R. d'Arge, R. de Groot, S. Farber, M. Grasso, B. Hannon, K. Limburg, et al. 1997. The value of the world's ecosystem services and natural capital. *Nature* 387:253–260.

Cunningham, V. Prakash, D. Pain G.R. Ghalsasi, G.A.H. Wells, G.N. Kolte, P. Nighot, M.S. Goudar, S. Kshirsagar, and A. Rahmani 2003. Indian vultures: Victims of an infectious disease epidemic? *Animal Conservation* 6:189–197.

Cuthbert, R., J. Parry-Jones, R.E. Green, and D.J. Pain. 2006. NSAIDs and scavenging birds: Potential impacts beyond Asia's critically endangered vultures. *Biology Letters* 3:90–93. doi:10.1098/rsbl.2006.0554.

Davidson, C., H.B. Shaffer, and M.R. Jennings. 2001. Declines of the California red-legged frog: Climate, UV-B, habitat, and pesticides hypotheses. *Ecological Applications* 14:464–479.

Davidson, C., H.B. Shaffer, and M.R. Jennings. 2002. Spatial tests of the pesticide drift, habitat destruction, UV-B, and climate-change hypotheses for California amphibian declines. *Conservation Biology* 16:1588–1601.

Davidson, C. 2004. Declining downwind: Amphibian population declines in California and historical pesticide use. *Ecological Applications* 14:1892–1902.

Donald, P.F., R.E. Green, and M.F. Heath. 2001. Agricultural intensification and the collapse of Europe's farmland bird populations. *Proceedings of the Royal Society of London Series B–Biological Sciences* 268:25–29.

EMEA 2004. Committee for Veterinary Medicinal Products: Diclofenac. Summary Report. EMEA/MRL/885/03-FINAL.

Engelman, C. 2008. Ecotoxicological simulation modeling: The effects of agricultural chemical exposure on borrowing owls wintering in south Texas cotton fields. MS Thesis, Texas A&M University.

Firbank, L.G., S. Petit, S. Smart, A. Blain, and R.J. Fuller. 2008. Assessing the impacts of agricultural intensification on biodiversity: A British perspective. *Philosophical Transactions of the Royal Society B – Biological Sciences* 363:777–787.

Fleishman, E., D.D. Murphy, and P.F. Brussard. 2000. A new method for selection of umbrella species for conservation planning. *Ecological Applications* 10:569–579.

Forbes, V.E., A. Palmqvist, and L. Bach. 2006. The use and misuse of biomarkers in ecotoxicology. *Environmental Toxicology & Chemistry*. 25:272–280.

Freda, J., and D.H. Taylor. 1992. Behavioral response of amphibian larvae to acidic water. *Journal of Herpetology* 26:429–433.

Freemark, K., and C. Boutin. 1995. Impacts of agricultural herbicide use on terrestrial wildlife in temperate landscapes: A review with special reference to North America. *Agriculture Ecosystems & Environment* 52:67–91.

Freemark K.E., and D.A. Kirk. 2001. Birds on organic and conventional farms in Ontario: Partitioning effects of habitat and practices on species composition and abundance. *Biological Conservation* 101:337–350.

Fuller, R. J., R.D. Gregory, D.W. Gibbons, J.H. Marchant, J.D. Wilson, S.R. Baillie, and N. Carter. 1995. Population declines and range contractions among lowland farmland birds in Britain. *Conservation Biology* 9:1425–1441.

Gamfeldt, L., H. Hillebrand, and P.R. Jonsson. 2008. Multiple functions increase the importance of biodiversity for overall ecosystem functioning. *Ecology* 89:1223–1231.

Genghini, M., S. Gellini, and M. Gustin. 2006. Organic and integrated agriculture: The effects on bird communities in orchard farms in northern Italy. *Biodiversity and Conservation* 15:3077–3094.

Gilbert, M., M.Z. Virani, R.T. Watson, J.L. Oaks, P.C. Benson, A.A Khan, S. Ahmed, et al. 2002. Breeding and mortality of oriental white-backed vulture *Gyps bengalensis* in Punjab Province, Pakistan. *Bird Conservation International,* 12:311–326.

Gilbert. M., R.T. Watson, S. Ahmed, M. Asim, and J.A. Johnson. 2007. Vulture restaurants and their role in reducing diclofenac exposure in Asian vultures. *Bird Conservation International* 17:63–77.

Gilbertson, M.K., G.D. Haffner, K.G. Drouillard, A. Albert, and B. Dixon. 2003. Immunosupression in the northern leopard frog (*Rana pipiens*) induced by pesticide exposure. *Environmental Toxicology & Chemistry* 22:101–110.

Goldstein M.I., T.E. Lacher, Jr., B. Woodbridge, M.J. Bechard, S.B. Canavelli, M.E. Zaccagnini, G.P. Cobb, E.J. Scollon, R. Tribolet, and M.J. Hooper. 1999. Monocrotophos-induced mass mortality of Swainson's hawks in Argentina, 1995–96. *Ecotoxicology* 8:201–214.

Goleman, W.L., L.J. Urquidi, T.A. Anderson, E.E. Smith, R.J. Kendall, and J.A. Carr. 2002. Environmentally relevant concentrations of ammonium perchlorate inhibit development and metamorphosis in *Xenopus laevis*. *Environmental Toxicology & Chemistry* 21:424–430.

Gotelli, N.J. and R.K. Colwell. 2001. Quantifying biodiversity: Procedures and pitfalls in the measurement and comparison of species richness. *Ecology Letters* 4:379–391.

Green, R.E., I. Newton, S. Shultz, A.A. Cunningham, M. Gilbert, D.J. Pain, and V. Prakash. 2004. Diclofenac poisoning as a cause of vulture population declines across the Indian subcontinent. *Journal of Applied Ecology* 41:793–800.

Green, R.E., M.A. Taggart, D. Das, D.J. Pain, C. Sashikumar, A.A. Cunningham, and R. Cuthbert. 2006. Collapse of Asian vulture populations: risk of mortality from residues of the veterinary drug diclofenac in carcasses of treated cattle. *Journal of Applied Ecology,* 43:949–956

Green, R.E., M.A. Taggart, K.R. Senacha, D.J. Pain, Y. Jhala, and R. Cuthbert. 2007. Rate of decline of the oriental white-backed vulture *Gyps bengalensis* population in India estimated from measurements of diclofenac in carcasses of domesticated ungulates. *PloS One* 2:e686. doi:10.1371/journal.pone.0000686.

Harfenist, A., T. Power, K.L. Clark, and D.B. Peakall. 1989. A review and evaluation of the amphibian toxicological literature. *Technical Report Series*, No. 61, Canadian Wildlife Service, Ottawa.

Hayes, T., K. Haston, M. Tsui, A. Hoang, C. Haeffele, and A. Vonk. 2003. Atrazine-induced hermaphroditism at 0.2 ppb in American leopard frogs (*Rana pipiens*): Laboratory and field evidence. *Environmental Health Perspectives* 111:568–575.

Henriques, W., R.D. Jeffers, T.E. Lacher, Jr., and R.J. Kendall. 1997. Agrochemical use on banana plantations in Latin America: perspectives on ecological risk. Annual Review Issue on Tropical Ecotoxicology. *Environmental Toxicology & Chemistry* 16:100–111.

Hoffman, D.J., B.A. Rattner, G.A. Burton, Jr., and J. Cairns, Jr. 2002. *Handbook of Ecotoxicology,* 2nd edition. Boca Raton, FL: CRC Press.

Hooper, D.U. et al. 2005. Effects of biodiversity on ecosystem functioning: a consensus of current knowledge. *Ecological Monographs* 75:3–35.

Hooper, M. J., P. Mineau, M.E. Zaccagnini, G.W. Winegrad, and B. Woodbridge. 1999. Monocrotophos and the Swainson's hawk. *Pesticide Outlook* 10:97–102.

Hopkins, W.A. 2007. Amphibians as models for studying environmental change. *Journal of the Institute for Laboratory Animal Research* 48:270–277.

Huber, W. 1993. Ecotoxicological relevance of atrazine in aquatic systems. *Environmental Toxicology & Chemistry*. 1993:1865–1881.

Hurlbert, S.H. 1971. The nonconcept of species diversity: A critique and alternative parameters. *Ecology* 52:577–586.

ISARPW 2004. Report on the international South Asian recovery plan workshop. *Buceros* 9:1–48.

James, S. M., E.E. Little, and R.D. Semlitsch. 2004. The effect of soil composition and hydration on the bioavailability and toxicity of cadmium to hibernating juvenile American toads (*Bufo americanus*). *Environmental Pollution* 132:523–532.

Kendall, R.J., and T.E. Lacher, Jr. 1991. Ecological modeling, population ecology, and wildlife toxicology: A team approach to environmental toxicology. *Environmental Toxicology & Chemistry* 10:297–299.

Kendall, R.J., and T.E. Lacher, Jr. 1994. *Wildlife Toxicology and Population Modeling: Integrated Studies of Agroecosystems*. Boca Raton, FL: Lewis Publishers.

Kendall, R.J., C.M. Bens, G.P. Cobb III, R.L. Dickerson, K.R. Dixon, S.J. Klaine, T.E. Lacher, Jr., et al. 1996. Aquatic and terrestrial ecotoxicology. In *Casarett and Doull's Toxicology: The Basic Science of Poisons,* 5th edition, C.D. Klassen et al., eds, 883–905. New York: McGraw Hill.

Kiesecker, J.M. 2002. Synergism between trematode infection and pesticide exposure: A link to amphibian limb deformities in nature? *Proceedings of the National Academy of Sciences USA* 99:9900–9904.

Kiely, T., D. Donaldson, and A. Grube. 2004. *Pesticides Industry Sales and Usages: 2000 and 2001 Market Estimates*. USEPA Office of Prevention, Pesticides, and Toxic Substances.

Killeen, T.J. 2007. *A Perfect Storm in the Amazon Wilderness: Development and Conservation in the Context of the Initiative for the Integration of the Regional Infrastructure of South America (IIRSA)*. Advances in Applied Biodiversity Science Number 7. Arlington, VA: Center for Applied Biodiversity Science, Conservation International.

Kunin, W.E. 1998. Extrapolating species abundance across spatial scales. *Science* 281:1513–1515.

Lacher, T.E., Jr. 1994. Theoretical ecology and research on agroecosystems. In *Wildlife toxicology and Population Modeling: Integrated Studies of Agroecosystems*, R.J. Kendall and T.E. Lacher, Jr., eds., 429–433. Chelsea, MI: Lewis Publishers.

Lacher, T.E., Jr., and M.I. Goldstein. 1997. Tropical ecotoxicology: status and needs. Annual Review Issue on Tropical Ecotoxicology. *Environmental Toxicology & Chemistry* 16:91–99.

Lacher, T.E., Jr., S. Mortensen, K. Johnson, and R.J. Kendall. 1997. Environmental aspects of pesticide use in banana plantations. *Pesticide Outlook* 8:24–28.

Lamoreux, J., H.R. Akçakaya, L. Bennun, N.J. Collar, L. Boitani, D. Brackett, A. Bräutigam, et al. 2003. Value of the IUCN Red List. *Trends in Ecology and Evolution* 18:214–215.

La Sorte, F.A., and W.J. Boecklen. 2005. Changes in the diversity structure of avian assemblages in North America. *Global Ecology and Biogeography* 14:367–378.

Laurance, W.F. 2008. Global warming and amphibian extinctions in eastern Australia. *Australian Ecology* 33:1–9.

Linder, G., S.K. Krest, and D.W. Sparling. 2003. *Amphibian Decline: An Integrated Analysis of Multiple Stressor Effects*. Pensacola, FL: SETAC Press.

Lips, K.R., J. Diffendorfer, J.R. Mendelson III, and M.W. Sears. 2008. Riding the wave: Reconciling the roles of disease and climate change in amphibian declines. *PLoS Biology* 6(3):e72. doi:10.1371/journal.pbio.0060072.

Little, E.E., R. Calfee, L. Cleveland, R. Skinker, A. Zaga-Parkhurst, and M.G. Barron. 2000. Photo-enhanced toxicity in amphibians: Synergistic interactions of solar ultraviolet radiation and aquatic contaminants. *Journal of Iowa Academy of Sciences* 107:67–71.

Magurran, A.E. 1988. *Ecological Diversity and Its Measurement.* Princeton, NJ: Princeton University Press.

Markandya, A., T. Taylor, A. Longo, M.N. Murty, S. Murty, and K. Dhavala. 2008. Counting the cost of vulture decline: An appraisal of the human health and other benefits of vultures in India. *Ecological Economics* 67:194–204.

Matson, C.W., M.M. Lambert, T.J. McDonald, R L. Autenrieth, K.C. Donnelly, A. Islamzadeh, D.I. Politov, and J.W. Bickham. 2006. Evolutionary toxicology and population genetic effects of chronic contaminant exposure on marsh frogs (*Rana ridibunda*) in Sumgayit, Azerbaijan. *Environmental Health Perspectives* 114:547–552.

Matson, C.W., G. Palatnikov, T.J. McDonald, R.L. Autenrieth, K.C. Donnelly, T.A. Anderson, J.E. Canas, A. Islamzadeh, and J.W. Bickham. 2005a. Patterns of genotoxicity and contaminant exposure: evidence of genomic instability in the marsh frogs (*Rana ridibunda*) of Sumgayit, Azerbaijan. *Environmental Toxicology & Chemistry* 24:2055–2064.

Matson, C.W., G. Palatnikov, A. Islamzadeh, T.J. McDonald, R.L. Autenrieth, K.C. Donnelly, and J.W. Bickham. 2005b. Chromosomal damage in two species of aquatic turtles (*Emys orbicularis* and *Mauremys caspica*) inhabiting contaminated sites in Azerbaijan. *Ecotoxicology* 14:1–13.

Meteyer, C.U., B.A. Rideout, M. Gilbert, H.L. Shivaprasad, and J.L. Oaks. 2005. Pathology and pathophysiology of diclofenac poisoning in free-living and experimentally exposed oriental white-backed vultures (*Gyps bengalensis*). *Journal of Wildlife Diseases* 41:707–716.

Metts, B.S., W.A. Hopkins, and J.P. Nestor. 2005. Density-dependent effects of an insecticide on a pond-breeding salamander assemblage. *Freshwater Biology* 50:685–696.

Mills, N.E. and R.D. Semlitsch. 2004. Competition and predation mediate indirect effects of an insecticide on southern leopard frogs. *Ecological Applications* 14:1041–1054.

Mineau P., M.R. Fletcher, L.C. Glaser, N.J. Thomas, C. Brassard, L.K. Wilson, J.E. Elliott, et al. 1999. Poisoning of raptors with organophosphorus and carbamate pesticides with emphasis on Canada, US and UK. *Journal of Raptor Research* 33:1–37.

Mineau, P., C. M. Downes, D. A. Kirk, E. Bayne, and M. Csizy. 2005. Patterns of bird species abundance in relation to granular insecticide use in the Canadian prairies. *Ecoscience* 12:267–278.

Mineau, P., and M. Whiteside. 2006. Lethal risk to birds from insecticide use in the United States: A spatial and temporal analysis. *Environmental Toxicology & Chemistry* 25:1214–1222.

Mittermeier, R.A., C.F. Kormos, C.G. Mittermeier, P. Robles Gil, T. Sandwith, and C. Besançon. 2005. *Transboundary Conservation: A New Vision for Protected Areas.* CEMEX – Agrupación Sierra Madre – Conservation International, Mexico City, Mexico.

Mortensen, S.R., K.A. Johnson, C.P. Weisskopf, M.J. Hooper, T.E. Lacher, Jr., and R.J. Kendall. 1998. Avian exposure to pesticides in Costa Rican banana plantations. *Bulletin of Environmental Contamination and Toxicology* 60:562–568.

Myers, N., R.A. Mittermeier, C.G. Mittermeier, G.A.B. da Fonseca, and J. Kent. 2000. Biodiversity hotspots for conservation priorities. *Nature* 403:853–858.

Nevo, E., R. Noy, B. Lavie, A. Beiles, and S. Muchtar. 1986. Genetic diversity and resistance to marine pollution. *Biological Journal of the Linnean Society* 29:139–144.

Newman, M.C., and R.H. Jagoe. 1998. Allozymes reflect the population level effect of mercury: Simulations of the mosquitofish (*Gambusia holbrooki* girard) GPI-2 response. *Ecotoxicology* 7:141–150.

Newton, I. 2004. The recent declines of farmland bird populations in Britain: An appraisal of causal factors and conservation actions. *Ibis* 146:579–600.

Oaks, J.L., B.A. Rideout, M. Gilbert, R. Watson, M. Virani, and A.A. Khan. 2001. (Abstract) Summary of diagnostic investigations into vulture mortality: Punjab Province, Pakistan, 2000–2001. 4th Eurasian Congress on Raptors, 25–29 September 2001. Seville, Spain: Raptor Research Foundation and Estacion Biologica Doñana.

Oaks, J.L., S.L. Donahoe, F.R. Rurangirwa, B.A. Rideout, M. Gilbert, and M.Z. Virani. 2004. Identification of a novel mycoplasma species from an oriental white-backed vulture (*Gyps bengalensis*). *Journal of Clinical Microbiology* 42:5909–5912.

Oaks, J.L., M. Gilbert, M.Z. Virani, R.T. Watson, C.U. Meteyer, B.A. Rideout, H.L. Shivaprasad, et al. 2004. Diclofenac residues as the cause of vulture population decline in Pakistan. *Nature* 427:630–633.

Pain, D.J., A.A. Cunningham, P.F. Donald, J.W. Duckworth, D.C. Houston, T. Katzner, J. Parry-Jones, et al. 2003. Gyps vulture declines in Asia; temporospatial trends, causes and impacts. *Conservation Biology* 17:661–671.

Pain, D.J., C.G.R. Bowden, A.A. Cunningham, R. Cuthbert, D. Das, M. Gilbert, R.D. Jakati, et al. 2008. The race to prevent the extinction of South Asian vultures. *Bird Conservation International* 18:S30–S48.

Pakendorf B., M. Stoneking. 2005. Mitochondrial DNA and human evolution. *Annual Review of Genomics and Human Genetics* 6:165–183.

Peterjohn, B.G., J.R. Sauer. 1999. Population status of North American grassland birds from the North American Breeding Bird Survey, 1966–1996. *Studies in Avian Biology* 19:27–44.

Pimentel, D. 2005. Environmental and economic costs of the application of pesticides primarily in the United States. *Environment, Development, and Sustainability* 7:229–252.

Pisani, J.M., W.E. Grant and M.A. Mora. 2008. Simulating the impact of cholinesterase-inhibiting pesticides on non-target wildlife in irrigated crops. *Ecological Modeling* 210:179–192.

Potts, G.R. 1986. *The Partridge: Pesticides, Predation and Conservation.* London: Collins.

Pounds, J.A., M.R. Bustamante, L.A. Coloma, J.A. Consuegra, M.P.L. Fogden, P.N. Foster, E. La Marca, et al. 2006. *Nature* 439:161–176.

Prakash, V., R.E. Green, D.J. Pain, S.P. Ranade, S. Saravanan, N. Prakash, R. Venkitachalam, R. Cuthbert, A.R. Rahmani, and A.A. Cunningham. 2007. Recent changes in populations of resident *Gyps* vultures in India. *Journal of the Bombay Natural History Society* 104:129–135.

Relyea, R.A., and N.M. Mills. 2001. Predator-induced stress makes the pesticide carbaryl more deadly to gray treefrog tadpoles (*Hyla versicolor*). *Proceedings of the National Academy of Sciences USA* 98:2491–2496.

Relyea, R.A. 2003. Predator cues and pesticides: A double dose of danger for amphibians. *Ecological Applications* 13:1515–1521.

Relyea, R.A. 2004. Growth and survival of five amphibian species exposed to combinations of pesticides. *Environmental Toxicology & Chemistry* 23:1737–1742.

Relyea, R.A. 2005. The impact of insecticides and herbicides on the biodiversity and productivity of aquatic communities. *Ecological Applications* 15:618–627.

Relyea, R.A., N.M. Schoeppner, and J. Hoverman. 2005. Pesticides and amphibians: The importance of community context. *Ecological Applications* 15:1125–1134.

Relyea, R.A., and J. Hoverman. 2006. Assessing the ecology in ecotoxicology: A review and synthesis in freshwater systems. *Ecology Letters* 9:1157–1171.

Rodrigues, A.S.L., and T.M. Brooks. 2007. Shortcuts for biodiversity conservation planning: The effectiveness of surrogates. *Annual Review of Ecology, Evolution and Systematics* 38:713–737.

Roe, J.H., W.A. Hopkins, S.E. Durant, and J. M. Unrine. 2006. Effects of competition and coal combustion wastes on recruitment and life history characteristics of salamanders in temporary wetlands. *Aquatic Toxicology* 79:176–184.

Rohr, J.R., and B.D. Palmer. 2005. Aquatic herbicide exposure increases salamander desiccation risk eight months later in a terrestrial environment. *Environmental Toxicology & Chemistry* 24:1253–1258.

Rohr, J.R., T.R. Raffel, J.M. Romansic, H. McCallum, and P.J. Hudson. 2008. Evaluating the links between climate, disease spread, and amphibian declines. *PNAS* 105:17436–17441.

Rowe, C.L., and W.A. Dunson. 1994. The value of simulated pond communities in mesocosms for studies of amphibian ecology and ecotoxicology. *Journal of Herpetology* 28:346–356.

Sala, O.E., F.S. Chapin, J.J. Armesto, E. Berlow, J. Bloomfield, R. Dirzo, E. Huber-Sanwald, et al. 2000. Biodiversity: Global biodiversity scenarios for the year 2100. *Science* 287:1770–1774.

Sauer, J.R., J.E. Hines, I. Thomas, J. Fallon, G. Gough. 2000. *The North American Breeding Bird Survey, Results and Analysis 1966–1999*. Version 98.1, Laurel, MD: USGS, Patuxent Wildlife Research Center.

Schipper, J., J. Chanson, F. Chiozza, N. Cox, M. Hoffmann, V. Katariya, J. Lamoreux, et al. 2008. The status of the world's land and marine mammals: Diversity, threat, and knowledge. *Science* 322:225–230.

Scott, D.E. 1994. The effect of larval density on adult demographic traits in *Ambystoma opacum*. *Ecology* 75:1383–1396.

Secretariat of the Convention on Biological Diversity. 2005. *Handbook of the Convention on Biological Diversity Including Its Cartegena Protocol on Biosafety,* 3rd edition. Montreal, Canada.

Secretariat of the Convention on Biological Diversity. 2006. *Global Biodiversity Outlook 2*. Montreal, Canada.

Semlitsch, R.D., D.E. Scott, and J.H.K. Pechmann. 1988. Time and size at metamorphosis related to adult fitness in *Ambystoma talpoideum*. *Ecology* 69:184–192.

Shultz, S., H.S. Baral, S. Charman, A.A. Cunningham, D. Das, D.R. Ghalsasi, M.S. Goudar, et al. 2004. Diclofenac poisoning is widespread in declining vulture populations across the Indian subcontinent. *Proceedings of the Royal Society of London B (Suppl.)* 271:S458–S460.

Siriwardena, G.M., S.R. Baillie, S.T. Buckland, R.M. Fewster, J.H. Marchant, and J.D. Wilson. 1998. Trends in the abundance of farmland birds: a quantitative comparison of smoothed Common Birds Census indices. *Journal of Applied Ecology* 35:24–43.

Siriwardena, G.M., H.Q.P. Crick, S.R. Baillie, and J.D. Wilson. 2000. Agricultural land-use and the spatial distribution of granivorous lowland farmland birds. *Ecography* 23:702–719.

Skelly, D.K. 1994. Activity level and the susceptibility of anuran larvae to predation. *Animal Behavior* 47:465–468.

Smith, D.C. 1987. Adult recruitment in chorus frogs: Effects of size and date at metamorphosis. *Ecology* 68:344–350.

Storrs, S.I., and J.M. Kiesecker. 2004. Survivorship patterns of larval amphibians exposed to low concentrations of atrazine. *Environmental Health Perspectives* 112:1054–1057.

Stuart, S.N., J.S. Chanson, N.A. Cox, B.E. Young, A.S.L. Rodrigues, D.L. Fischman, and R.W. Waller. 2004. Status and trends of amphibian declines and extinctions worldwide. *Science* 306:1783–1786.

Swan, G.E., R. Cuthbert, M. Quevedo, R.E. Green, D.J. Pain, P. Bartels, A.A. Cunningham, et al. 2006. Toxicity of diclofenac to *Gyps* vultures. *Biology Letters* 2:279–282.

Swan, G., V. Naidoo, R. Cuthbert, R.E. Green, D.J. Pain, D. Swarup, V. Prakash, et al. 2006. Removing the threat of diclofenac to critically endangered Asian vultures. *PLoS Biology* 4:396–402.

Swartz, C.D., K.C. Donnelly, A. Islamzadey, G.T. Rowe, W. J. Rogers, G.M. Palatnikov, R. Kasimov, T.J. McDonald, J.K. Wickliffe, and J.W. Bickham. 2003. Chemical contaminants and their effects in fish and wildlife from the industrial zone of Sumgayit, Republic of Azerbaijan. *Ecotoxicology* 12:511–523.

Swarup, D., R.C. Patra, V. Prakash, R. Cuthbert, D. Das, P. Avari, D.J. Pain, et al. 2006. The safety of meloxicam to critically endangered *Gyps* vultures and other scavenging birds in India. *Animal Conservation* 10:192–198.

Taggart, M.A., R. Cuthbert, D. Das, D.J. Pain, R.E. Green, S. Shultz, A.A. Cunningham, and A.A. Meharg. 2006. Diclofenac disposition in Indian cow and goat with reference to *Gyps* vulture population declines. *Environmental Pollution* 147:60–65.

Taggart, M.A., K. Senacha, R.G. Green, Y.V. Jhala, B. Ragathan, A.R. Rahmani, R. Cuthbert, D.J. Pain, and A.A. Meharg. 2007. Diclofenac residues in carcasses of domestic ungulates available to vultures in India. *Environment International* 33:759–765.

Thomas, C.D., A. Cameron, R.E. Green, M. Bakkenes, L.J. Beaumont, Y.C. Collingham, B.F.N. Erasmus, et al. 2004. Extinction risk from climate change. *Nature* 427:145–148.

Tilman, D., J. Fargione, B. Wolff, C. D'Antonio, A. Dobson, R. Howarth, D. Schindler, W.H. Schlesinger, D. Simberloff, and D. Swackhamer. 2001. Forecasting agriculturally driven global environmental change. *Science* 292:281–284.

Turner, W.R., K. Brandon, T.M. Brooks, R. Costanza, G. A.B. da Fonseca, and R. Portela. 2007. Global conservation of biodiversity and ecosystem services. *BioScience* 57:868–873.

Truhaut, R. 1977. Ecotoxicology: Objectives, principles and perspectives. *Ecotoxicology and Environmental Safety* 1:151–173.

Unrine, J.M., W.A. Hopkins, C.S. Romanek, and B.P. Jackson. 2007. Bioaccumulation of trace elements in omnivorous amphibian larvae: Implications for amphibian health and contaminant transport. *Environmental Pollution* 149:182–192.

U.S. EPA. 1992. *Risk Assessment Forum: Framework for Ecological Risk Assessment.* EPA/630/R-92/001. Washington, D.C.: U.S. EPA.

Wilson, E.O., and F.M. Peter. 1988. *Biodiversity*. Washington, DC: National Academy Press.

Wu, T. 2009. Integrating biodiversity conservation with poverty reduction: A socioeconomic perspective. *Bulletin of the Ecological Society of America* 2009:80–85.

Yu, M.-H. 2005. *Environmental Toxicology: Biological and Health Effects of Pollutants*, 2nd edition. Boca Raton, FL: CRC Press.

6 Impacts of Anthropogenic CO_2 and Climate Change on the Biology of Terrestrial and Marine Systems

Lee Hannah

CONTENTS

6.1 INTRODUCTION

Toxicology traditionally deals with contaminants released locally or regionally that impact one or several species. Even though chemical application and direct habitat alteration have long been recognized as impacting regional or global ecosystems, we now realize that less obviously harmful human activities are affecting not only individual species, but also the entire planet. Chief among the planetary concerns are the effects, direct and indirect, of the large amounts of fossil fuel pollution being disposed of in Earth's atmosphere. Combustion of fossil fuels releases CO_2, which has been measurably increasing in the atmosphere for over half a century. Impacts of this CO_2 buildup may affect all wildlife on Earth, both marine and terrestrial, and are the subject of this chapter.

It has been known for over a century that accumulating CO_2 in the atmosphere could retain heat and might lead to warming of the surface of the planet and associated changes in climate (Keeling and Rakestraw 1960). For almost as long, it has been recognized that increasing CO_2 has direct effects on plant life. More recently, it has become clear that CO_2 buildup influences ocean chemistry, with perhaps profound consequences for marine life. While there are other greenhouse gases contributing to climate change, CO_2 is unique in being the greatest contributor to climate change, as well as driving large-scale changes in ocean chemistry and plant growth.

Appreciation of the impacts of climate change on wildlife began with a landmark book by Rob Peters and Thomas E. Lovejoy, as well as a series of journal publications by Peters and colleagues (Peters and Darling 1985; Peters 1991; Peters and Lovejoy 1992). These authors recognized that species ranges had shifted in response to climate change in the past, and would likely do so again in response to human-driven climate change. They recognized that changing species range boundaries relative to fixed protected area's boundaries might create more, or less, protection for species. They also saw the relevance to wildlife of myriad climate changes associated with planetary warming and the direct effects of CO_2.

The field has matured over the intervening two decades. New research is revealing the complexity of climate change and its direct effects on a weekly basis. Marine impacts that were only hinted at in the early publications are now not only recognized, but have begun (e.g., Glynn 2001). Reefs around the world have died in episodes of extremely high water temperatures. Anticipated terrestrial impacts are unfolding as well, with species range shifts, changes in phenology, and changes in species interactions all the focus of a substantial and increasing literature (Parmesan and Yohe 2003; Root et al. 2003).

Wildlife biologists need to be aware of these changes and how they may influence species in even remote locations (Hannah, Midgley et al. 2002). There is no pristine wilderness immune to these effects, and they will compound many existing stresses in heavily impacted landscapes. A new era of wildlife toxicology has begun, one in which CO_2 and associated climate change must take their places among other contaminants as major concerns. Understanding the impacts associated with rising CO_2 will help wildlife managers everywhere anticipate interactions with other contaminants and stressors and, ultimately, be more effective managers in the face of change.

6.2 RISING CO_2 IN THE ATMOSPHERE

CO_2 enters the atmosphere from many combustion-related sources. The burning of fossil fuels is the number one human-caused source of CO_2, with the burning of tropical forests as the second leading cause (IPCC 2007). Tropical forest burning is a surprisingly large source, accounting for about one-fifth of all CO_2 emissions, or as much as the entire transportation sector.

CO_2 from forest burning may peak in the first half of this century, as forests are depleted and efforts to abate forest emissions are introduced. Reducing emissions from deforestation is now a focus of international policy on reducing greenhouse gas emissions (da Fonseca et al. 2007). It is one of the easiest ways to reduce CO_2

emissions, because the economic gains associated with forest burning are often low and the means to reduce the destruction are technically simple. For wildlife, there is the double benefit of reducing greenhouse gas emissions and reducing habitat destruction simultaneously when forest burning is curtailed. A strong effort to control CO_2 emissions therefore logically begins with reducing deforestation and burning of forests.

The CO_2 levels from fossil fuel emissions are rapidly rising and expected to continue to peak in mid- to late century (Pacala and Socolow 2004). The major focus of international efforts to curtail greenhouse gas emissions is on these industrial and transport emissions from fossil fuel. Stabilizing greenhouse gases in the atmosphere will require an almost complete cessation of fossil fuel emissions, representing a complete energy transformation to renewable sources. Because of the magnitude of this task, fossil fuel emissions are expected to continue to increase for a decade or more. Optimistic scenarios indicate that stabilization by mid-century is possible with existing technologies (Pacala and Socolow 2004). Past mid-century, however, large amounts of new renewable or carbon-emissions-free energy sources must come online to ensure that human drivers of climate change are minimized.

The effects of emissions from fossil fuel use and deforestation are clearly visible in atmospheric chemistry. Atmospheric CO_2 concentrations have been monitored since the 1950s from the Mauna Loa Observatory in Hawaii (Keeling and Rakestraw 1960) (Figure 6.1). Charles Keeling of the Scripps Institute established monitoring for atmospheric CO_2 at Mauna Loa in 1958, and the record he initiated shows a clear trace of increasing CO_2 ever since. Pre-industrial CO_2 concentrations are usually accepted to have been 280 ppm, while current CO_2 concentration is about 380 ppm and rising (IPCC 2007).

Within the rising CO_2 trace is a small sawtooth rise and fall associated with the seasons. In the northern summer, as plants begin to grow and take up CO_2, global CO_2 concentrations fall (Keeling et al. 1996). They are not balanced by fall decay in the southern hemisphere because there is little landmass at temperate latitudes in the south. Similarly, in the northern autumn, leaves fall and decay, releasing CO_2 into the atmosphere and causing a small but perceptible rise in CO_2 concentration in the Mauna Loa record. Other observations of rising CO_2, such as gas bubbles trapped in Antarctic glacial ice, confirm a strong and increasing rise in atmospheric concentration (IPCC 2007).

The evidence of CO_2 buildup in the atmosphere is therefore clear and highly resolved. The CO_2 that has been released into the atmosphere since the dawn of the industrial revolution is accumulating, with attendant effects on global chemistry and climate. Effects of the CO_2 buildup on wildlife are both direct and indirect. The direct effects include changes in plant growth and competitiveness due to increased photosynthetic precursors, changes in seawater chemistry that impact organisms that build calcium carbonate shells, and changes related to decreasing seawater pH. Indirect effects stem from the effects of CO_2 as a greenhouse gas on climate, with attendant changes in almost all organisms that respond to climate to determine their range, timing of biological events, and interactions with other species.

Future CO_2 concentration will continue to rise, at least for several decades. The Intergovernmental Panel on Climate Change (IPCC) has established several scenarios for future increase (IPCC 2001). These scenarios are logically consistent stories

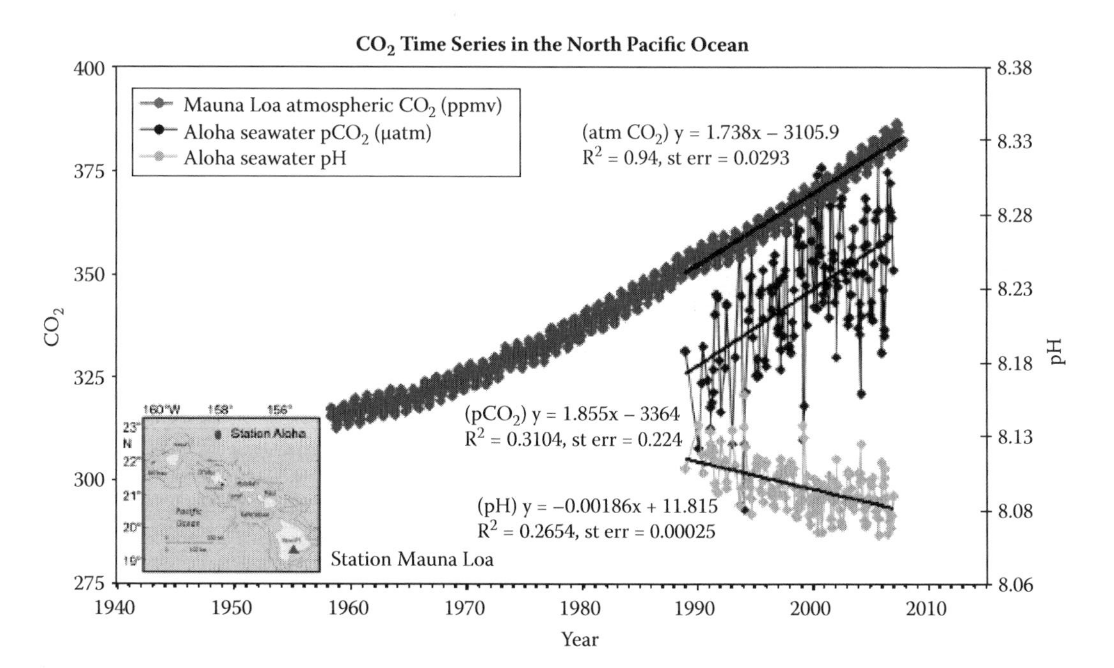

FIGURE 6.1 Rising CO_2 records in the atmosphere and ocean. CO_2 increase in the atmosphere (upper trace) has been monitored from the Mauna Loa observatory (map, triangle) on the island of Hawaii. Increasing CO_2 in seawater (middle trace) and decreasing pH of seawater (bottom trace) have been monitored more recently from Station Aloha, north of Oahu (map, filled circle).

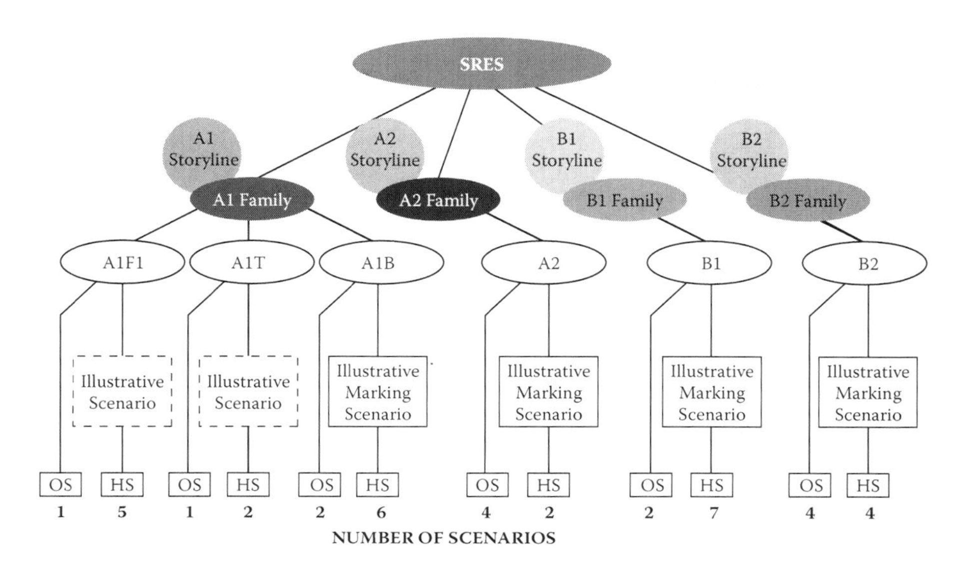

FIGURE 6.2 IPCC emissions scenarios. The A1 family includes the scenarios with the highest near-term emissions rates, while the A2 family has the highest overall emissions. The B1 and B2 families are lower emissions trajectory scenarios, but do not include active greenhouse gas reduction policies. Current actual emissions exceed all IPCC scenarios.

of how global economies, industry, and emissions may evolve. Some assume continuing globalization and rapidly expanding economic growth and fossil fuel use. Others assume a more technological society, regionalization, or other factors that may lower emissions. The full family of IPCC scenarios is represented in Figure 6.2.

None of the IPCC scenarios includes specific policy actions to reduce emissions. Global greenhouse gas reduction policy could, therefore, lower emissions much more than in any of the IPCC scenarios. The international treaty designed to produce global emissions reductions is the United Nations Framework Convention on Climate Change (UNFCCC). Most nations of the world, including the United States, are signatories to the UNFCCC. The Kyoto Protocol is the first set of rules for emissions reductions under the UNFCCC. The Kyoto Protocol has been ratified by enough nations to come into force, but the United States is notably absent from the list of countries that has ratified Kyoto. This means that the world's largest CO_2 polluter is not participating in the first round of efforts to control the problem. The next round of UNFCCC implementation is expected to aim for much more ambitious reductions, likely with a target of keeping global mean temperature increase from rising above 2°C.

Despite the efforts of the IPCC and UNFCCC, global CO_2 emissions are accelerating. The current emissions trend is higher than even the highest IPCC scenario (Raupach et al. 2007). This is possible because the targets set in the Kyoto Protocol were modest, the largest polluter (the United States) did not participate, and emissions have rapidly increased in countries such as China and India, which were exempted from Kyoto in order to promote economic development and to focus early reductions on countries that had caused most of the pollution in the past.

This leaves the world with large and rapidly increasing CO_2 emissions, ambitious targets for reduction, and a highly uncertain trajectory of CO_2 concentration in the atmosphere as a result. The implications for wildlife are profound, as will be explored below. The combination of rapidly rising emissions and low ultimate targets may even mean that emissions will undergo an "overshoot"—a rise in greenhouse gases above target levels, followed by gradual reductions back to the target (Parry et al. 2009). This would put wildlife through a double transition—a warming *and* a cooling—with consequences that are difficult to predict. In the remainder of this chapter, we explore known and expected consequences of rising CO_2 warming, as well as possible endpoints for wildlife.

6.3 DIRECT EFFECTS OF CO_2 ON TERRESTRIAL SYSTEMS

Physiological theory predicts that rising CO_2 will have a direct effect on plants, resulting from an alteration of the concentration of this important precursor to photosynthesis (Woodward 1998). This theoretical prediction has now been borne out in both model simulations and experimental treatments. The experimental record reveals effects not initially recognized by theorists, resulting in complexities that are not yet fully resolved.

Theory suggests that when the atmospheric partial pressure of CO_2 is increased, photosynthesis is stimulated (Beerling and Woodward 1993). This stimulation should facilitate faster plant growth and possibly increased biomass. In addition, plants using different photosynthetic pathways may be affected differently. In particular, plants using the C3 photosynthetic pathway are generally favored in high CO_2 environments, meaning that C4 plants may lose dominance as human CO_2 contributions to the atmosphere continue. In C3 plants, photosynthesis is catalyzed by the enzyme Rubisco (Drake et al. 2005). Rubisco is affected directly by changes in CO_2 since it is subsaturated at current atmospheric CO_2 levels (280–380 ppm). Increases in atmospheric CO_2 concentration may increase saturation of Rubisco, thereby enhancing photosynthetic rates, increasing competitiveness of C3 plants.

Experiments support these expectations. Both in laboratory and whole plot experiments, elevated CO_2 enhances plant growth (Lin et al. 1998; Reich et al. 2001). Enhancement in C3 plants is higher than in C4 plants in laboratory experiments (Korner 2000). Total biomass of plants grown in elevated CO_2 is often higher than in control, ambient CO_2 experiments (Lawton 1995). C3 species in laboratory experiments were found to average a 45% enhancement of biomass under elevated CO_2 of between 500 and 750 ppm (Poorter and Navas 2003). C4 plants showed a lower, but still significant biomass enhancement of 12%. CAM species were intermediate at 23% (Poorter and Navas 2003). Laboratory experiments are generally conducted on individual plants in greenhouses or small enclosures in which CO_2 is added (Lawton 1995). Whole plot experiments may have CO_2 pumped into greenhouses, or, in the most advanced experiments, piped into whole fields in situ. The in situ field enhancement of CO_2 experiments are known as Free-Air CO_2 Enrichment, or FACE experiments (Smith et al. 2000). Natural seeps of CO_2 have also been manipulated to study effects on vegetation.

The physiological response observed in experiments is reduced over time, resulting from an apparent physiological adjustment to elevated CO_2 (Idso 1999). This down-regulation occurs in many types of plants in many experimental environments, resulting in less long-term net increase in productivity and biomass increase than might be predicted from the results of short-term experiments. Conversely, increased air temperature accentuates CO_2 stimulation, so the combined effects of increased CO_2 and temperature may be more complex than suggested by CO_2 enrichment alone (Idso et al. 1987) (Figure 6.3).

Effects of elevated CO_2 are further complicated in whole ecosystem settings. In some systems, down-regulation seems to reduce aboveground biomass production in the first few years of the experiment, while belowground (root) biomass is stimulated (Korner 2000). Eventually, the stimulation of root biomass facilitates greater growth aboveground, and the initial down-regulation is overcome, with a net increase in biomass observed (Drake 2005). In other settings, down-regulation seems to dominate, leaving less overall increase in biomass from an elevated CO_2. Competition may be affected, but overall evidence for reduction in C4 dominance is limited (Korner 2000). C4 plants may continue to be more competitive in drought-limiting conditions, despite some diminution of their competitive advantage, at least up to double pre-industrial CO_2 conditions.

There is little literature on the direct effect of changing CO_2 on terrestrial vertebrates at concentrations expected from anthropogenic pollution. CO_2 toxicity in humans occurs at concentrations above 1%. Since CO_2 concentrations in the atmosphere are low relative to partial pressure in the lungs, there is little theoretical reason to expect an effect at the low levels of increase associated with fossil fuel burning. Increased CO_2 in the ppm range may have effects on soil microbes and processes, however. Nutrient uptake mediated by mycorrhizae has been found to be stimulated by elevated CO_2, and the effect is more pronounced in C4 plants (Wilson and Hartnett 1998).

Simulation of elevated CO_2 in models of global vegetation suggests an important direct effect of CO_2, though smaller than the effect of climate change (Cramer et al. 2001; Sitch et al. 2008). These computer simulations calculate photosynthesis and biomass from photosynthetic and carbon cycle equations. They show reasonable agreement with free air enrichment (FACE) experiments for stimulation of net primary productivity. NPP enhancement in FACE experiments averages 23%, while global vegetation simulations show increases of 18–34% (Sitch et al. 2008).

6.4 DIRECT EFFECTS OF CO_2 ON MARINE SYSTEMS

CO_2 has two main direct effects on marine systems, one mediated by the calcium carbonate chemistry of seawater, the other mediated by pH (Fabry et al. 2008). The calcium carbonate effect impedes the shell-building ability of marine plankton, corals, and mollusks, while the pH effect has wide-ranging implications for physiological changes in marine organisms.

Marine creatures that build shells or tests from calcium carbonate do so using one of two variants of calcium carbonate: aragonite or calcite (Feely et al. 2004). The saturation of seawater with these variants strongly influences the ability of organisms to build shells, exoskeletons. and other structural features. The aragonite saturation of seawater, for instance, determines whether or not corals are able to secrete their skeletons (Hoegh-Guldberg et al. 2007). Where the aragonite saturation state of seawater

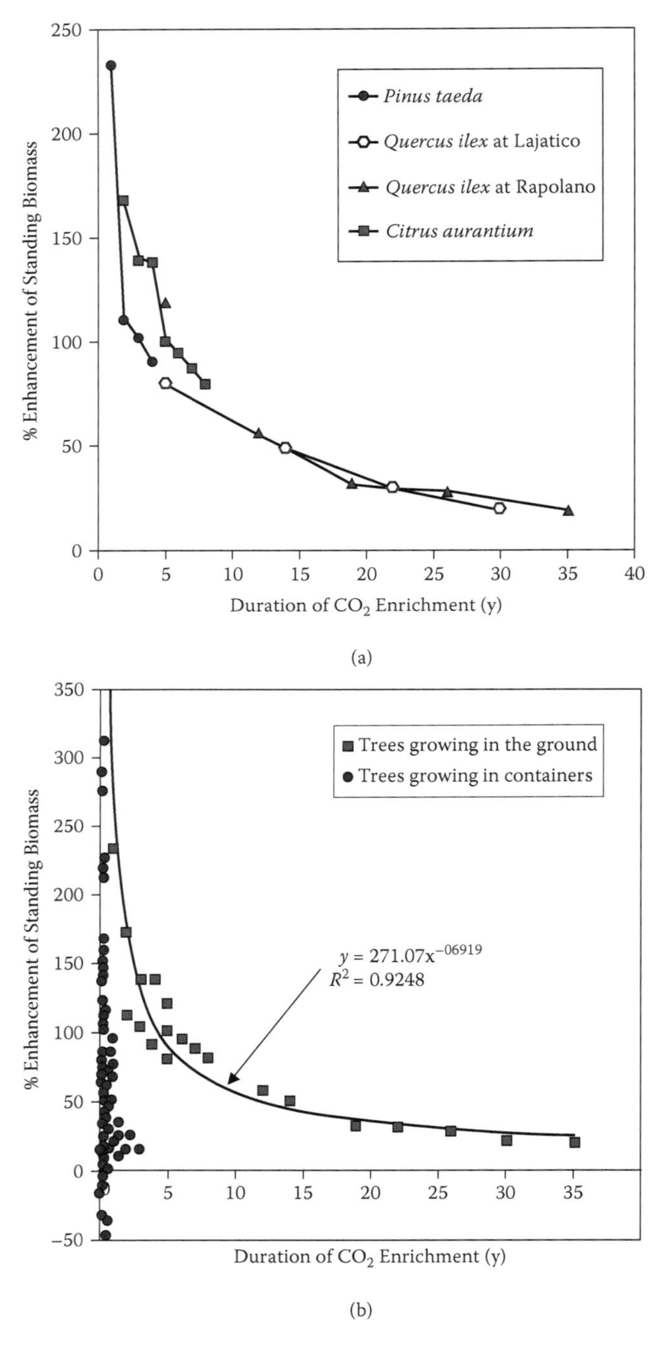

FIGURE 6.3 Attenuation of CO_2 stimulation. CO_2 stimulation is reduced over time in multiple species (a) and is more pronounced in in situ experiments (b).

is too low, corals are unable to exist. Other shell-building organisms, such as clams and test-building plankton, are similarly affected. Plankton grown in low-saturation-state seawater have reduced or deformed tests (Feely et al. 2004).

CO_2 influences calcium carbonate chemistry of seawater because CO_2 dissolves in seawater to form a carbonic acid, which dissociates to hydrogen ion and bicarbonate (Orr et al. 2005). The bicarbonate in turn liberates another hydrogen ion to form calcium carbonate. The net effect of these reactions is to lower pH by releasing hydrogen ions and to shift the balance of the calcium carbonate reaction in seawater, reducing the saturation state of both aragonite and calcite, thereby decreasing the driving force for formation of both minerals (Figure 6.4).

Experiments in elevated CO_2 seawater show serious impacts on calcification (shell building) in a variety of marine organisms. At double pre-industrial CO_2 levels, calcification falls by 5–50% (Feely et al. 2006). This makes it more difficult for corals to build reefs that can keep pace with erosion, for instance. The algae and mollusks that compose the base of many marine food chains are strongly affected. Lag times in CO_2 mixing into the oceans and subsequent impact on saturation states are significant, so effects on reefs, on the base of food webs, and on other critical marine systems from changes in ocean calcium carbonate chemistry are in effect time bombs that have already been set in motion (Fabry et al. 2008).

The direct effect of the hydrogen ions released in seawater when CO_2 dissociates is to lower the pH of seawater. This is known as ocean acidification, even though seawater is basic and will remain so even with the influx of acid (Fabry et al. 2008). Acidification may have effects on organisms from shallow subtidal areas to the deepest ocean. As with all organisms, pH is a fundamental cellular property that is closely regulated in marine organisms. When pH is altered, the organism must expend energy to maintain preferred pH levels, reducing the energy available for other life functions. Shutdown of protein synthesis may result (Seibel and Fabry 2003). For instance, in squid a decrease in arterial pH of 0.15—a 50% increase in hydrogen ions—expected this century, is enough to impair metabolism and limit mobility (Drake 2005).

The effects of acidification are delayed by the long mixing times of the oceans. Gas exchange and acidification at the surface reaches the depths of the oceans only slowly. Deep mixing may take centuries, so the full effect of acidification on sensitive bottom-dwelling organisms, for instance, will not be known for many years.

6.5 INDIRECT EFFECTS OF CO_2— ATMOSPHERIC WARMING AND CLIMATE CHANGE

As serious as are the direct effects of CO_2 on wildlife, perhaps of even greater concern are the impacts from the indirect effects of climate change. The physical habitat changes that result vary from the loss of sea ice in the Far North to loss of tropical glaciers, intensification of storms, and change in mean temperature and precipitation.

Climate change results from CO_2 buildup because CO_2 is a greenhouse gas: it in effect traps outgoing long-wave radiation, warming the surface of the planet

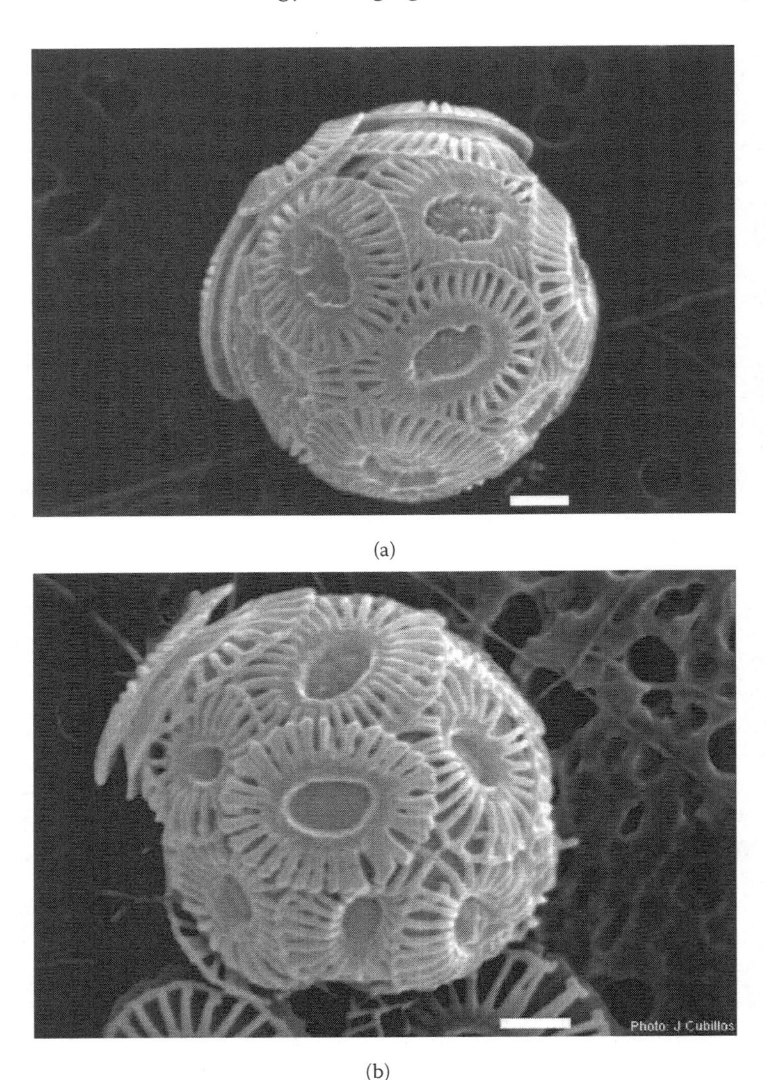

(a)

(b)

FIGURE 6.4 (a) Healthy shell structure of the coccolithophorid *Emiliania huxleyi*; (b) deformed shell formed under high CO_2 conditions. (From Cubillos, J.C., Wright, S.W., Nash, G., de Salas, M.F., Griffiths, B., Tilbrook, B., Poisson, A., and Hallegraeff, G.M. 2007. Calcification morphotypes of the coccolithophorid *Emiliania huxleyi* in the Southern Ocean: Changes in 2001 to 2006 compared to historical data. *Marine Ecology–Progress Series*, 348: 47–54.)

(IPCC 2007). This warming increases global mean temperature, but also sets in motion a series of related climatic changes. Alterations in precipitation and circulation take place as the warming alters the moisture-holding capacity of the atmosphere and changes pressure gradients, especially between continental interiors and the sea.

Warming has already been observed, and global mean temperature has increased about 0.74°C over the past century (IPCC 2007). There is strong regional variation in this global pattern, with some regions warming markedly and others only slightly or not at all. Precipitation patterns have altered significantly as well, with even more regional variation.

One of the most striking physical alterations due to warming is the loss of sea ice in the Arctic. More than 50% of peak season sea ice has been lost in the past 30 years. This loss represents a major habitat change for polar bears, walruses, spectacled eider, and other ice-dependent species (Grebmeier et al. 2006).

Future changes are projected to include further warming, large and difficult to predict precipitation changes, almost total loss of summer arctic sea ice, and increases in sea level. The variety of impacts this will have on the species and ecosystems of the planet is impossible to fully gauge, but evidence from past changes, computer simulations, and current observations strongly suggest what may lie ahead for wildlife.

6.6 RESPONSE OF PLANTS AND ANIMALS TO CLIMATE CHANGE

The fossil record suggests that one of the dominant responses of species to past climate change was changing ranges (Huntley and Birks 1983). For instance, tree species in North America underwent a dramatic northward range shift as the continent emerged from the last ice age (Overpeck et al. 2003). But beyond ice ages, many species from all parts of the world have responded to many types of climate change (warming, cooling, drying, increased rainfall) by adjusting their physical location to match their preferred climate tolerances. This leads us to expect that a major response to human-induced climate change should be species' range shifts (Hannah, Lovejoy et al. 2005). And indeed, many such range shifts are now being observed.

6.6.1 Range and Population Effects

One of the first demonstrations of a range shift resulting from climate change is from a butterfly found in western North America. Checkerspot butterflies (*Euphydras*) had been known for some time to be vulnerable to population crashes or booms resulting from weather conditions. Edith's checkerspot butterfly (*Euphydras* spp.) is responding to changing climate by undergoing a major range shift (Parmesan 1996). The entire range of Edith's checkerspot has been studied—one of the few species for which such a complete range shift has been established (Figure 6.5).

Populations of Edith's checkerspot in the south (Mexico) and in the lowlands are disappearing faster than populations in the north and in the uplands (Parmesan 1996). Continuation of this trend would lead to loss of the lowland and southern ranges, and increases in range at upper elevations and poleward—exactly the pattern expected with climate change. Other causes for the shift have been eliminated, clearly pointing to climate change as the cause of the shift (Parmesan 1996).

Evidence of climate-change-related range shifts has mounted rapidly since the pioneering studies in the 1990s. Butterflies and birds are particularly well represented, as are impacts at the poles (Parmesan et al. 1999). But changes have been

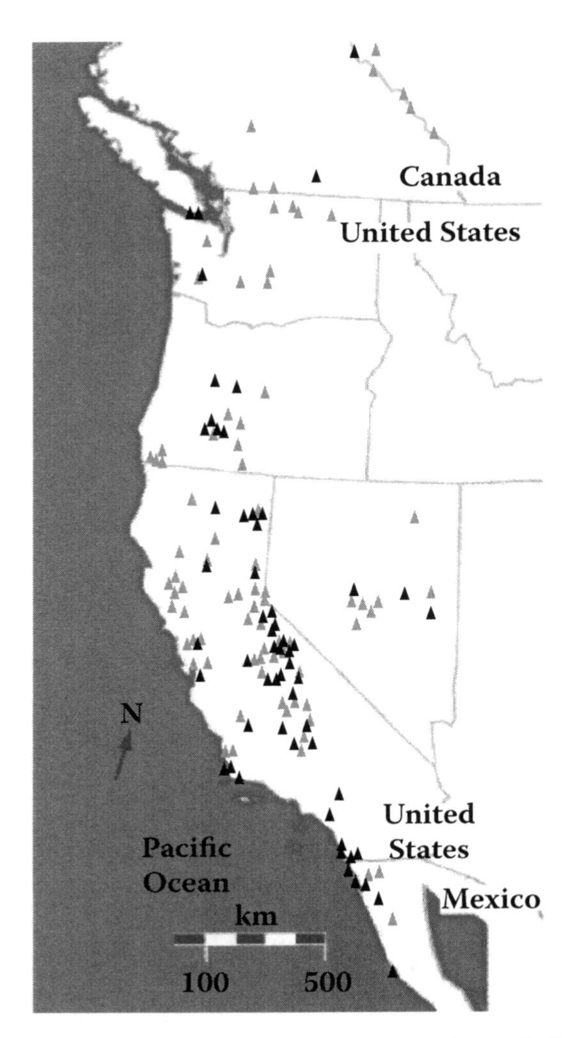

FIGURE 6.5 Map of Edith's checkerspot range shift. Light triangles indicate extinct populations; dark triangles indicate persistent populations.

recorded in a variety of species and many latitudes (Parmesan and Yohe 2003; Root et al. 2003).

The Arctic is the scene of some of the most dramatic changes. The northern polar region is rapidly losing sea ice, with severe impacts on species that depend on sea ice. Polar bears, dependent on sea ice in their hunting, are declining in numbers in some populations, and those that do survive sometimes have to swim long distances to shore as sea ice retreats (Norris et al. 2002). Large-scale range changes have been seen in the Arctic fox, which is retreating northward as its more competitive cousin the red fox expands its range with warming (Hansell et al. 1998).

In the Bering Sea, walrus and spectacled eider populations have shifted northward and declined precipitously (Grebmeier et al. 2006). Both species rest on sea ice and dive to eat clams in waters less than 100 meters deep. Warming ocean temperatures have shifted benthic fauna from clams to sea stars and other less palatable species, while ice melt has reduced areas for resting. Ice is melting so fast that walrus pups left on sea ice by their mothers have the ice literally melt out from under them, making it impossible for their mothers to relocate them. Lost pups have been found hundreds of miles from walrus range in waters too deep for survival. Spectacled eider populations have decreased more than 90% because of similar habitat changes and deterioration of food quality (Grebmeier et al. 2006).

In the Antarctic, some penguin populations have declined in parallel with changes in sea ice (Barbraud and Weimerskirch 2001). Emperor penguin populations have undergone population declines as high as 50%, and Adelie penguins 70% in some locations (Moline et al. 2004). Populations farthest from the pole have been hardest hit, causing a poleward range shift.

In temperate climates, studies involving large numbers of species indicate a broad biological response to climate change. In Europe, of 35 butterfly species studied, 63% have undergone northward range shifts, as would be expected with global warming, with only 3% southward shifts (Parmesan et al. 1999). The range shifts were large—between 35 and 240 kilometers. Similarly, among 59 species of birds in Great Britain, an average northward range shift of 20 kilometers has been observed (Thomas and Lennon 1999).

Dragonflies in Great Britain have expanded northward, with 23 of 24 well-documented species showing a northward shift, with a mean shift of 88 kilometers (Parmesan 2006). Seventy-seven lichen species have expanded their ranges northward into the Netherlands. Alpine plants are moving upslope in Swiss mountains. The pika (*Ochotona princeps*), a montane rodent, is disappearing from lower elevation sites in the United States (Grayson 2005). Many species in less well-studied taxa are likely to be shifting ranges in the temperate zone, and the number of documented cases is rising steadily (Parmesan and Yohe 2003; Root et al. 2003)

Treelines have shifted northward and upslope in a wide variety of settings and regions, including Russia, eastern Canada, Sweden, and New Zealand. In many other cases, it has been difficult to demonstrate a treeline effect of climate change. For example, increased growth at treeline has been observed in the southwestern United States, but in Alaska, response is mixed (Wilmking et al. 2004). In California, significant changes in mortality and population density of trees have all been observed without changes in treeline (Millar). Increases in tree mortality attributable to climate change have also been noted in western U.S. forests (van Mantgem et al. 2009).

A range shift of the quiver tree (*Aloe dichotoma*) has been demonstrated across its entire range from South Africa into Namibia (Foden et al. 2007). This is one of the few plant range shifts that have been documented across the entirety of a range. Quiver tree populations are declining in the north and at lower elevations, with northernmost populations all at high elevation.

Range shifts in the tropics are less well documented, partly because species in the tropics are less well known and partly because there are fewer scientific institutions and professional biologists in the tropics. Nonetheless, there is evidence of range shifts in the tropics. In the Monteverde cloud forest of Costa Rica, toucans and other birds are moving upslope in response to warming (Pounds et al. 1999). Butterflies from North Africa are expanding their ranges into southern Europe (Parmesan et al. 1999). The rufous hummingbird has undergone a large population shift northward in its winter range, moving from Mexico into the southern United States.

There have now been so many documented cases of range shifts that it is possible to look for patterns. Can biological responses be used to prove that global climate is warming? Two studies published in 2003 looked at a large dataset of biological evidence compiled for the IPCC and concluded that a distinct "fingerprint" of climate change was discernable (Parmesan and Yohe 2003; Root et al. 2003). One study looked on a species-by-species basis, while the other aggregated based on individual research studies. Both concluded that the pattern of range shifts was so strong and unidirectional that it could result only from a change in climate. In the species-by-species study, the overwhelming majority of species showed the poleward and upslope shifts expected with warming. In 1,700 species studied, poleward range shifts averaged 6 kilometers per decade (Parmesan and Yohe 2003). The second study looked at 143 research publications and discovered that most found shifts in the direction expected by climate change (Root et al. 2003).

In the marine realm, the most severe and wide-ranging impact of climate change has been coral bleaching (Hoegh-Guldberg et al. 2007). Bleaching occurs when sea surface temperature rises more than 1° or 2°C above normal summer maximal temperatures. Corals have microscopic symbionts known as zooxanthellae inhabiting their cells. The zooxanthellae photosynthesize and pass nutrients to the coral while the coral host provides a physical structure that keeps the algae near enough to the surface to get adequate sunlight for photosynthesis. However, when corals are exposed to very warm water, they expel their zooxanthellae. Without photosynthetic pigments from the algae, the corals lose their color. The coral appears "bleached."

Mortality due to bleaching may be severe enough to wipe out entire species over large areas. It is therefore a strong driver of range shifts in corals. For example, in the central lagoon of Belize, staghorn coral (*Acropora cervicornis*) was the dominant species until the 1980s, when it was wiped out by a combination of disease and rising water temperatures (Aronson et al. 2000). The scroll-like coral *Agaricia tenuifolia* took over as the dominant coral, only to be wiped out itself in the high water temperatures of the 1998 El Niño event. These massive mortalities were the worst in at least 3,000 years, resulting in range changes over large areas of the Caribbean for staghorn and other corals.

Coral bleaching was undescribed in the scientific literature 50 years ago, yet is so common today that reefs in all parts of the world have been affected by bleaching within the past two decades (Hoegh-Guldberg 1999). There have been seven major coral bleaching events, affecting reefs in all parts of the world, between 1979 and 2002, all associated with El Niño conditions (Baker et al. 2008). The 1997–1998 El Niño was the worst of the last century for coral bleaching. In that event, reefs around the world were affected, many experiencing record damage (Baker et al. 2008). Over 10% of all the world's corals died in that event, with mortality in some regions, such as the Indian Ocean, going as high as 46%.

6.6.2 Changes in Timing

As climate warms, it is expected that biological events cued to temperature will change. The study of the influences on timing of biological events is *phenology* (Root 1997). Many phenological events are advancing in spring and retarding in fall, characteristic of a warming climate (Corlett and Lafrankie 1998; Inouye et al. 2000; Menzel 2000; Chmielewski and Roetzer 2001; Penuelas and Filella 2001). The overwhelming majority of taxa that have been studied are showing these responses—in plants and vertebrates, and in terrestrial, marine, and freshwater systems (Root et al. 2003).

The earlier arrival of spring is perhaps the best-documented change over the past two centuries because of the strong interest people have in the breakup of winter in the colder climates of Europe, Asia, and North America. In these regions, spring means the beginning of the planting season, the advent of ice-free conditions on rivers and lakes that permits navigation, and, in the past, the beginning of the end of a period of food scarcity. As a result, northern cultures have a long history of recording the first signs of spring: budburst on trees, arrival of first migrants, and break-up of ice and other signs of impending warmth. Many of these records are of biological phenomena and of sufficient accuracy to serve as a historical baseline against which the effects of warming may be judged (Menzel 2000).

Some of the best records come from Europe and involve changes in plant species that have been domesticated as food crops and are sensitive to climate. Wine grapes are sensitive to climate, and harvest conditions are critical to wine quality, so meticulous records of wine grape phenology have been kept over centuries in wine-growing regions of Europe (Meier et al. 2007). The heatwave of 2003 brought on the earliest harvest in a 500-year record (Menzel 2005). A strong correlation in these data exists between harvest date and summer temperature, with April–August temperatures explaining 84% of the harvest-time variation.

In birds and butterflies, spring migrants are arriving earlier, and the date of first flight for species in diapause is advancing. California birds are arriving earlier, though not departing significantly later (Macmynowski et al. 2007). Migrant birds of the North Sea have arrived several days earlier every decade since 1960. In butterflies, advances in emergence and first arrival of spring migrants have been identified in 26 species in the United Kingdom (Roy and Sparks 2000).

Birds are nesting and laying eggs earlier as well. An analysis of over 70,000 nest records in Britain revealed an 8-day advance in first egg laying in 20 species from the 1970s to the mid-1990s (Crick et al. 1997). Mexican jays (*Aphelocoma ultramarina*) are nesting earlier in Arizona, as are tree swallows (*Tachycineta bicolor*) across the United States and Canada and pied flycatchers (*Ficedula hypoleuca*) in Europe (Brown et al. 1999; Dunn and Winkler 1999; Both and te Marvelde 2007). Brunnich's guillemot (*Uria lomvia*) is breeding earlier around Hudson Bay in response to decreasing sea ice as a result of warming (Gaston et al. 2005).

All of these present effects of climate change suggest that the lesson of the past—that species ranges and timing will shift with climate change—are being borne out and will intensify in the future. Examining future patterns of wildlife response to climate change require computer models that simulate responses.

The species distribution model (SDM) is perhaps the most widely implemented of all climate change biology models (Guisan and Zimmermann 2000). SDMs are statistical models that establish a relationship between current climate and known occurrences of a species, and then extrapolate that relationship to all suitable current climates or to changed future climates. SDMs are also sometimes referred to in the literature as "envelope models," "bioclimatic models," and "range shift models" (Elith et al. 2006).

SDM simulations of species range shifts with climate change have now been conducted for many regions and thousands of species. These studies support expectations from theory that species will move toward the poles and upslope as the planet warms (Thomas et al. 2004). They also support the notion that species respond individualistically to climate change (Midgley et al. 2003). This means that species move independently of one another as climate changes, resulting in the tearing apart of vegetation "communities" and the reassembly of species into new communities (Figure 6.6).

Europe and North America have been the focus of the greatest number of SDM studies. In these north temperate settings, they have projected northward movement in a wide range of species, and upslope movement in the Alps, Rocky Mountains, Cape Fold Mountains, and other ranges (Bugmann and Fischlin 1996; Peterson 2004; Peterson et al. 2004; Hannah, Midgley et al. 2005). Studies in the Alps have demonstrated significant loss of range in alpine plants, due to the effect of decreasing

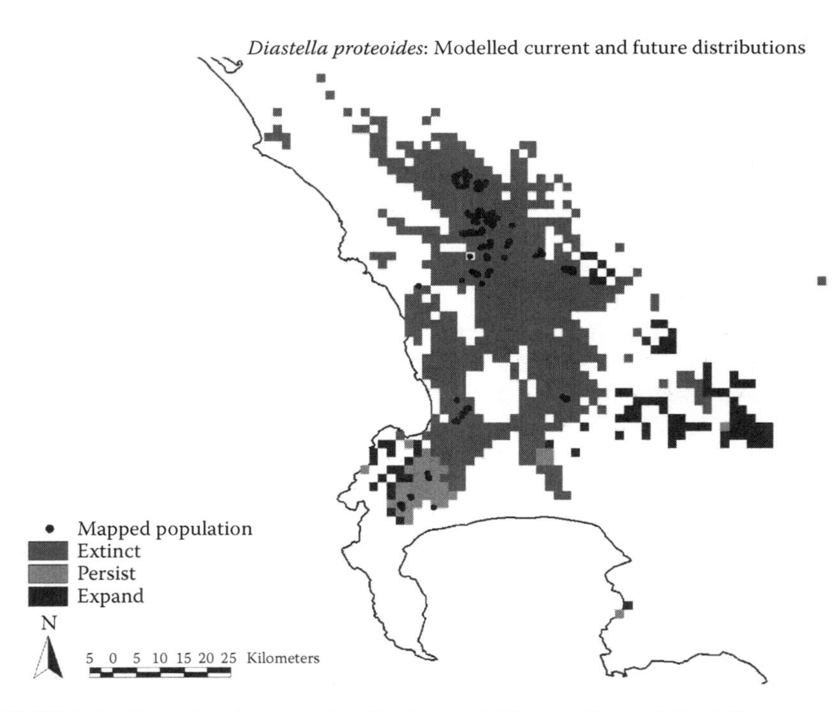

FIGURE 6.6 Example of species distribution model for the Cape of Good Hope region of South Africa.

area as species move upslope, just as there is less area at the tip of a cone than at the base (Gottfried et al. 1999).

Results from Australia have shown simulated range losses in mountains and across a range of species. In a study of 819 eucalypts, 53% were found to have a temperature range under 3°C (Hughes et al. 1996). Of 92 endemic plant species, most were found to lose range and 28% lost all range with only a 0.5°C warming (Hughes 2003). Large amounts of range loss have been demonstrated in numerous tropical montane species in Queensland (Williams et al. 2004).

Mexico and South Africa are other regions in which active research programs have produced SDM results for thousands of species (Erasmus et al. 2002; Peterson et al. 2002; Midgley et al. 2003). These studies have been used to estimate the need for connectivity and new protected areas in these regions, applications that will be explored in detail in later sections (Hannah, Midgley et al. 2005; Williams et al. 2005).

The Dynamic Global Vegetation Model (DGVM) is another modeling tool used in future simulations. DGVMs use equations describing photosynthesis, carbon cycles, and other physical processes to define vegetation types likely to exist in geographical space (Sitch et al. 2008). DGVMs are often run for the globe, but increasingly are finding regional applications (Lenihan et al. 2003). A limitation of DGVMs is that their output is denominated in vegetation type or plant functional type, not species, so they are unable to simulate the individualistic range shifts expected of many species.

For future, warming climates, DGVMs show northward expansion of boreal forest, expansion of some temperate southern forests toward the pole, expansion of several temperate forest types, and contraction of some tropical forest types, most notably in the Amazon (Cramer et al. 2001; Cox et al. 2004; Sitch et al. 2008).

DGVM results show the wet forests of the Amazon shrinking in response to climate change (Cox et al. 2004). The mechanism driving this effect is moisture recycling within the basin. Moisture driven into the basin from the Atlantic Ocean is taken up by the wet tropical forest of the Amazon and then transpired. As the vegetation respires, water is lost through the stomata of leaves. This moisture enters the atmosphere, rises, condenses, and falls again as rain. Thus forests in the eastern Amazon are the source of rainfall for forests farther west, especially in the central Amazon. If climate change results in the drying of these eastern forests, there is less water and less rainfall for the central Amazon, and it dries too—in something akin to a domino effect for forest water. Models that incorporate feedbacks from vegetation to climate show an even stronger effect.

6.7 EXTINCTION RISK FROM CLIMATE CHANGE

Summing SDM results across various regions can allow estimation of extinction risk from climate change. SDM simulations in most regions indicate range losses for many species, including complete loss of suitable climate for some species, even for conservative, mid-century climate projections. Thomas et al. (2004) used these range losses to estimate extinction risk across six geographically and taxonomically diverse regions. These authors used the Species Area Relationship (SAR) to estimate extinctions associated with modeled levels of range loss.

The use of the Species Area Relationship depends on the assumption that loss of suitable climatic space is analogous to loss of habitat or land area. On islands, or in areas of habitat loss to human uses, smaller areas carry smaller numbers of species. This effect is mathematically defined as the SAR, generally in the form of a power law for areas that have been empirically sampled (Kinzig and Harte 2000). If climate is assumed to be a controlling habitat variable, then loss of suitable climatic space should result in fewer species surviving in an area—the SAR can simply be applied to the range losses calculated in the species models. This assumption is not without controversy (Harte et al. 2004), but has been useful in allowing the first estimates of extinction risk from climate change.

Thomas et al. (2004) calculate that 18% to 32% of the species in the regions they sampled are at risk of extinction from climate change under midrange scenarios. If these results are typical of other regions, then upward of a million species may be at risk of extinction from climate change, using a low-end figure of 5 million species for the number of terrestrial species in the world. If the percentages held for the marine realm as well, the number of species at risk would climb to over 2 million. Since estimates of the total number of species on Earth range as high as 100 million, these initial estimates may be low.

One means of corroborating these early extinction risk estimates is by looking to the record of past climate changes. Evidence from the past is difficult to interpret, because no past climate is precisely analogous to the climate change that will unfold as a result of human CO_2 pollution (Overpeck et al. 2003). Few extinctions have been associated with rapid climate changes coming in and out of ice ages over the past 2 million years, perhaps because these changes were greatest in the Northern Hemisphere, where species were already adapted to ice ages (Jackson and Johnson 2000). But the glacial-interglacial changes have been predominantly warming in cool climates or cooling in warm ones. The coming climate change will be warming on an already warm, interglacial climate. The last time Earth experienced a large warming event in an already warm climate was at the Paleocene-Eocene Thermal Maximum, 56 million years ago (Gingerich 2006). That event saw massive extinctions and adaptive radiation in very short geological time spans. However, that event was set in a global temperature much warmer than even today's climate, so again the analogy is very imprecise. In general, major extinction episodes in the fossil record are associated only rarely with climate. Response to major past climate change took place in totally natural landscapes, so it may not adequately represent the consequences of combining climate-change-driven range shifts across heavily fragmented, human-dominated landscapes (Hannah, Lovejoy et al. 2005).

More recent modeling approaches have tried to reproduce the results of Thomas et al. (2004) and arrived at broadly comparable conclusions. Malcolm et al. (2007) used DGVMs to assess loss of characteristic vegetation in global biodiversity hotspots and found extinction risk of an order of magnitude similar to the SDM-based estimates.

6.8 CONSERVATION RESPONSES TO CLIMATE CHANGE

Conservation responses to climate change must take into account species range dynamics usually ignored in traditional conservation strategies (Bush 1996; Hannah, Midgley, Bond et al. 2002; Hannah, Midgley, Millar et al. 2002). Many of the elements of climate change conservation strategies overlap with, or are adaptations of, current conservation tools, however. The most important of these climate change responses are

- Protected areas and corridors
- Management of human systems
- Assisted migration and ex situ management
- Management of interactions of human adaptation and conservation

New protected areas are needed to cope with climate change, because current protection was not formulated with climate change in mind (Hannah, Midgley et al. 2002; Hannah, Midgley et al. 2005; Hannah et al. 2007). Species range-shift dynamics means that more protected area is needed to capture current and future suitable climates for a number of species. Hannah et al. (2007) explored the magnitude of needed new protection in three regions, and found that 1% to 30% additional protected area would be needed to maintain minimal species representation. This study used a selection process that captured all species in "chains" of protection (Williams et al. 2005) from present to future suitable climates.

One of the conclusions of this study is that population-level connectivity is needed to ensure species' survival as they move in response to climate. Rather than connecting protected areas, the population-connectivity approach emphasized maintaining populations. Existing protected areas became important destinations in areas that would become climatically suitable in the future. The most cost-effective approach to maintaining population-level connectivity is then to ensure that present populations are able to reach these future destinations. This can be accomplished by adding protected area adjacent to existing protected area or by working with landholders to conserve species on lands adjacent to conserved areas. Thus contract conservation or conservation easements are an attractive alternative to land purchase and protection.

For some species, additional protection will not be possible, and rescue or assisted migration may be necessary (McLachlan et al. 2007). Species that lose all suitable climate space in the future may need to be captive-propagated to avoid total extinction. Species whose suitable climate shifts outpace their dispersal ability may need assistance in the form of translocation of current populations to sites that will hold suitable climate in the future. Questions of ethics, establishing baseline conditions, and ensuring that artificial translocation is based on sound science are currently being explored in the literature.

Human land-use shifts must be anticipated and managed alongside species range shifts (da Fonseca 2005). Anticipating changing patterns of human use as the result of climate change is important in ensuring that species range shifts can be accommodated in appropriately conserved lands into the future, in areas that are secure from loss to changes in human land use. For example, where agricultural suitability changes, joint land use planning can help ensure that agricultural needs and conservation goals are simultaneously met. In South Africa, where wine production is highly sensitive to climate change and to consumer preference for environmentally sound production, vineyards and conservationists are working to mutually plan future vineyard expansion into uplands to minimize loss of native habitat important to species shifting upslope (see www.bwi.co.za).

The cost of these conservation measures may be substantial relative to current conservation investments (da Fonseca 2005). Many connectivity needs will require re-establishment of habitat. Restoration is typically many times more expensive than conserving intact habitat, so the cost per unit area of conservation will increase substantially. Initial estimates from Madagascar suggest that restoration of a fraction of a current protected area may cost on the same order of magnitude as current conservation expenditures (Hannah 2008). Early estimates from other regions indicate that climate change conservation costs may be frontloaded (for example, to acquire land needed as future habitat before it is converted to human uses) and a major expense relative to current investment. Establishing adaptation-funding mechanisms adequate to meet this need is a major challenge.

6.9 POLICIES TO CONTROL HUMAN GREENHOUSE GAS EMISSIONS

Limiting damage to wildlife and ecosystems from climate change requires response on two fronts: limiting greenhouse gas emissions and improving conservation strategies to deal with climate change. While wildlife biologists will be primarily concerned with the second of these essential responses, they must also be aware of, and often involved in, efforts to reduce greenhouse gas emissions. This is because the success of policies to combat human causes of climate change will determine the stresses wildlife biologists must expect and manage, and because the direct response to the problem will often be foremost in the public's mind—enlisting help for wildlife may also require demonstrating concern for the problem as a whole, for its root causes, and for the solutions in which we all must participate. Wildlife managers may be effective advocates for public policies to reduce greenhouse gas emissions because we see so clearly the damages to natural systems that may result.

The primary international vehicles for combating greenhouse gas emissions are the United Nations Framework Convention on Climate Change (UNFCCC) and the Kyoto Protocol. The UNFCCC is the overarching international agreement to address climate change. The United States is a signatory to the UNFCCC. The Kyoto Protocol is the first set of implementation rules under the UNFCCC. The United States has, until recently, resisted ratification of the Kyoto Protocol. The Kyoto Protocol will expire and a new set of rules will take effect in 2012. Therefore, policy discussions often reference post-2012 agreements in which the United States is much more likely to participate.

The goal of the UNFCCC is to "avoid dangerous anthropogenic interference in the climate system." The three benchmarks for success are avoiding impeding agriculture, allowing development to proceed, and allowing ecosystems to adapt naturally. Although it may be difficult to determine what natural adaptation means for ecosystems in totally novel climates, the important part of the UNFCCC is that it acknowledges impacts on nature and wildlife as one of its key benchmarks for success. It is the job of ecologists and wildlife biologists to help interpret that language and let policymakers know when the limits of natural adaptation in ecosystems may be exceeded.

A key part of the UNFCCC and Kyoto Protocol responses is the treatment of loss of natural forests. Deforestation accounts for about one-fifth of global greenhouse gas emissions. Most of this deforestation occurs in developing countries. Since most of the greenhouse gas buildup to date has resulted from fossil fuel burning in developed countries, the Kyoto Protocol initially excluded deforestation from greenhouse gas reduction rules. The idea was that developed countries that had caused most of the problem and had the greatest ability to pay should make the first cuts in emissions. However, as the Kyoto Protocol took effect, developing countries saw a carbon market emerge worth billions of dollars. These markets allowed reductions in emissions to be achieved in the most economically efficient (cheapest) way, by allowing reductions in CO_2 to be traded. Developing countries realized that keeping deforestation out of the Kyoto rules meant that they were excluded from participating in the carbon markets, potentially costing them tens or hundreds of billions of dollars in aggregate. As a result, in 2009, new rules were approved that allowed reductions in deforestation to count toward Kyoto goals and be traded in carbon markets.

The emerging trade in deforestation carbon has important benefits for wildlife. To sell deforestation carbon, a country must demonstrate that it is reducing its CO_2 emissions from deforestation. That reduction in deforestation benefits wildlife living in forests that would otherwise be destroyed. Helping countries participate in carbon markets and steering efforts to combat deforestation toward biologically valuable forests are therefore important roles for wildlife biologists. Local communities will often be involved in avoided-deforestation arrangements, so the link between wildlife biologists and national policy-makers designing avoided-deforestation programs is not as remote as it might seem. For instance, a partnership between an indigenous group and conservationists could help ensure the survival of a forest that might otherwise be lost, and therefore make it eligible for carbon payments. The division of the payments would be determined by the rules in effect in the particular country, but usually would involve some revenue staying with national and state governments and some revenue accruing to local communities and forest managers.

Other efforts to combat greenhouse gas emissions may not be so beneficial for wildlife. Replacing fossil fuel burning requires development of renewable energy sources, some of which may require extensive land area. Where this renewable energy development cannot be co-located in cities (for instance, through rooftop solar panels) or agriculture (for instance, through location of wind turbines in crop fields), natural areas may come under pressure for energy development. For example, in California, much of the state's projected renewable energy production will come from wind and solar farm developments in the Mojave Desert, on federal government lands previously in low-intensity uses compatible with conservation. Major environmental and social impacts have already been felt from the development of biofuels, including conversion of cropland to fuel production, rising prices for food staples such as corn (in Mexico), conversion of land in conservation set-asides to production of biofuel (in the United States), conversion of natural habitat for biofuel production (in Brazil, Indonesia, and other tropical countries), and increases in pollution from fertilizer and pesticide applications on biofuel crops (see Chapter 3, this book, for more detailed descriptions).

Wildlife impacts from renewable energy development such as these will mount as conversion to fossil-free energy sources intensifies. If not properly managed, the cure (renewable energy development) may be worse than the disease (climate change). Wildlife managers have a large stake in alternate energy development planning that includes strong consideration of environmental and social impacts. Where markets are used for economic efficiency in renewable energy development, new regulation may be required to constrain unwanted impacts. Wildlife managers in state and federal agencies and NGOs may have important roles to play in molding such legislation.

6.10 CONCLUSION

The impacts of greenhouse gas pollution in the atmosphere on wildlife are both direct and indirect. Marine and terrestrial systems are both affected. Responses include direct effects of CO_2 on plant growth and metabolism, direct effects of pH change in the oceans, and changing calcium carbonate chemistry on corals and other organisms with calcium carbonate skeletons or shells. The indirect effects of climate change include range shifts in both plants and animals, changes in phenology (biological timing), and disassembly and re-arrangement of communities as species shift ranges relative to one another. The challenges posed by these changes are great, leaving wildlife biologists with a legacy of dynamics that must be addressed this century. At the same time, policies to lessen these impacts have their own consequences for wildlife, requiring managers and researchers to play multiple roles to protect ecosystems and species.

As emissions of CO_2 and other greenhouse gases intensify this century, these impacts will intensify. International action to reduce emissions seems likely to result in declining emissions by the later half of the century. Earlier reductions may be required to avoid the most negative impacts on wildlife. At the same time, rapid deployment of renewable energy will have its own impacts that will vary from location to location. Sound energy development planning, strong greenhouse gas reduction policies, and conservation management that incorporates climate change are all needed to see wildlife and nature thrive in this century and beyond.

REFERENCES

Aronson, R. B., W. F. Precht, et al. 2000. Ecosystems: Coral bleach-out in Belize. *Nature* 405 (6782): 36–36.

Baker, A. C., P. W. Glynn, et al. 2008. Climate change and coral reef bleaching: An ecological assessment of long-term impacts, recovery trends and future outlook. *Estuarine Coastal and Shelf Science* 80 (4): 435–471.

Barbraud, C. and H. Weimerskirch. 2001. Emperor penguins and climate change. *Nature London* 411 (6834): 183–186.

Beerling, D. J. and F. I. Woodward. 1993. Ecophysiological responses of plants to global environmental change since the Last Glacial Maximum. *New Phytologist* 125 (3): 641–648.

Both, C. and L. te Marvelde. 2007. Climate change and timing of avian breeding and migration throughout Europe. *Climate Research* 35 (1–2): 93–105.

Brown, J. L., S. H. Li, et al. 1999. Long-term trend toward earlier breeding in an American bird: A response to global warming? *Proceedings of the National Academy of Sciences of the United States of America* 96 (10): 5565–5569.

Bugmann, H. and A. Fischlin. 1996. Simulating forest dynamics in a complex topography using gridded climatic data. *Climatic Change* 34 (2): 201–211.

Bush, M. B. 1996. Amazonian conservation in a changing world. *Biological Conservation* 76 (3): 219–228.

Chmielewski, F. M. and T. Roetzer. 2001. Response of tree phenology to climate change across Europe. *Agricultural and Forest Meteorology* 108 (2): 101–112.

Corlett, R. T. and J. V. Lafrankie, Jr. 1998. Potential impacts of climate change on tropical Asian forests through an influence on phenology. *Climatic Change* 39 (2–3): 439–453.

Cox, P. M., R. A. Betts, et al. 2004. Amazonian forest dieback under climate-carbon cycle projections for the 21st century. *Theoretical and Applied Climatology* 78 (1–3): 137–156.

Cramer, W., A. Bondeau, et al. 2001. Global response of terrestrial ecosystem structure and function to CO_2 and climate change: Results from six dynamic global vegetation models. *Global Change Biology* 7: 357–373.

Crick, H. Q. P., C. Dudley, et al. 1997. UK birds are laying eggs earlier. *Nature* 388:526.

da Fonseca, G. A. B. 2005. Managing the matrix. In *Climate Change and Biodiversity*, T. Lovejoy and L. Hannah, editors. New Haven, CT: Yale University Press.

da Fonseca, G. A. B., C. M. Rodriguez, et al. 2007. No forest left behind. *Plos Biology* 5 (8): 1645–1646.

Drake, B.G., L. Hughs, E. Johnson, B. Siebel, M. Cochrane, V. Fabry, D. Rasse, and L. Hannah. 2005. Synergistic effects. In *Climate Change and Biodiversity*, T. Lovejoy and L. Hannah, editors. New Haven, CT: Yale University Press.

Dunn, P. O. and D. W. Winkler. 1999. Climate change has affected the breeding date of tree swallows throughout North America. *Proceedings of the Royal Society Biological Sciences Series B* 266 (1437): 2487–2490.

Elith, J., C. H. Graham, et al. 2006. Novel methods improve prediction of species' distributions from occurrence data. *Ecography* 29 (2): 129–151.

Erasmus, B. F. N., A. S. VanJaarsveld, et al. 2002. Vulnerability of South African animal taxa to climate change. *Global Change Biology* 8: 679–693.

Fabry, V. J., B. A. Seibel, et al. 2008. Impacts of ocean acidification on marine fauna and ecosystem processes. *ICES Journal of Marine Science* 65(3): 414–432.

Feely, R. A., C. L. Sabine, et al. 2004. Impact of anthropogenic CO_2 on the $CaCO_3$ system in the oceans. *Science* 305 (5682): 362–366.

Foden, W., G. F. Midgley, et al. 2007. A changing climate is eroding the geographical range of the Namib Desert tree Aloe through population declines and dispersal lags. *Diversity and Distributions* 13 (5): 645–653.

Gaston, A. J., H. G. Gilchrist, et al. 2005. Climate change, ice conditions and reproduction in an Arctic nesting marine bird: Brunnich's guillemot (*Uria lomvia* L.). *Journal of Animal Ecology* 74 (5): 832–841.

Gingerich, P. D. 2006. Environment and evolution through the Paleocene-Eocene thermal maximum. *Trends in Ecology & Evolution* 21 (5): 246–253.

Glynn, P. W. 2001. History of significant coral bleaching events and insights regarding amelioration. In *Coral Bleaching and Marine Protected Areas*, R. V. Salm and S. L. Coles, editors. Honolulu: The Nature Conservancy.

Gottfried, M., H. Pauli, et al. 1999. A fine-scaled predictive model for changes in species distribution patterns of high mountain plants induced by climate warming. *Diversity and Distributions* 5: 241–251.

Grayson, D. K. 2005. A brief history of Great Basin pikas. *Journal of Biogeography* 32 (12): 2103–2111.

Grebmeier, J. M., J. E. Overland, et al. 2006. A major ecosystem shift in the northern Bering Sea. *Science* 311 (5766): 1461–1464.

Guisan, A. and N. E. Zimmermann. 2000. Predictive habitat distribution models in ecology. *Ecological Modelling* 135 (2–3): 147–186.

Hannah, L., T. E. Lovejoy, et al. 2005. Biodiversity and climate change in context. In *Climate Change and Biodiversity*, T. E. Lovejoy and L. Hannah, editors. New Haven, CT: Yale University Press.

Hannah, L., G. F. Midgley, G. Bond, et al. 2002. Conservation of biodiversity in a changing climate. *Conservation Biology* 16(1): 11–15.

Hannah, L., G. F. Midgley, T. Lovejoy, et al. 2002. Conservation of biodiversity in a changing climate. *Conservation Biology* 16(1): 264–268.

Hannah, L., G. F. Midgley, D. Millar, et al. 2002. Climate change-integrated conservation strategies. *Global Ecology & Biogeography* 11 (66): 485–495.

Hannah, L., G. F. Midgley, et al. 2005. The view from the cape: Extinction risk, protected areas and climate change. *BioScience* 55 (3): 231–242.

Hannah, L., G. Midgley, et al. 2007. Protected area needs in a changing climate. *Frontiers in Ecology and the Environment* 5 (3): 131–138.

Hannah, L., R. Dave, P. P. Lowry II, S. Andelman, M. Andrianarisata, L. Andriamaro, A. Cameron, et al. 2008. Climate change adaptation for conservation in Madagascar. *Biology Letters* 4(5): 590–594.

Hansell, R. I. C., J. R. Malcolm, et al. 1998. Atmospheric change and biodiversity in the arctic. *Environmental Monitoring and Assessment* 49 (2–3): 303–325.

Harte, J., A. Ostling, et al. 2004. Climate change and extinction. *Nature* (online).

Hoegh-Guldberg, O. 1999. Climate change, coral bleaching and the future of the world's coral reefs. *Marine and Freshwater Research.* 50:839–866.

Hoegh-Guldberg, O., P. J. Mumby, et al. 2007. Coral reefs under rapid climate change and ocean acidification. *Science* 318 (5857): 1737–1742.

Hughes, L. 2003. Climate change and Australia: Trends, projections and impacts. *Austral Ecology* 28 (4): 423–443.

Hughes, L., E. M. Cawsey, et al. 1996. Climatic range sizes of Eucalyptus species in relation to future climate change. *Global Ecology and Biogeography Letters* 5:23–29.

Huntley, B. and H. J. B. Birks. 1983. *An Atlas of Past and Present Pollen Maps for Europe, 0–13000 B.P.* Cambridge: Cambridge University Press.

Idso, S. B. 1999. The long-term response of trees to atmospheric CO_2 enrichment. Global *Change Biology* 5 (4): 493–495.

Idso, S. B., B. A. Kimball, et al. 1987. Effects of atmospheric CO_2 enrichment on plant-growth: the interactive role of air-temperature. *Agriculture Ecosystems & Environment* 20 (1): 1–10.

Inouye, D. W., B. Barr, et al. 2000. Climate change is affecting altitudinal migrants and hibernating species. *Proceedings of the National Academy of Sciences of the United States of America* 97 (4): 1630–1633.

IPCC. 2001. *Climate Change 2001: The Scientific Basis. Contribution of Working Group I to the Third Assessment Report of the Intergovernmental Panel on Climate Change.* Port Chester, NY: Cambridge University Press.

IPCC. 2007. *Climate Change 2007: The Scientific Basis. Contribution of Working Group I to the Fourth Assessment Report of the Intergovernmental Panel on Climate Change.* Cambridge UK: Cambridge University Press.

Jackson, J. B. C. and K. G. Johnson. 2000. Life in the last few million years. *Paleobiology* 26 (4): 221–235.

Keeling, C. D. and N. W. Rakestraw. 1960. The concentration of carbon dioxide in the atmosphere. *Journal of Geophysical Research* 65 (8): 2502–2502.

Keeling, C. D., J. F. S. Chin, et al. 1996. Increased activity of northern vegetation inferred from atmospheric CO_2 measurements. *Nature London* 382 (6587): 146–149.

Kinzig, A. P. and J. Harte. 2000. Implications of endemics-area relationships for estimates of species extinctions. *Ecology* 81 (12): 3305–3311.

Korner, C. 2000. Biosphere responses to CO_2 enrichment. *Ecological Applications* 10 (6): 1590–1619.

Lawton, J. H. 1995. Ecological experiments with model systems. *Science* 269 (5222): 328–331.

Lenihan, J. M., R. Drapek, et al. 2003. Climate change effects on vegetation distribution, carbon, and fire in California. *Ecological Applications* 13 (6): 1667–1681.

Lin, G., B. D. V. Marino, et al. 1998. An experimental and modeling study of responses in ecosystems carbon exchanges to increasing CO_2 concentrations using a tropical rainforest mesocosm. *Australian Journal of Plant Physiology* 25 (5): 547–556.

Lovejoy, T. E. and L. Hannah. 2005. *Climate Change and Biodiversity*. New Haven, CT: Yale University Press.

Macmynowski, D. P., T. L. Root, et al. 2007. Changes in spring arrival of Nearctic-Neotropical migrants attributed to multiscalar climate. *Global Change Biology* 13 (11): 2239–2251.

McLachlan, J. S., J. J. Hellmann, et al. 2007. A framework for debate of assisted migration in an era of climate change. *Conservation Biology* 21 (2): 297–302.

Meier, N., T. Rutishauser, et al. 2007. Grape harvest dates as a proxy for Swiss April to August temperature reconstructions back to AD 1480. *Geophysical Research Letters* 34: L20705. DOI: 10.1029/2007GL031381.

Menzel, A. 2000. Phenology as global change bioindicator. *Annalen der Meteorologie* 39:41–43.

Menzel, A. 2000. Trends in phenological phases in Europe between 1951 and 1996. *International Journal of Biometeorology* 44:76–81.

Menzel, A. 2005. A 500 year pheno-climatological view on the 2003 heatwave in Europe assessed by grape harvest dates. *Meteorologische Zeitschrift* 14 (1): 75–77.

Midgley, G. F., L. Hannah, et al. 2003. Developing regional and species-level assessments of climatic change impacts on biodiversity in the Cape Floristic Region. *Biological Conservation* 112:87–97.

Moline, M. A., H. Claustre, et al. 2004. Alteration of the food web along the Antarctic Peninsula in response to a regional warming trend. *Global Change Biology* 10 (12): 1973–1980.

Norris, S., L. Rosentrater, et al. 2002. Polar bears at risk. 1–27. Gland, Switzerland: World Wildlife Fund World Wide Fund for Nature.

Orr, J. C., V. J. Fabry, et al. 2005. Anthropogenic ocean acidification over the twenty-first century and its impact on calcifying organisms. *Nature* 437 (7059): 681–686.

Overpeck, J., C. Whitlock, et al. 2003. Terrestrial biosphere dynamics in the climate system: past and future. In *Paleoclimate, global change, and the future*. K. D. Alverson, R. S. Bradley and T. F. Pederson, editors. Berlin: Springer Verlag.

Pacala, S. and R. Socolow. 2004. Stabilization wedges: solving the climate problem for the next 50 years with current technologies. *Science* 305 (5686): 968–972.

Parmesan, C. 1996. Climate and species' range. *Nature* 382: 765–766.

Parmesan, C., N. Ryrholm, et al. 1999. Poleward shifts in geographical ranges of butterfly species associated with regional warming. *Nature London* 399 (6736): 579–583.

Parmesan, C. and G. Yohe. 2003. A globally coherent fingerprint of climate change impacts across natural systems. *Nature* 421: 37–42.

Parry, M., J. Lowe, et al. 2009. Overshoot, adapt and recover. *Nature* 458 (7242): 1102–1103.

Penuelas, J. and I. Filella. 2001. Phenology: Responses to a warming world. *Science* 294 (5543): 793–795.

Peters, R. L. 1991. Consequences of global warming for biological diversity. In *Global Climate Change and Life on Earth.* R. L. Wyman, editor. New York: Routledge, Chapman & Hall.

Peters, R. L. and J. D. Darling. 1985. The green-house effect and nature reserves. *BioScience* 35 (1): 707–717.

Peters, R. L. and T. E. Lovejoy. 1992. *Global Warming and Biological Diversity.* London: Yale University Press.

Peterson, A. T. 2004. Projected climate change effects on Rocky Mountain and Great Plains birds: Generalities of biodiversity consequences. *Global Change Biology* 9: 647–655.

Peterson, A. T., E. Martinez-Meyer, et al. 2004. Modeled climate change effects on distributions of Canadian butterfly species. *Canadian Journal of Zoology* 82: 851–858.

Peterson, A. T., M. A. Ortega-Huerta, et al. 2002. Future projections for Mexican faunas under global climate change scenarios. *Nature* 416 (6881): 626–629.

Poorter, H. and M. L. Navas. 2003. Plant growth and competition at elevated CO_2: On winners, losers and functional groups. *New Phytologist* 157 (2): 175–198.

Pounds, J. A., M. P. L. Fogden, et al. 1999. Biological response to climate change on a tropical mountain. *Nature London* 398 (6728): 611–615.

Raupach, M. R., G. Marland, et al. 2007. Global and regional drivers of accelerating CO_2 emissions. *Proceedings of the National Academy of Sciences* of the United States of America 104 (24): 10288–10293.

Reich, P. B., J. Knops, et al. 2001. Plant diversity enhances ecosystem responses to elevated CO2 and nitrogen deposition. *Nature London* 410 (6830): 809–812.

Root, T., J. T. Price, et al. 2003. Fingerprints of global warming on wild animals and plants. *Nature* 421: 57–60.

Root, T. L. 1997. Changes over 30 years in spring arrival dates of migrating birds: Is spring arriving three weeks earlier? Boulder, Co: Birdlife International, WWF.

Roy, D. B. and T. H. Sparks 2000. Phenology of British butterflies and climate change. *Global Change Biology* 6:407–416.

Sitch, S., C. Huntingford, et al. 2008. Evaluation of the terrestrial carbon cycle, future plant geography and climate-carbon cycle feedbacks using five Dynamic Global Vegetation Models (DGVMs). *Global Change Biology* 14 (9): 2015–2039.

Smith, S. D., T. E. Huxman, et al. 2000. Elevated CO_2 increases productivity and invasive species success in an arid ecosystem. *Nature London* 408 (6808): 79–82.

Thomas, C. D., A. Cameron, et al. 2004. Extinction risk from climate change. *Nature* 427 (8 January 2004): 145–148.

van Mantgem, P. J., N. L. Stephenson, et al. 2009. Widespread increase of tree mortality rates in the western United States. *Science* 323 (5913): 521–524.

Williams, P., L. Hannah, et al. 2005. Planning for climate change: Identifying minimum-dispersal corridors for the Cape proteaceae. *Conservation Biology* 19 (4): 1063–1074.

Williams, S. E., E. E. Bolitho, et al. 2004. Climate change in Australian tropical rainforests: An impending environmental catastrophe. *Proceedings of the Royal Society of London, Series B* 270: 1887–1892.

Wilmking, M., G. P. Juday, et al. 2004. Recent climate warming forces contrasting growth responses of white spruce at treeline in Alaska through temperature thresholds. *Global Change Biology* 10 (10): 1724–1736.

Wilson, G. W. T. and D. C. Hartnett. 1998. Interspecific variation in plant responses to mycorrhizal colonization in tallgrass prairie. *American Journal of Botany* 85 (12): 1732–1738.

Woodward, F. I. 1998. Do plants really need stomata? *Journal of Experimental Botany* 49: 471–480.

7 Statistical Models in Wildlife Toxicology

Stephen B. Cox

CONTENTS

7.1 INTRODUCTION

Within the biological sciences, there is a significant shift taking place regarding the way data and other observations are being used to make inferences about biological phenomena and processes of interest. This shift is beginning to spill over into environmental toxicology and chemistry (Newman 2008). However, perhaps in no other field is this more apparent than in the wildlife sciences, where information theoretic approaches to model selection (Burnham and Anderson 2002; Johnson and Omland 2004) have become widely adopted. In other arenas, Bayesian approaches are being promoted as alternative frameworks that perhaps provide a more informative understanding of the data in hand (Clark 2005; Link and Albers 2007; Ellison 2004). However, these novel approaches are not merely alternative statistical "tests"; rather, they represent entirely different philosophical approaches, a fact that may not be entirely grasped by many practicing ecologists (Ellison 2004; Dennis 1996). Thus, to really understand the issues associated with the use of these novel approaches, one must investigate them in the context of the philosophy of science. In fact, it has been argued that much misunderstanding about statistics among graduate students arises not from a lack of technical knowledge about when and how to apply certain tests, but rather from a lack of training on "how to conduct science as an integrated process from hypothesis creation through statistical analysis" (Boyles et al. 2008). Similarly, I would argue that to effectively incorporate novel quantitative approaches into their own toolbox, practicing wildlife toxicologists need to be fully aware of how such approaches fit within their overall approaches to conducting science, how they relate to the actual biological hypotheses of interest, and how they can be reported in such a way as to promote the generation of new knowledge—not just "statistical ritual" (Guthery 2008).

This chapter will discuss recent developments in statistics (and more generally, the application of models to scientific data) that are altering the way scientists approach data analysis and attempt to validate model output. To lay a foundation for understanding the philosophical differences inherent in many of the approaches, a brief discussion of the role of models in science and the scientific method will first be presented. There is no possible way for this topic to be thoroughly reviewed in an entire book, much less in one book chapter! However, the goal is to introduce wildlife toxicologists to the key issues and provide some key references for where to explore these issues further.

7.2 THE ROLE OF MODELS IN THE SCIENTIFIC METHOD

The ultimate goal of science is to increase understanding of the natural world. Although there is not one exact way to accomplish this goal, if asked to define the "scientific method," most scientists would describe some form of the hypothetico-deductive method (Popper 1972). In its simplest form, it contains some variation of the following steps.

$$theory \rightarrow hypothesis \rightarrow experiment \rightarrow evaluation$$

In practice, biological hypotheses typically are translated into statistical hypotheses that are evaluated in light of experimental (or observational) data. The emphasis in this evaluation is on being able to falsify a statistical null hypothesis, and there has been much debate about how this evaluation should be conducted (Berger 2003) (see Salsburg (2001) for an especially entertaining discussion of this debate). Although founded on the principles of deductive inference and falsificationism,* modern application of the scientific method actually involves elements of both inductive and deductive reasoning (Gauch 2003; Quinn and Dunham 1983; Platt 1964; Crane and Newman 1996) (see Figure 7.1). Theory is advanced not only by proposing theory-based hypotheses that are evaluated in light of observations, but also by generating hypotheses based on the recognition of patterns observed in a series of observations.

So, where do models come in? Whether we explicitly realize it or not, models are used extensively throughout this process. What is a model? Simply put, it is a conceptual construct that represents a simplified view of nature (Pickett et al. 1994). This construct can take on a variety of forms. It may be verbal or quantitative.

---

* Simply put, falsificationism recognizes that, although a hypothesis logically can never be proven true, it can logically be proven false.

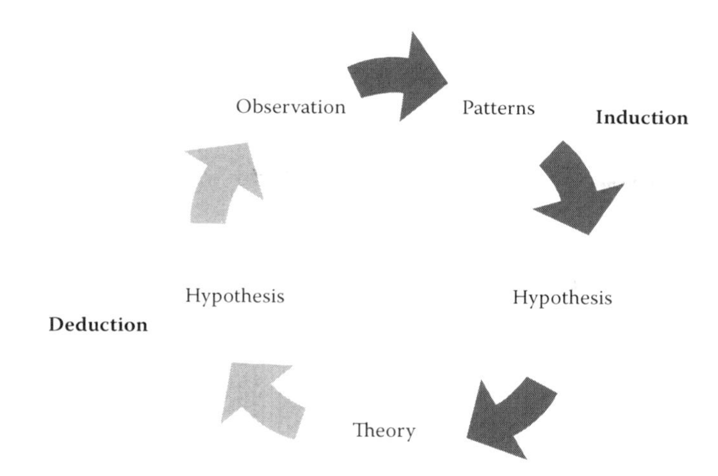

FIGURE 7.1 A conceptual model of inductive and deductive reasoning. In inductive reasoning, a series of confirmatory observations are used to lend evidence to the credibility of a hypothesis, which may eventually become part of a body of theory. In deductive reasoning, hypotheses are derived from theory, and observations are used to evaluate the merit of hypotheses. Early work in the philosophy of science focused on the use of induction; however, later work emphasized the use of deduction and falsification as hallmarks of the scientific method.

James and McCulloch (1985) further classify quantitative models as either empirical (derived from data) or theoretical (derived from theory) (Quinn and Keough 2002). For example, we may be interested in the effect of the concentration of chemical x on species y. Thus, a working verbal model is

$$x \rightarrow y$$

In fact, this conceptual model is really the simple hypothesis "chemical x affects species y."

Both the empirical and theoretical modeling approaches could be used to evaluate this hypothesis. The empirical modeling approach would first gather observations on the response of species y when it was exposed to chemical x. These observations could be derived from actual occurrences of exposure in nature, or they could be generated via lab and field experiments. Statistical tools would then be used to evaluate the hypothesis in light of the data. I will refer to this as a "top down" approach. In the second approach—the theoretical modeling approach—current knowledge about how chemical x operates would be utilized to develop an expectation about how species y will respond. Frequently, this expectation would be cast in terms of a mathematical model that links x and y. Various modeling tools (validation, simulation, etc.) would then be used to study the potential effects of x on y. I will refer to this approach as a "bottom up" approach.

As an example of a top down approach, consider the statistical evaluation of data that have been gathered during an experiment to "test" for an effect of chemical x.

A common approach to analyzing data from a dose-response experiment is to utilize probit or logit regression. In logit regression, we fit the model

$$p_i = \frac{e(\beta_0 + \beta_1 x_i)}{1 + e(\beta_0 + \beta_1 x_i)} + \varepsilon_i \tag{7.1}$$

to the experimental data using an algorithm to estimate the parameters β_0 and β_1 (p represents the proportion of individuals that survive). The model makes certain assumptions about the relationship, and, because it is a statistical model, it recognizes that there is some degree of variability (i.e., stochasticity) in survival that will not be accounted for (the ε_i). As long as this leftover variability adheres to certain assumptions (namely, for logit regression the assumption that the tenets of the binomial probability density function hold), the model is valid. After the model is fit, a variety of statistical hypothesis tests are used to discern whether or not chemical x had an effect on species y.

As an example of the bottom up approach, we could ascribe a specific mathematical relationship to the above working conceptual model based on knowledge about how chemical x operates. Suppose we expect some sort of a logistic dose-response relationship. A possible model is

$$y = c + \frac{d - c}{1 + \exp\{b[\ln(x) - \ln(e)]\}}, \tag{7.2}$$

where we have defined the logistic relationship between x and y by the parameters b, c, d, and e (Cedergreen et al. 2005). By using the formal language of mathematics, we are now forced to explicitly specify certain aspects of the relationship, including any assumptions we hold. For example, here we have assumed that the change in survival that occurs as a consequence of increasing concentration (i.e., the slope of the relationship) is not consistent (Figure 7.2). At low concentrations, that change is small, but it quickly becomes larger. At high concentrations, that change becomes small again until survival finally levels off at some minimum (denoted by the parameter c). We know that this simplified version of reality is not perfect. In fact, the model is completely deterministic. It states that, given a certain concentration of x, you will get a specific response every time. "Essentially, all models are wrong, but some are useful" (Box and Draper 1987). Nevertheless, we can use this formalized version of the above conceptual model to test hypotheses, generate new hypotheses, and otherwise investigate the potential impacts of chemical x on species y.

Thus, we use models in a variety of ways. Both approaches outlined above attempt to model the relationship between x and y. In the empirical modeling approach, we translate a working conceptual model (a biological hypothesis) into a series of statistical hypotheses that can be addressed using statistical models. In theoretical modeling, we cast biological hypotheses in terms of mathematical

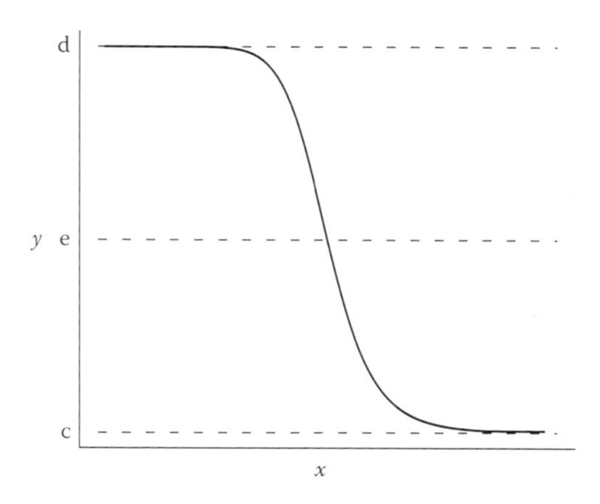

FIGURE 7.2 A model of the hypothetical relationship between the concentration of contaminant x and survival of species y. Survival at control is represented by the parameter d, survival at infinite dose is represented by c, and survival that is midway between c and d (i.e., the LC_{50}) is denoted by e.

equations that can be analytically or numerically evaluated to allow for testing and prediction.

7.3 STATISTICAL MODELS

7.3.1 Null Hypothesis Testing

We will begin our discussion of statistical models with a brief overview of null hypothesis testing and the models that underlie the statistical tests wildlife toxicologists frequently use. Classical null hypothesis testing (NHT) is familiar territory for most wildlife toxicologists. It is founded on the principle of falsificationism, and is a primary tool in the hypothetico-deductive method. In fact, most graduate education in applied statistics usually focuses on NHT. The basic steps of NHT are

1. generate a biological hypothesis
2. translate the biological hypothesis into a statistical hypothesis
3. gather data
4. calculate the test statistic
5. calculate the p-value
6. make a decision about the statistical hypothesis
7. draw biological conclusions with respect to the original biological hypothesis.

These basic steps are based on the work of Sir Ronald Fisher (Fisher 1954, 1956), who also noted that, in order to adhere to falsificationism, the statistical hypothesis should really be stated in the form of a "null hypothesis"—that is, the hypothesis of "no effect," labeled H_0. If we can falsify this hypothesis, then we feel confident that there is indeed an effect to consider further. Fisher regarded the p-value as a measure of the evidence against this H_0. Jerzy Neyman and Egon Pearson (1928, 1933) extended this idea by also including an "alternative hypothesis" and an a priori criterion (called α) that represented a critical p-value. If the observed p-value is less than α, then we reject H_0 in favor of the alternative (see Quinn and Keough (2002) for a succinct, yet useful discussion of these ideas).

To really understand NHT, however, you have to recognize that it is based on the frequentist notion of probability. To a frequentist, probability represents the relative frequency with which you expect an outcome to occur, or the "long-run average frequency of that outcome in a sequence of repeated experiments" (Bolker 2008). Thus, frequentists really interpret experimental results with an implicit assumption that the experiments are repeated over and over. The p-value of NHT, then, represents the probability of the observed outcome (the observed test statistic) or anything more extreme, given a null hypothesis. Philosophically speaking, the p-value is providing information on the likelihood of the data, given the null hypothesis, or

$$p = Pr(data|H_0) \tag{7.3}$$

With this in mind, statistical tests are viewed, at least implicitly, in the context of a series of repeated experiments. Scientific progress is made because, in the long run, the chances of making decision errors are controlled. As a result, NHT historically has been a very useful tool in the biological sciences, and continues to be especially useful in experimental settings. It also provides a useful tool when performed in the context of "strong inference" (Platt 1964), where a series of multiple alternative hypotheses are narrowed down via a series of well-designed experiments.

However, NHT has also been criticized (Johnson 1999; Newman 2008; Sterne and Smith 2001; Ellison 1996). Part of the criticism arises because of inappropriate application and a lack of understanding about the true meaning of the p-value (Newman 2008; Robinson and Wainer 2002). But, even when appropriately applied and interpreted, NHT still has weaknesses. For example, the implicit goal of NHT is to reject H_0, which is the hypothesis of "no effect." However, it is difficult, if not impossible, to really guess what patterns would emerge in the absence of biological processes of interest (Quinn and Dunham 1983). In addition, H_0 is usually some statement about a parameter being equal to zero or being equal to the same parameter from another population, etc. Such hypotheses are rather strange and contrived entities, and before we even start the whole testing procedure, we know that they are rarely, if ever, true. Failure to reject them usually has more to do with sample sizes and statistical power than the fact that they may, in fact, be true. This, in turn, has led to another related problem. Because the implicit goal of NHT is to reject H_0, it is difficult to publish studies that have failed to find "significant" results. Thus, the published literature may be biased (Ioannidis 2005; Newman 2008). More generally, this bias could impact theory development; Loehle (1987) provides an interesting discussion of the role of NHT in theory development within ecology. Finally, when using NHT, it is tempting to forget that statistical significance is not the same thing as biological significance (see Ellison

1996). Care must be taken to ensure that the statistical hypothesis being tested is appropriately tied to the actual biological hypothesis of interest.

7.3.2 Modeling Data

7.3.2.1 The General Linear Model

Most wildlife toxicologists are familiar with using the NHT paradigm to evaluate the "significance" of effects of toxicants on wildlife organisms. What many do not realize, however, is that the assumptions (normality, equal variances, etc.) we frequently worry about for statistical tests arise from an underlying statistical model that describes the process that gave rise to the data in hand. Even though we may teach a series of traditional hypothesis tests, we are really relying on a series of statistical models. In particular, most of the widely applied statistical tests (e.g., *t*-tests, ANOVA, regression, ANCOVA) rely on the general linear model

$$Y_i = \beta_0 + \beta_1 X_1 + \beta_2 X_2 + \cdots + \beta_p X_p + \varepsilon_i, \tag{7.4}$$

where we try to predict Y as a linear function of a series of predictor variables (X_i). The predictors can be continuous or categorical (categorical predictors get converted to a series of binary "dummy variables"), and each of the commonly used statistical tests is a special case of this general model. For example, having one categorical predictor means ANOVA, having one continuous predictor means regression, and so on. However, all share a suite of common assumptions about the response variable, namely that the residuals (ε_i) are normally distributed and have a well-behaved variance.

An underlying assumption in the use of the general linear model is that "true" values of the parameters (the ε_i) exist and are fixed. Thus, the goal of the modeler is to use observed data to develop adequate estimates of those parameters. For the general linear model, the method of ordinary least squares (OLS) is used to obtain those estimates. Simply put, OLS estimation chooses parameter values such that the difference between observations and model predictions is minimized. As long as the assumptions pertaining to residuals hold, analytic formulas for these parameter estimates exist.

The value of the general linear model, however, goes beyond the NHTs that are derived from it. In fact, examination of the model structure can provide additional insight into the data. For example, parameter estimates often can be interpreted as measures of effect size. "Whereas NHST (null hypothesis significance testing) is useful in determining statistical significance, effect sizes are useful in determining practical importance" (Robinson and Wainer 2002). For example, in simple linear regression, the slope, β_1, is a measure of the unit change in the response variable as a result of a unit change in the predictor. When coupled with a confidence interval, it provides more insight into the relationship between response and predictor than a significance test alone would provide.

In ANOVA, categorical predictor variables get coded into a series of dummy variables, and examination of the parameter estimates for these dummy variables provides useful information above and beyond the ANOVA hypothesis test. The number of dummy variables required to code for a particular factor is one less than the number of groups, and the exact way in which the coding gets done varies by statistical package. In R (Team 2008), for example, the default is "treatment contrasts." If you have three dosing groups (control, low, and high, for example), two dummy variables are created according to the following:

Dose Group	Dummy Variable 1 (D_1)	Dummy Variable 2 (D_2)
Control	0	0
Low	1	0
High	0	1

The statistical model becomes

$$Y_i = \beta_0 + \beta_1 D_1 + \beta_2 D_2 + \varepsilon_i \,. \tag{7.5}$$

Observations in the control get zeros for D_1 and D_2, observations in the low group get a one and a zero, and observations in the high group get a zero and then a one. With this coding in place, the estimate for β_1 is an estimate of the difference in mean response between observations from the control and observations in the low dose group. Similarly, β_2 provides insight into the difference in mean response between control and high dose group. Thus, these parameter estimates, and their associated confidence intervals, provide meaningful biological information on the effects of dose. This example focuses on treatment contrasts, but there are a variety of other ways that the dummy variable coding can be done—each allowing for additional insight into the biological questions of interest.

As an example, consider data on tobacco budworms exposed to *trans*-cypermethrin (Figure 7.3). Although this is a rather simple dataset on a species that is not exactly charismatic megafauna, it provides a useful starting place for illustrating the

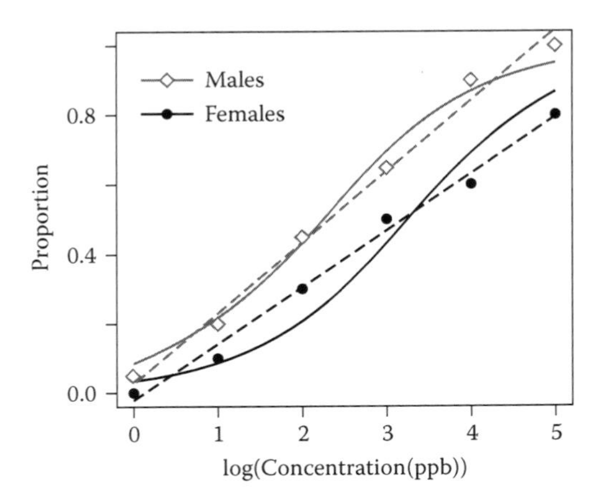

FIGURE 7.3 Relationship between the proportion of individual tobacco budworms that died and the log (base 2) of exposure (to *trans*-cypermethrin) concentration. Data are from Collett (1991). Predictions from the linear model (dashed lines) and logit model (solid lines) are shown for both sexes.

approaches to be discussed. The data were first collected by Collett (1991), and were later discussed in Venables and Ripley (2002). Twenty individuals were exposed to a series of concentrations, and the response variable of interest was the proportion of individuals that died as a result of exposure. Both males and females were examined, and one approach to modeling the data is to allow both the slope and intercept of the linear model to vary between sexes. Using treatment contrasts, the resulting model is

$$p_i = \beta_0 + \beta_1 conc + (\beta_2 + \beta_2 conc)sex, \quad (7.6)$$

where p_i represents the predicted proportion of individuals that die, *conc* is the $\log_2$ of the exposure concentration, and *sex* is a dummy variable denoting sex; 0 = females, 1 = males.

Standard model output is shown in Table 7.1, and parameter estimates can be viewed as a measure of effect size. For females, increasing $\log_2$(concentration) by one unit increases the proportion of individuals dying by 0.16 (95% confidence interval, 0.142 to 0.184). Furthermore, the parameters β_2 and β_3 represent the change in intercept and slope, respectively, when going from females to males. Thus, for males, the change in slope (i.e., effect size) is 0.04 (95% confidence interval, 0.009 to 0.0678). In other words, males experienced an increase in the proportion of individuals dying of 0.20 for every unit increase in exposure concentration. Furthermore, the hypothesis test associated with β_3 indicates that this change in slope is significant. As a result, when comparing males to females, the concentration at which the comparison is made will influence the outcome.

Thus, by understanding the way that the general linear model is used to "model" the data being analyzed, we can glean much more useful information from the data than just the rejection of a null hypothesis. In particular, examination of model structure can provide additional insight into aspects of effect size, thus providing a clearer linkage to the biological questions of interest (Robinson and Wainer 2002; Newman 2008). However, using a general linear model with proportions as a response variable has two limitations. First, proportions are not normally distributed. Second, the linear models make predictions that are not realistic for proportions (e.g., values less than 0 and greater than 1). Methods for dealing with these limitations will be discussed in the next section.

TABLE 7.1
Standard Output from a General Linear Model of the Effects of *trans*-Cypermethrin on Tobacco Budworms

Parameter	Estimate	Std. Error	*t* value	Pr(>|*t*|)
β_0	−0.023810	0.027126	−0.878	0.4057
β_1	0.162857	0.008959	18.177	8.62e-08 ***
β_2	0.061905	0.038362	1.614	0.1453
β_3	0.038571	0.012671	3.044	0.0160 *

Each parameter estimate has an associated hypothesis test (*t*-test) of H_o: $\beta_i = 0$.

7.3.2.2 Generalized Linear Models

Wildlife toxicologists frequently deal with data that do not adhere to the assumptions of the general linear model. In particular, they are frequently interested in measures of response that do not adhere to a Gaussian (i.e., normal) probability density function. Examples include counts (e.g., number of abnormal cells in some sort of histological study) or binary outcomes (e.g., presence/absence of response). In fact, one of the most ubiquitous examples occurs when individual organisms are exposed to a series of contaminant concentrations, and the lethality of the various concentrations is tracked. Generalized* linear models (McCullagh and Nelder 1983), represented as

$$Y_i = f(\beta_0 + \beta_1 X_1 + \beta_2 X_2 + \cdots + \beta_p X_p) + \varepsilon_i, \tag{7.7}$$

allow for the examination of such response variables by extending the general linear model in two specific ways. First, a specific function is applied to the linear component of the right hand side of the equation. The inverse of this function, when applied to the Y_i, is referred to as the "link" function. Second, given a certain link function, the distribution of the errors (the ε_i) can take on a variety of forms other than the Gaussian distribution.

For dose-response data, two generalized models have been applied. The first is probit regression, which uses the model,

$$p_i = \phi(\beta_0 + \beta_1 X_1) + \varepsilon_i, \tag{7.8}$$

where ϕ is a function based on the normal cumulative distribution function. It can also be represented using the probit link function,

$$\phi^{-1}(p_i) = \beta_0 + \beta_1 X_1, \tag{7.9}$$

where the probit, ϕ^{-1}, is the inverse of the normal cumulative distribution function. It is sometimes referred to as a Normal Equivalent Deviant (NED).

The second model, sometimes referred to as logistic or logit regression, is

$$p_i = \frac{e^{(\beta_0 + \beta_1 X_i)}}{1 + e^{(\beta_0 + \beta_1 X_i)}} + \varepsilon_i, \tag{7.10}$$

where p_i is the probability of death at a series of concentrations (X_i). This is exactly equivalent to using the logit link function as the response variable,

$$\log\left(\frac{p_i}{1 - p_i}\right) = \beta_0 + \beta_1 X_i. \tag{7.11}$$

* Both the general linear model and generalized linear models are frequently referred to as GLM, and, unfortunately, software packages differ in their usage of the term. SAS uses GLM to refer to the general linear model, whereas R/S-plus users tend to think of generalized linear models when they see GLM. Some users use GLiMs to refer to generalized linear models.

When using either the probit or the logit link function, we now have a linear model that is conceptually very similar to the general linear model. Categorical predictors can be represented as a series of dummy variables. Those dummy variables can be coded using a variety of contrasts (just like in the general linear model), and the parameter estimates of the β_i can be interpreted as measures of effect size. Other covariates also can be added into the model. Thus, NHTs that are similar to ANOVA, ANCOVA, and regression of the general linear model can be addressed.

The choice of when to use logit versus probit has received some attention in the ecotoxicology literature (Newman 1995; Newman and Clements 2008). In practice, results from the two models are frequently almost identical. However, if the data tend to have more observations at low and high concentrations, with relatively fewer observations in the middle, the results of logit regression tend to be more robust (Venables and Ripley 2002). In addition, because the response variable of logit regression is the natural log of the "odds" that an individual will die, the parameters from logit regression are the natural log of odds ratios. As a result, they can be interpreted in a manner that provides a relevant biological interpretation.

As an example, we will revisit the data on tobacco budworms. Again, we will consider a model in which both the slope and intercept (of the relationship between the proportion of individuals dying and exposure concentration) are allowed to vary between sexes. If p_i is the probability of dying at the ith exposure concentration, we assume

$$r_i \sim Binomial(p_i, n_i)$$
$$\ln\left(\frac{p_i}{1-p_i}\right) = \beta_0 + \beta_1 conc + (\beta_2 + \beta_3 conc)sex, \tag{7.12}$$

where r_i is the actual number of individuals that died at the ith exposure concentration. It is assumed to be distributed as a binomial variate with n_i trials. The parameters, β_i, can be interpreted as in the general linear model, this time linking the *logit(p)* to *log*$_2$*(concentration)* and *sex*. However, the method we used to obtain parameter estimates in the general linear model (i.e., the method of OLS) is not appropriate for estimating the parameters in this model. Instead, maximum likelihood estimators are frequently used (discussion of the concept of the "likelihood" will be left for the section on Bayesian techniques). The results are shown in Table 7.2, and the p-value for β_3 indicates that, now, we do not have convincing evidence that the slope of the relationship differs between males and females. Thus, we will consider the reduced model

$$\ln\left(\frac{p_i}{1-p_i}\right) = \beta_0 + \beta_1 conc + \beta_2 sex. \tag{7.13}$$

In this version of the model, β_1 represents the slope for both males and females. Moreover, because this is a logit model, parameters can be interpreted as the natural log of odds ratios. Taking the exponential of the parameter estimates will result in an odds ratio. Thus, we see that if you increase the $\log_2$(concentration) by one unit, the odds of death are 2.9 (that is, $e^{1.06}$, where 1.06 is the slope estimate of the reduced model) times higher. In addition, after controlling for concentration, the odds of death for males are

TABLE 7.2
Standard Output from Generalized Linear Models (Logit Model) of the Effects of *trans*-Cypermethrin on Tobacco Budworms

Parameter	Estimate	Std. Error	z-value	Pr(>|z|)
Full model:				
β_0	−2.9935	0.5527	−5.416	6.09e-08 ***
β_1	0.9060	0.1671	5.422	5.89e-08 ***
β_2	0.1750	0.7783	0.225	0.822
β_3	0.3529	0.2700	1.307	0.191
Reduced model:				
β_0	−3.4732	0.4685	−7.413	1.23e-13 ***
β_1	1.0642	0.1311	8.119	4.70e-16 ***
β_2	1.1007	0.3558	3.093	0.00198 **

Each parameter estimate has an associated hypothesis test (z-test) of H_o: $\beta_i = 0$.

3.01 (i.e., $e^{1.1}$) times the odds for females. This is the same thing as saying that *for a given concentration*, the odds of death for males are 3.01 times the odds for females.

Thus, by explicitly considering details about the response variable being considered, generalized linear models are a robust tool for examining data within wildlife toxicology. In the above example, we see that when the response of interest (number of individuals dying) is explicitly modeled as a binomial variate, the overall conclusions of the analysis change. In particular, the logit model indicates that the data are consistent with the hypothesis that the slope (i.e., the effect of exposure concentration on the likelihood of death) is the same for males and females. Thus, it is meaningful to discuss the likelihood of death for males relative to females, *controlling* for the effects of exposure concentration.

7.3.2.3 Mixed-Effects Models

In all of the above linear models, the effects that are being considered (and hence, the parameters representing those effects) are assumed to be fixed. That is, there is an implicit assumption that true parametric values of the parameters being estimated actually exist. The goal of the model is to estimate those values, and, as a result, get an estimate of effect size. Such effects typically are represented as an estimate of the change in mean response. However, in many cases, we may be interested in factors that are not fixed. Instead, they may actually represent a potential source of variability that the researcher has limited or no control over, perhaps because of experimental limitations. In such cases, the effect of interest is the amount of variability (represented as an estimate of a standard deviation or variance) in response that can be attributed to the factor of interest. Such factors are termed *random effects factors*.

The following example is familiar to most wildlife toxicologists and will serve as a useful illustration. A researcher wishes to conduct a dose-response experiment for a bird species, and a series of dosing media is prepared using five concentrations of a contaminant of interest. The researcher wishes to use 10 individuals per treatment, but, because of size limitations, she can place only 5 individuals within each dosing cage. Thus, she is left with two cages per treatment. The researcher may choose to analyze the resulting data using the ANOVA model*

$$Y_{ij} = \mu + \alpha_i + \varepsilon_{ij}, \tag{7.14}$$

where μ is the grand mean of the Y_{ij}, and α_i represents the change in μ as a result of treatment i. One disadvantage of this model is that it does not account for a possible cage effect. That is, the ε_{ij}, in addition to representing random variability in the response, also contain some variability that is attributable to the different cages within each treatment. Stated another way, the assumption of independence has been violated; individuals within the same cage are not independent. One practical consequence of this is that the test for treatment effects loses statistical power.

A more powerful approach is to explicitly model this variability using

$$Y_{ijk} = \mu + \alpha_i + b_j + \varepsilon_{ijk}, \tag{7.15}$$

where b_j represents the random effect of being placed into cage j. The added assumption of this model is that the b_j are random draws from a normal distribution with a constant variance. Thus, the mixed effects model allows for an explicit examination of both the fixed effects attributable to dose and the random effects attributable to cage.

Methods for dealing with specific examples of mixed-effects models have been around for some time. For example, Model II ANOVA, nested ANOVA (such as the above example), and simple repeated measures are all examples of mixed-effects models. In these examples, traditional methods are based on partitioning sum of squares with a primary focus on developing NHTs (Sokal and Rohlf 1995). However, approaches for estimating parameters based on likelihood and restricted likelihood and the computational methods for dealing with such models have been developed and now allow for examination of a wider variety of linear and nonlinear mixed-effects models. Even when dealing with classic examples, these current methods have some distinct advantages over the sums of squares approaches. For example, they are less sensitive to unbalanced data. In addition, they allow other aspects of the potential non-independence of observations to be modeled. Data that are heteroscedastic (i.e., exhibit non-constant variances) can be modeled. Moreover, data that are correlated (e.g., in space or time) can be modeled. Thus, mixed-effects models provide a very powerful and flexible tool for analyzing wildlife toxicology data. Popular tools for examining mixed-effects models are SAS's Proc Mixed (Littell 1996) and R's nlme and lme4 packages (Bates et al. 2008; Pinheiro and Bates 2000).

To illustrate the use of a mixed-effects model, consider the data in Figure 7.4. In an experimental setting, lizards were chronically exposed to five concentrations of a toxicant. Lizard weight was measured over time as an endpoint of interest. One approach to modeling these data is to consider the slopes and intercepts of each

---

* Note that this ANOVA model is equivalent to the general linear model, using four dummy variables to code for the five treatment groups, and where β_0 represents the grand mean.

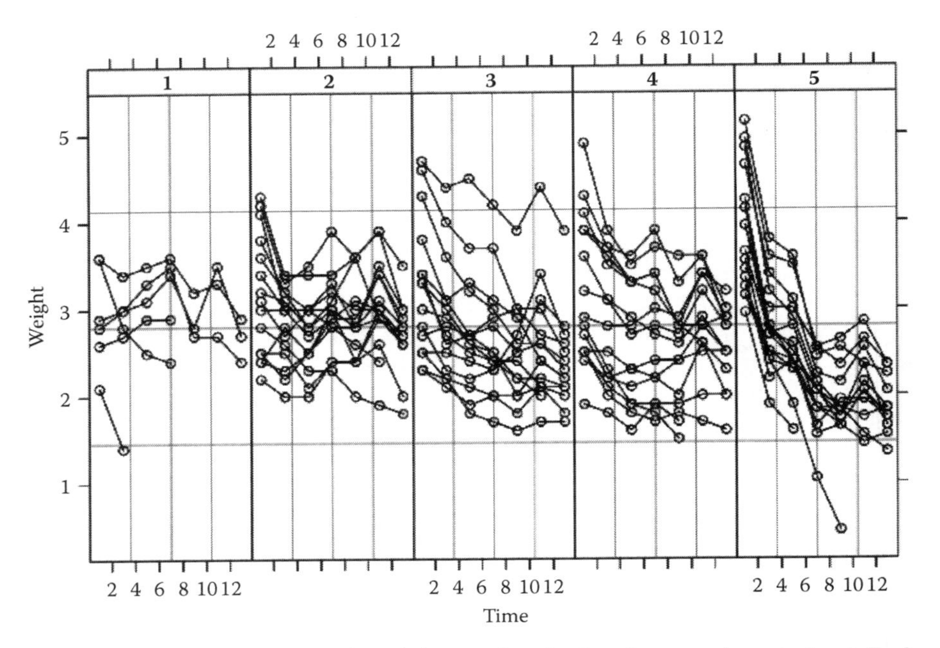

FIGURE 7.4 Plot of the change in weight over time for lizards exposed to a toxicant. Each line represents an individual, and each of the five panels represents a different toxicant concentration. Note that some individuals did not survive the entire duration of the experiment.

individual as random effects.* In effect, this accounts for the lack of independence between observations from the same individual. The model is

$$Y_{ij} = \beta_1 + \beta_2 G + \beta_3 T_{ij} + \beta_4 GT + b_{i1} + b_{i2} T_{ij} + \varepsilon_{ij}, \tag{7.16}$$

where b_{i1} and b_{i2} are the random intercepts and slopes, respectively. The β_i represent the fixed effects; however, for the sake of simplicity, the dummy variables representing the five groups are not explicitly listed.

Results of the model fit are in Table 7.3. Here, instead of labeling each of the parameter values using βs, fixed effects are labeled by the variable names. Because treatment contrast coding for "Group" was used, the estimates labeled "Intercept" and "Time" represent the predicted intercept and slope, respectively, for Group 1 (controls). Other estimates represent the change in intercept or slope for the different groups. For this analysis, we anticipate that if dosing effects are present, they will manifest as changes in the slope of the relationship between weight and time. That is, our hypothesis is that dosing will increase the rate of weight loss above what we observe in the control group (Group 1). Thus, the relevant measures of effect size are the fixed effects estimates of the changes in slope. In Group 5 (the highest-dosing group), the slope was significantly more negative than in the controls. In fact, it is

* This is analogous to the multi-level linear modeling approach (sometimes called hierarchical modeling or derived variable analysis in other contexts) to longitudinal data.

TABLE 7.3
Results of a Mixed-Effects Model of the Effects of Dosing Group and Time on Lizard Weight

Parameters	Value	Std. Error	*t*-value	Pr(>|*t*|)
Fixed effects:				
(Intercept)	3.1095599	0.23826805	13.050679	0.0000
Time	–0.0320488	0.02642850	–1.212662	0.2260
Group 2	–0.1008429	0.30474550	–0.330909	0.7418
Group 3	0.0502338	0.30474550	0.164839	0.8696
Group 4	–0.0534563	0.30490477	–0.175321	0.8614
Group 5	0.3997986	0.23563826	1.696663	0.0906
Time: Group 2	0.0157515	0.03024569	0.520785	0.6028
Time: Group 3	–0.0324457	0.03024569	–1.072738	0.2841
Time: Group 4	–0.0184610	0.03035741	–0.608123	0.5435
Time: Group 5	–0.1346196	0.02932312	–4.590905	0.0000
Random effects:				
Intercept	0.69887835			
Slope	0.04908608			

Fixed effects parameter estimates are based on treatment contrast coding of the categorical variable "Group" and are interpreted as in other linear models. The random effects values are estimates of standard deviation.

–.135 (95% CI, –0.192 to –0.077) g/week lower than the control slope, which is –0.032 g/week (95% CI, –0.084 to 0.020). Other groups were not significantly different from the controls.

7.3.3 Information Theoretic Approaches

Up to this point, we have seen that the general linear, generalized linear, and mixed-effects models can be used for the purposes of developing NHTs and estimating effect sizes. The information theoretic approach (Burnham and Anderson 2002), also referred to as model or multimodel selection, explicitly recognizes that models represent hypotheses about the biological processes that gave rise to the data in hand. The basic steps of the approach are

1. develop a suite of a priori hypotheses
2. represent the hypotheses as a set of candidate models*
3. collect relevant data
4. determine the likelihood of each of the candidate models, given the data
5. identify the model or models that appear to have the most support

* Although most application of the IT within the wildlife sciences has focused on statistical models, the principles of model selection can be applied to both empirical (i.e., statistical) and theoretical (i.e., mechanistic) models.

The metric that is commonly used to compare models is Akaike's Information Criterion, which, when corrected for small sample sizes, is

$$AIC_c = -2\log[L(\hat{\theta}_K|\text{data})] + 2K\left(\frac{n}{n-K-1}\right), \quad (7.17)$$

where K is the number of parameters in the model and n is the sample size. The term $L(\hat{\theta}_K|\text{ data})$ represents the "likelihood" of the K parameter estimates, given the data. We will discuss the meaning of the likelihood in more detail in a later section. In a regression model, for example, it is analogous to the concept of the residual sum of squares.

Conceptually, this is a major shift away from null hypothesis testing. For example, in NHT, the p-value represents the probability of our data, given a specific model (i.e., hypothesis). In contrast, model selection considers the likelihood of models, given the data. As a result, studies that choose an information theoretic approach should not mix in language (for example, "significant," "non-significant," or "rejected") or other aspects of NHT (e.g., p-values) (Anderson and Burnham 2002; Burnham and Anderson 2002).

Moreover, the IT approach emphasizes "careful, a priori consideration of alternative models" (Burnham and Anderson 2002). This is really the most important step of model selection approaches. They are intended to highlight only the best model with respect to the set of candidate models being considered. If all the candidate models are poor, results from model selection will be poor. Therefore, "we cannot overstate the importance of the scientific issues [rather than statistical issues], the careful formulation of multiple working hypotheses, and the building of a small set of models to clearly and uniquely represent these hypotheses" (Burnham and Anderson 2002). Thus, model selection is not a test of hypotheses. Rather, model selection aligns more closely with confirmation and inductive reasoning than it does with falsification and deductive reasoning (Figure 7.1). The objective of model selection is not to reject any specific hypothesis but to ascertain which hypotheses appear to be most consistent with data. As a result, it is perhaps better suited to observational studies, rather than controlled, experimental settings where traditional hypothesis testing along with quantification of effect size and precision have proven valuable.

To illustrate the application of the IT approach, and to facilitate comparison with the other approaches, we will once again consider the data on tobacco budworms.* The first step of the approach is to develop a priori hypotheses. For the sake of illustration, assume the following three hypotheses are of interest:

- H1: mortality is determined solely by exposure concentration.
- H2: mortality is determined solely by the sex of the organism.
- H3: mortality is determined by both exposure concentration and sex, each operating independently.

---

* Again, experimental data are perhaps better suited to traditional hypothesis testing. By considering the IT approach to these data, the goal is to facilitate comparison and conceptual understanding of the differences between IT and NHT approaches.

TABLE 7.4
Information Theoretic Assessment of Three Models Regarding Determinants of Mortality in Tobacco Budworms

Model	Δ_i	ω_i
1: $\log_2$(conc)	0.990	0.379
2: sex	102.805	~ 0.0
3: $\log_2$(conc), sex	0	0.621

Models were selected a priori and are denoted by the predictors they included. The Δ_i represent the change in Akaike's Information Criterion, corrected for small sample size. The model weights, ω_i, denote the relative likelihood of various models, represented as proportions.

We will represent each of these hypotheses using a logit model and examine the resulting AIC_c values. In particular, we will examine two metrics. First, the Δ_i represent the change in AIC_c for the ith model relative the model with the lowest AIC_c. In addition, the model weights, ω_i, calculated as $\exp(-0.5 * \Delta_i)$, represent the relative likelihood of the ith model, expressed as a proportion. Results are shown in Table 7.4. As a rule of thumb, models with a $\Delta_i < 2$ are supported by the data in hand and deserve additional consideration. Thus, hypothesis 3 has the most support, although hypothesis 1 cannot be ruled out. However, the data do not support the hypothesis that mortality is determined solely by the sex of the organism (a somewhat dubious hypothesis to begin with considering the true nature of these data!).

7.3.4 Bayesian Statistics

By abandoning *p*-values and using the concept of likelihood, the model selection approach takes one step away from the traditional hypothetico-deductive framework and the frequentist notion of probability. Bayesian approaches, on the other hand, take a giant leap away. Indeed, Bayesian approaches represent a complete abandonment of traditional approaches (Dennis 1996). To understand the differences, one must first understand the concept of likelihood. Up to now, we have somewhat sloppily used probability and likelihood to be synonymous. Technically, they have different meanings.

Fisher developed the notion of the likelihood when devising alternative methods to estimate parameter values. In all of the previous modeling frameworks we have discussed, there is an assumption that one true value of the parameter exists in nature and the goal of statistical modeling is to develop point estimates and confidence statements about those estimates using data. These confidence statements implicitly assume (under the frequentist notion of probability) that an infinite number of samples are collected, and that the parameter value is fixed. Ordinary least-squares methods, where analytical formulas give parameter estimates that minimize distances between model predictions and observed data, do not work in some cases (for example, in probit or logit generalized linear models). As an alternative, Fisher developed methods based on the notion that the observed data are held fixed, and it is the parameter estimates that are allowed to vary. Because this concept turned traditional interpretations of probability around, he coined the term *likelihood* to distinguish it from probability (Fisher 1954). In addition,

$$L(\theta|data) = \alpha P(data|\theta) \tag{7.18}$$

or, the likelihood of a parameter value (i.e., a hypothesis), given the data, is equal to the probability of the data, given the parameter value, multiplied by the constant α. Calculus is then used to find parameter estimates that maximize this likelihood,* thus creating "maximum likelihood" methods. Bayesians fully embrace this notion and take it a step further. In the Bayesian approach, parameters explicitly are considered random variables, and the data are considered fixed. The basic steps, modified from Gotelli and Ellison (2004), are

1. develop a hypothesis or hypotheses
2. represent the hypothesis or hypotheses as models
3. specify the parameters as random variables
4. specify a prior probability distribution
5. calculate the likelihood
6. calculate the posterior probability distribution
7. interpret the results

Intuitively, the Bayesian approach is rather simple. It is founded on the following relation, derived from Bayes' theorem. We will use θ to represent a hypothesized parameter value. Conceptually, however, it also could be interpreted as a "model" or "hypothesis" of interest.

$$P(\theta \mid data) \propto P(\theta)P(data \mid \theta) \tag{7.19}$$

The posterior probability distribution (usually called the "posterior" and denoted $P(\theta|data)$), is the outcome of interest. It is the probability distribution of our parameter (our hypothesis), given the observed data, and is proportional to the prior probability distribution (usually called the "prior" and denoted $P(\theta)$) multiplied by $P(data|\theta)$. As described by Ellison (1996), the prior can be interpreted in three different ways: a frequency distribution based on a synthesis of previous data, an objective statement about the distribution based on ignorance (i.e., a noninformative prior), or a subjective statement of belief based on previous experience. The likelihood (denoted $P(data|\theta)$) is derived from Fisher's likelihood methods as described above.

Thus, the Bayesian approach abandons the frequentist notion of probability altogether by incorporating notions of belief. Although this can be a bit alarming to many scientists, it is actually more in line with common usage of the term *probability*.

* In practice, it is the log-likelihood that is maximized.

Moreover, instead of a p-value (and, at least superficially, a nicely packaged decision to reject/fail to reject an arbitrary hypothesis), the result in a Bayesian analysis is a posterior probability distribution. This can roughly be translated as representing the probability of different values of the parameter of interest, given the observed data. This language may be better suited to communication with decision makers in a policy context because it is more closely aligned with common usage and it more directly addresses the actual question of interest (Ellison 1996). In addition, because it includes the prior, it contains a mechanism to incorporate prior knowledge. In fact, one potential benefit of the Bayesian approach is the manner in which knowledge about a tangible biological phenomenon (e.g., an effect size) can be continually updated as new information or observations are discovered. As a result, some have argued that it is ideally suited to environmental sciences, especially where adaptive management is being considered (Ellison 1996).

Consider a simple experiment where two groups of animals are being compared: one group was dosed, while the other was a control. The endpoint Y was measured, and the parameter of interest is the change in Y as a result of being dosed. For the sake of argument, let's assume that the observed difference in Y between the groups was 5 units with a 95% confidence interval (CI) from 3 to 7. Compare the following results statements:

1. Assuming that there is no effect of dose on Y, the probability that we would have observed a difference between the two groups as large as or larger than the observed difference, had we repeated the experiment an infinite number of times, was less than 5%. Therefore, we conclude that there is an effect of dosing on Y. (This is the interpretation of a p-value under classical NHT.)
2. The observed difference in Y between the groups was 5 ± 2 units. Had we repeated the experiment an infinite number of times, 95% of those CIs would have included the true change in Y as a result of dose. (This is the correct interpretation of effect size and confidence interval. Note that it is *not* correct to say that there is a 95% chance that the true effect size lies in the interval 3 to 7, a common interpretation.)
3. Based on current knowledge, these data suggest that 95% of all possible changes in Y as a result of dosing will lie between a and b. (A valid interpretation assuming a Bayesian approach. Here, we have used a and b to represent the 95% CI of the posterior distribution. When referring to the posterior, this CI is sometimes referred to as the "credibility" or "belief" interval. We have only used the interval as a convenient summary of the entire distribution, which could be described in a variety of ways.)

Thus, Bayesian proponents assert that it is a straightforward way to explicitly state starting assumptions, quantify pre-existing information, and allow for communication of results, and to do so in a manner that is more intuitive to decision makers (Ellison 1996). In fact, Clark (2005) argues that the "emergence of modern Bayes has little to do with philosophy, but comes rather from pragmatism." Nevertheless, although Bayes' theorem is conceptually simple to understand, the calculation of the posterior distribution based on the prior and likelihood is difficult and beyond the capabilities

TABLE 7.5
Summary of the Posterior Distributions of Parameters in the Full Logit Model Examining the Effects of Exposure Concentration and Sex on Mortality of Tobacco Budworms

Parameter	Posterior Mean	95% CI of Posterior
β_0	−3.085	−4.276 to −2.045
β_1	0.933	0.610 to 1.293
β_2	0.149	−1.443 to 1.706
β_3	0.379	−0.151 to 0.945

of most widely used statistical packages.* Moreover, this difficulty is compounded when classical distributions such as the normal do not adequately describe the distribution of the parameter of interest. This fact is probably a major reason we have yet to see widespread adoption of Bayesian methods in wildlife toxicology.

As a final illustration, let's once again consider the data on tobacco budworm mortality. Any modeling approach can be used; however, in the Bayesian framework, we now consider the data to be fixed, and the estimated parameters to be random variates. The goal of the analysis is to estimate the likely distribution of those parameters, given the observed data. For the purposes of illustration, we will use the full logit model (equation 12), and assign each of the parameters minimally informative priors based on the normal distribution. Because of the difficulties in determining posterior distributions in practice, they typically are approximated using Markov chain simulation (Gelman 2004). Posterior means and confidence intervals for the parameters are described in Table 7.5. The actual posterior distributions are shown in Figure 7.5.

Readers are referred to Ellison (1996) and Dennis (1996) for a useful introduction to, and discussion of, the potential strengths and weaknesses of the Bayesian approach. Analytical difficulties are not the only reason Bayesian methods have been criticized. Briefly, much of the debate has centered on the use of priors. While some argue that it is an effective means of including information and knowledge that has already been gained (many times at the cost of considerable resources and effort), others argue that it introduces a level of subjectivity that is unacceptable. On the other hand, Ellison points out that, as long as the likelihood dominates the prior (which is in line with sound scientific practice), the data have a much greater impact on the posterior than any potential subjective influence introduced via the prior.

In summary, Bayesian methods are distinctly different than traditional statistics in two ways. First, they abandon the frequentist notion of probability and include aspects of subjective belief. Second, they consider parameter values to be random variables, with the data being fixed. Although the discussion above focused on the estimation of parameter values, posteriors can be calculated for entire models, and then models can be compared in a fashion analogous to that in the information theoretic approach. For example, Link and Albers (2007) provide a thoughtful example

* WinBUGS is a freely available program for Windows. In addition, there are packages available for R that facilitate Bayesian analysis. An R lab manual for Clark (2007) is also available.

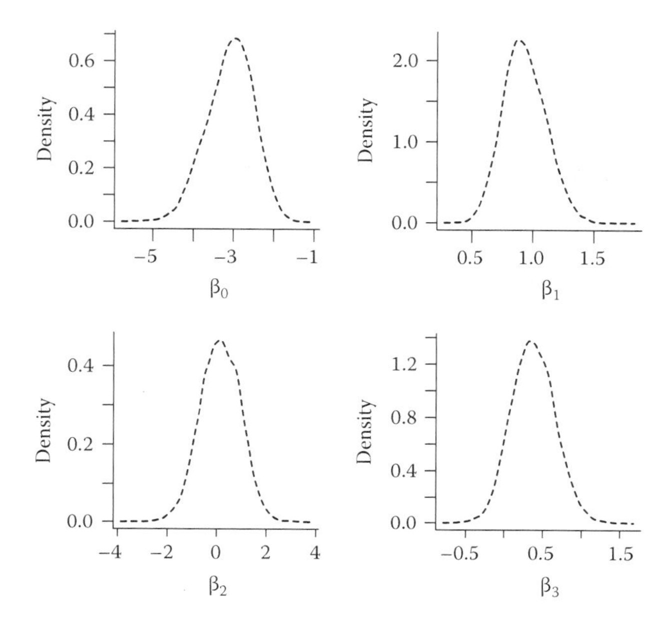

FIGURE 7.5 Plots of the posterior distributions of parameter values from the logit model examining the effects of exposure concentration and sex on mortality of tobacco budworms. Distributions represent the likelihood of possible parameter values, given observed data and prior knowledge.

of the application of Bayesian multimodel inference in the context of dose-response models using data on American kestrels exposed to methylmercury (Albers et al. 2007). Gelman (2004) provides a widely cited introduction to Bayesian methods, and, for perhaps the most thorough introduction to Bayesian methods in an ecological context, readers are referred to Clark (2007).

7.4 CONCLUSIONS

Novel methods are altering the way that biologists in general, and wildlife toxicologists in particular, are approaching data. By explicitly considering the statistical models underlying traditional hypothesis tests, many researchers are moving away from reliance on statistical significance as a primary means of making biological inferences. Instead, more attention is being placed on describing and reporting biologically meaningful estimates of effect size; and many times, such measures of effect size are found in model parameters. Moreover, current modeling approaches allow for researchers to explicitly model data that do not adhere to the assumptions of classical methods, thus freeing the researcher from arbitrarily transforming and otherwise manipulating data to fit relatively rigid models. Mixed-effects models, in particular, allow researchers to model correlation among non-independent observations and to model heteroscedastic and unbalanced data. Although there are still some areas of active research with respect to how degrees of freedom can be determined from such models (especially important when trying to assess ANOVA-like results), mixed-effects models are an especially useful tool for wildlife toxicologists.

In the larger picture, novel philosophical approaches to the use of such models are beginning to emerge. The use of the information theoretic approach has been readily adopted in the wildlife literature and is gaining popularity in ecology. It also may prove useful to wildlife toxicologists, who frequently deal with observational data from sites that have experienced a toxic release. In addition, Bayesian approaches are offering scientists the ability to quantitatively update expectations and inferences as new data and information become available.

REFERENCES

Albers, P. H., M. T. Koterba, R. Rossmann, W. A. Link, J. B. French, R. S. Bennett, and W. C. Bauer. 2007. Effects of methylmercury on reproduction in American kestrels. *Environmental Toxicology & Chemistry* 26 (9): 1856–1866.

Anderson, D., and K. Burnham. 2002. Avoiding pitfalls when using information-theoretic methods. *Journal of Wildlife Management* 66 (3): 912–918.

Bates, D., M. Maechler, and B. Dai. 2008. lme4: Linear mixed-effects models using S4 classes. R package version 0.999375-28.

Berger, J. O. 2003. Could Fisher, Jeffreys and Neyman have agreed on testing? *Statistical Science* 18 (1): 1–12.

Bolker, Benjamin M. 2008. *Ecological Models and Data in R*. Princeton, NJ: Princeton University Press.

Box, G. E. P., and Norman Richard Draper. 1987. *Empirical Model-Building and Response Surfaces*. Wiley Series in Probability and Mathematical Statistics: Applied Probability and Statistics. New York: Wiley.

Boyles, J. G., D. P. Aubrey, B. S. Cooper, J. G. Cox, D. R. Coyle, R. J. Fisher, J. D. Hoffman, and J. J. Storm. 2008. Statistical confusion among graduate students: Sickness or symptom? *Journal of Wildlife Management* 72 (8): 1869–1871.

Burnham, K. P., and D. R. Anderson. 2002. *Model Selection and Multimodel Inference: A Practical Information-Theoretic Approach*. 2nd ed. New York: Springer.

Cedergreen, N., C. Ritz, and J. C. Streibig. 2005. Improved empirical models describing hormesis. *Environmental Toxicology & Chem*istry 24: 3166–3172.

Clark, J. S. 2005. Why environmental scientists are becoming Bayesians. *Ecology Letters* 8 (1): 2–14.

Clark, J. S. 2007. *Models for Ecological Data: An Introduction*. Princeton, NJ: Princeton University Press.

Collett, D. 1991. *Modelling Binary Data*. 1st ed. London and New York: Chapman & Hall.

Crane, M., and M. C. Newman. 1996. Scientific method in environmental toxicology. *Environmental Reviews* 4:112–122.

Dennis, B. 1996. Discussion: Should ecologists become Bayesians? *Ecological Applications* 6 (4): 1095–1103.

Ellison, A. M. 1996. An introduction to Bayesian inference for ecological research and environmental decision-making. *Ecological Applications* 6 (4): 1036–1046.

———. 2004. Bayesian inference in ecology. *Ecology Letters* 7 (6): 509–520.

Fisher, R. A. 1954. *Statistical Methods for Research Workers*. 12th ed. Biological Monographs and Manuals, No. 5. Edinburgh: Oliver and Boyd.

———. 1956. *Statistical Methods and Scientific Inference*. New York: Hafner Pub. Co.

Gauch, H. G. 2003. *Scientific Method in Practice*. New York: Cambridge University Press.

Gelman, A. 2004. *Bayesian Data Analysis*. 2nd ed. Texts in Statistical Science. Boca Raton, FL: Chapman & Hall/CRC.

Gotelli, N. J., and A. M. Ellison. 2004. *A Primer of Ecological Statistics*. Sunderland, MA: Sinauer Associates Publishers.

Guthery, F. S. 2008. Statistical ritual versus knowledge accrual in wildlife science. *Journal of Wildlife Management* 72 (8): 1872–1875.

Ioannidis, J. P. A. 2005. Why most published research findings are false. *PLoS Medicine* 2 (8): 696–701.

James, F. C., and C. E. McCulloch. 1985. Data analysis and the design of experiments in ornithology. In *Current Ornithology*, edited by R. F. Johnston. New York: Plenum Press.

Johnson, D. H. 1999. The insignificance of statistical significance testing. *Journal of Wildlife Management* 63 (3): 763–772.

Johnson, J. B., and K. S. Omland. 2004. Model selection in ecology and evolution. *Trends in Ecology and Evolution* 19 (2): 101–108.

Link, W. A., and P. H. Albers. 2007. Bayesian multimodel inference for dose-response studies. *Environmental Toxicology & Chemistry* 26 (9): 1867–1872.

Littell, R. C. 1996. *SAS System for Mixed Models*. Cary, NC: SAS Institute Inc.

Loehle, C. 1987. Hypothesis-testing in ecology: Psychological-aspects and the importance of theory maturation. *Quarterly Review of Biology* 62 (4): 397–409.

McCullagh, P., and J. A. Nelder. 1983. *Generalized Linear Models*. Monographs on Statistics and Applied Probability. London and New York: Chapman & Hall.

Newman, M. C. 1995. *Quantitative Methods in Aquatic Ecotoxicology*. Advances in Trace Substances Research. Boca Raton, FL: Lewis Publishers.

Newman, M. C. 2008. What exactly are you inferring? A closer look at hypothesis testing. *Environmental Toxicology & Chemistry* 27 (5): 1013–1019.

Newman, M. C., and W. H. Clements. 2008. *Ecotoxicology: A Comprehensive Treatment*. Boca Raton, FL: Taylor & Francis.

Neyman, J., and E. Pearson. 1928. On the use and interpretation of certain test criteria for purposes of statistical inference. *Biometrika* 20A:175–240.

———. 1933. On the problem of the most efficient tests of statistical hypotheses. *Philosophical Transactions of the Royal Society of London*. 231:289–337.

Pickett, S. T., J. Kolasa, and C. G. Jones. 1994. *Ecological Understanding*. San Diego: Academic Press.

Pinheiro, J. C., and D. M. Bates. 2000. *Mixed-Effects Models in S and S-PLUS*. Statistics and Computing. New York: Springer.

Platt, J. R. 1964. Strong inference. *Science* 146 (3642): 347–353.

Popper, K. R. 1972. *The Logic of Scientific Discovery*. 6th impression revised ed. London: Hutchinson.

Quinn, G. P., and M. J. Keough. 2002. *Experimental Design and Data Analysis for Biologists*. Cambridge (UK) and New York: Cambridge University Press.

Quinn, J. F., and A. E. Dunham. 1983. On hypothesis-testing in ecology and evolution. *American Naturalist* 122 (5): 602–617.

R. Development Core Team. 2005. *R: A Language and Environment for Statistical Computing*. R Foundation for Statistical Computing. Vienna, Austria.

Robinson, D. H., and H. Wainer. 2002. On the past and future of null hypothesis significance testing. *Journal of Wildlife Management* 66 (2): 263–271.

Salsburg, D. 2001. *The Lady Tasting Tea: How Statistics Revolutionized Science in the Twentieth Century*. New York: W.H. Freeman.

Sokal, R. R., and F. J. Rohlf. 1995. *Biometry: The Principles and Practice of Statistics in Biological Research*. 3rd ed. New York: W.H. Freeman.

Sterne, J. A. C., and G. D. Smith. 2001. Sifting the evidence: What's wrong with significance tests? *British Medical Journal* 322 (7280): 226.

Venables, W. N., and B. D. Ripley. 2002. *Modern Applied Statistics with S*. 4th ed. Statistics and Computing. New York: Springer.

8 Global Perspectives on Wildlife Toxicology

Emerging Issues

Philip N. Smith, Mohamad Afzal, Redha Al-Hasan, Henk Bouwman, Luisa E. Castillo, Michael H. Depledge, Muralidharan Subramanian, Venugopal Dhananjayan, Cristina Fossi, Malsha Kitulagodage, Henrik Kylin, Robin Law, Letizia Marsili, Todd O'Hara, Manuel Spinola, Paul Story, and Céline Godard-Codding

CONTENTS

8.1 INTRODUCTION

The broad aim of this chapter was to assimilate and document global perspectives on emerging threats to wildlife populations. Unfortunately, sufficient funding, employer understanding, and domiciliary freedom were not granted for requisite travel, research, and documentation. Thus, expertise in wildlife conservation, wildlife toxicology, and ecotoxicology was sought throughout the planet. The process of identifying and soliciting collaborative input from some of the world's experts was less daunting than expected, and speaks to the passion with which these esteemed scientists pursue their life's work, and even more so to their collegiality. Coauthors were selected based on their scientific record, expertise, experience, and, in most cases, where they choose to live and work. The chapter is thus organized geographically, with subsections describing major anthropogenic threats to wildlife populations over large geographic regions. We have attempted to include perspectives on the topic from all corners of the globe, and with few exceptions, have succeeded.

Clearly, each individual coauthor has his or her own personal thoughts and interests related to wildlife toxicology issues. It follows that their perspectives may or may not reflect those of their countrymen, colleagues, or residents of the regions they are addressing. Therefore each geographic subsection should be considered from the context in which it was written. Though each coauthor is more than qualified to address the topic, readers should feel free to disagree with the perspectives that follow. Readers may feel strongly that there are other, more pressing anthropogenic (or otherwise) issues facing world wildlife populations. Undoubtedly there are other issues, many of them, and many acting in concert to exert negative pressures on wildlife. The global challenges facing humankind, let alone wildlife, are immense, and could not possibly be distilled into a single book chapter. Yet here we attempt to initiate a global dialogue on wildlife toxicology—a first (however meager) stab at documenting the emerging anthropogenic stressors that may impact wildlife populations in the future.

As such, chapter contributors were asked to address major concepts arising within their own geographic region (though Robin Law graciously addressed Asia/China since, surprisingly, no wildlife toxicology expertise could be recruited from that region). They were asked to provide a brief description of the region along with any relevant history or governance. In each geographic subsection, we have attempted to include an overview of regional biodiversity and the state of wildlife populations within the geographic region. Next, we asked collaborators to include a description of the major anthropogenic stressors of the region (e.g., pesticides, industrial chemicals, and petroleum), and then to provide a detailed description of what each contributor considered to be the highest priority emerging contaminant-related threat or threats to wildlife populations within the particular geographic region. Potential trans-regional stressors or effects on trans-regional wildlife populations were fair game, as were perspectives on the needs for future research and conservation efforts. Following are geographically based perspectives on emerging issues that may impact future generations of wildlife and humankind.

8.2 AFRICA

Africa is the world's second largest continent (~30,000,000 square kilometers), after Asia, with associated islands in both the Atlantic and Indian oceans, notably Madagascar (UNEP 2008). Recognized as the cradle of humankind, Africa has a unique history of human predecesors evolving side by side with wildlife for more than 6 million years. Although a number of hominid genera are recognized, only *Homo* has extended its range outside Africa, this in a series of waves with consequences for wildlife in other regions probably more severe than in Africa, especially for the larger fauna (Wilson 1999). Africa lies across the equator, and has several major climatic zones: bracketed between Mediterranean-type climates are various savannah zones, themselves bracketing tropical rainforests—another unique aspect of this continent. It is home to more than a billion people, 40% concentrated along the coasts and some inland areas, depending on water and other resources, leaving large tracts of lands thinly populated (UNEP 2008). Water is unevenly distributed across Africa, a feature that is of cardinal importance when considering wildlife toxicology (Osibanjo et al. 2002).

8.2.1 Regional Biodiversity

Africa is well known for its megafauna; in many areas they are free roaming, and large migrations still take place even across national borders (UNEP 2008). The faunal diversity in Africa remains largely intact, although many species have been under threat of extinction for 50 years or more, and many regions have seen severe contractions and local extinctions. The larger primates are under severe threat, and only isolated and small populations remain. The islands, with their unique biota, have experienced extinctions over the last four centuries, and many remaining island species lead a precarious existence (UNEP 2008).

The various climatic zones in Africa house a number of large and small biomes, from deserts to rain forests. Islands such as Madagascar are home to unique plants and animals, as are some of the isolated large lakes and mountains on the continent (UNEP 2008). There are many features in Africa explaining historical and current wildlife distributions, but, on the whole, and especially in the rural areas, wildlife and humans mostly co-exist in a continual state of stress and often conflict, although there are exceptions. To a greater or lesser extent, rural populations depend on biota for food, fiber, and shelter, thereby bringing pollution and pollutant-affected biota in much closer contact with humans than in more developed regions.

8.2.2 Major Chemical Stressors

With the possible exception of Antarctica, it is probable that Africa is the least known of all continents from a chemical stressor point of view (Osibanjo et al. 2002; Torto et al. 2007). Although some excellent analytical infrastructures exist, these are few and small, and most are severely constrained by continuity, capacity, funding, and logistics. The welcome initiatives and activities of the Stockholm Convention have added impetus to the gathering of information on persistent organic pollutants, or POPs (de Boer et al. 2008).

A very important consideration when looking at Africa from a chemical pollutant perspective is that the use and release patterns of pollutants and how they interact with biota (and humans) are often different from other regions. Insecticides (such as dichlorodiphenyltrichloroethane (DDT), organophosphates (OPs) and pyrethroids) are used to combat pests such as malaria, quelea finches, locusts, and tsetse flies outside agricultural regions over vast tracts of natural (and sometimes residential) land, requiring consideration of alternatives in addition to classical spray-drift, leaching, and run-off approaches (Bouwman 2004; Bouwman et al. 2006; Daka et al. 2006; Keith et al. 1994; Mbongwe et al. 2003; Mullie et al. 1999; Mullie and Keith 1993; Osibanjo et al. 2002; Torto et al. 2007; Wiktelius and Edwards 1997). In addition, many countries have received "donations" of pesticides, often close or beyond expiry date, and sometimes even unmarked. These donations are now found in dumps in remote regions, causing environmental pollution (Elfvendahl et al. 2004; Kishimba et al. 2004). Fortunately, this issue is being addressed, but contaminated soils and groundwater will remain for a long time (Figure 8.1).

Existing and developing industries in some countries are often less constrained in operations because of old processes, lack of investment, or lack of oversight and enforcement for various (mostly less acceptable) reasons. Major population centers are often located close by, but some industries, such as mining (soda ash, diamonds, gold, etc.), are located remotely in otherwise natural regions. It can be assumed (but with exceptions) that the locations and operations of some of these have little or no concern for ecology, biota, and pollutant releases, somehow reminiscent of Eastern Europe a couple of decades ago, and even of industry and operations in the era before environmental concern became important in Western Europe and other developed regions. Acid mine drainage, mercury and uranium releases, and the use of cyanide, mercury, and other chemicals associated with mining and other industries are some of the major problems that remain to be dealt with.

The nature of pollutant stressors in Africa includes many of the "legacy" compounds (such as DDT through malaria control and dieldrin from dumps) and the

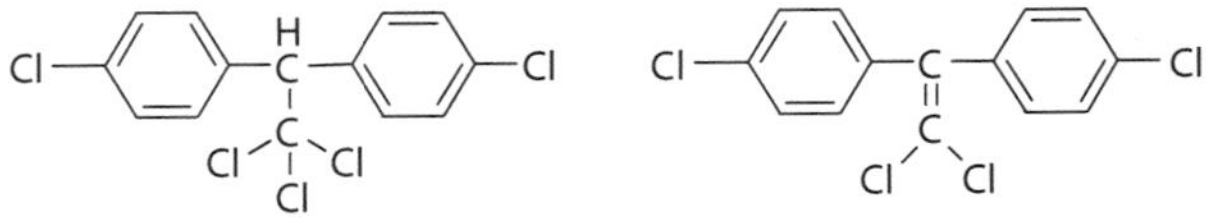

DDT

DDE

FIGURE 8.1 Chemical structures of dichlorodiphenyltrichloroethane (DDT) and dichlorodiphenyldichloroethylene (DDE).

newer emerging pollutants through imports, industry, and mining (Osibanjo et al. 2002). This mix of "old" and "new" pollutants is likely to remain a feature of the chemical residue landscape of Africa for the immediate future.

8.2.3 High Priority Emerging Contaminant Threats to Wildlife

Given the little available data, and taking into consideration that most releases occur in a few concentrated industrial regions, combined with generally much sunlight and high temperatures, it seems that pollution of biota, soils, and water in distant regions through long-range air transport results in detectable but generally lower levels when compared with more industrialized areas located in moderate to cold climates. This, however, remains to be confirmed. With an accelerating development of infrastructure, the types and amounts of pollutant released may well increase, as will the use of pesticides, possibly increasing the amount of air-transported pollutants.

It is in water, however, that serious warning signs are coming to the fore. Not only is there more demand for water and quality water, there is also a reduction in wetlands as a result of abstraction and demand for arable land, not only in the more arid regions, but also in the moister central, southern, and eastern regions. The remaining water has to supply the demands of humans and remaining biodiversity, causing conflict and often reduction in ecosystem function (Bouwman et al. 1990; Evans and Bouwman 2000; Everts 1997; Mbongwe et al. 2003; Ngabe and Bidleman 2006; vanderValk 1997). Overstressed aquatic systems visually signal their status through algal blooms and mass fish mortality, and recently also through a mass crocodile (*Crocodylus niloticus*) die-off in the well-known Kruger National Park in South Africa in 2008. This die-off of especially large crocodiles (~300) occurred in a remote part of the Olifants River, with crocodiles upstream seemingly healthy. A large array of chemical analysis has failed to identify an obvious single stressor, but heavy metals, POPs, and polybrominated diphenyl ethers (PBDEs) were detected in tissue and sediments. The current consensus is that the combination of a host of chemical stressors has reached a tipping point in the top-predator of that system, but many questions, including impact on human health, remain.

It is likely that most stress to wildlife caused by pollutants is currently and for the foreseeable future increasingly likely to manifest itself through aquatic systems. Dioxins, PBDEs, polychlorinated biphenyls (PCBs), pyrethroid insecticides, and many other pollutants from industrial and agricultural origin find their way into the remaining accessible water resources, bringing them into contact with wildlife and dependent communities. Pharmaceutical and household products will also find their way into water. Endocrine disruptive effects in fish, snails, and mammals have been detected in a nature reserve near Pretoria, South Africa (Barnhoorn et al. 2004; Bornman et al. 2007). The same dams also supply (treated) drinking water to the capital. Indications of eggshell thinning have also been seen in African darters (*Anhinga rufa*) at levels of detected pollutants an order of magnitude lower than expected, indicating either a sensitive biology or the presence of other unknown pollutants (Bouwman et al. 2008). Life history traits of animals (and possibly plants) in warmer climates differ from the more studied ones in temperate regions. Such traits might make organisms in warmer climates more susceptible to effects at ambient levels of pollution generally considered to be acceptable elsewhere (Bouwman et al. 2008). It is expected that more signs of pollutant stress on wildlife will become apparent when investigated, and more impacts will occur with an accelerated development and demand for consumer goods and services. Given the close association with and dependence of many people on their immediate environment, pollutant effects on wildlife may also translate into human health effects, if this has not already happened (Aneck-Hahn et al. 2007; Ntow 2001).

8.2.4 Trans-Regional Stressors and Effects

Since there is no known production of pesticides in Africa of note (UNEP 2008), all these products are shipped in from other continents, sometimes with less than adequate regard for regulations. Fortunately the dumping of expired products and wastes has diminished dramatically, but illegal shipments may still occur, and shipments into conflict territories with no government are likely to result in dumps with future consequences. Manufacturing of products that contain imported chemicals such as flame retardants, as well as the importation of manufactured products, will likely become (and may already be) a major pollution source once these products become waste. Scavenging of waste dumps by wildlife (but also harvesting by humans) will probably lead to elevated body burdens (Bouwman et al. 2008).

Little is known about long-range transport via air currents in Africa (Batterman et al. 2009; Karlsson et al. 2000), and this remains a research priority. Pollution of higher, cooler regions and mountains and subsequent catchment pollution are possible, and some indications have been found (Vosloo and Bouman 2005). Pollution plumes (N_2O) from coal-burning power stations in South Africa have been traced thousands of kilometers from their source (Wenig et al. 2003), and mercury and other pollutant transport is therefore likely.

The rivers and larger lakes of Africa might be the most important trans-regional transport conveyance for pollutants (Vosloo and Bouman 2005), but very little work has been done. Many of these rivers cross or form borders, are important transport routes, and are sources of water and food for many along the shores. These waters will also contribute toward coastal pollution of biota (Daby 2006; Marshall and Rajkumar 2003) and, consequently, of the many small and medium fisheries, as well as of marine mammals (Dekock et al. 1994; Vetter et al. 1999).

8.2.5 Research Needs and Conservation Issues

The data gap on the presence and levels of pollutants in many media in many regions of Africa needs to be addressed. The interaction of climate change, water scarcity, ecosystems, pests, disease vectors, industry, agriculture, urbanization, poverty, and human livelihoods with pollutants and their effects on wildlife and its conservation will provide major challenges to ecotoxicologists for decades to come.

8.3 MIDDLE EAST

The Middle East, an area of very diverse landscapes, is at the junction of trade routes connecting Europe and China, India, and Africa. Many of these routes have been documented from as early as 5,000 years ago. The Silk Route refers to historical trading routes stretching from China through Central Asia to the Middle East.

Middle East is a term derived from a European perspective. The 19th-century Europeans distinguished the Middle East from India and the Far East (Southeast Asia and China). Originally, the term *Near East* referred to areas under Ottoman control, from the Balkans to the border of Iran. The term *Middle East* was introduced in the early 20th century to include areas around the Persian Gulf, with *Near East* still being used to refer to the Ottoman Balkans. After World War II, *Middle East* became the dominant term for the whole region, minus the Balkans (Husain 1995).

Total population of the Middle East was approaching 200 million in 2008, which is about 2.9% of the world population. In a total area of 7.3 million square kilometers, there are only three major rivers in the Middle East; two of these rivers flow from Turkey into Iraq before entering the Arabian Gulf.

Farmers in North Africa and the Middle East plant more than 5 million hectares of barley each year for animal feed. The secret to barley's popularity among farmers lies in its adaptation to very harsh conditions and in its use as feed for sheep and goats, which are the main source of meat, milk, and milk products for rural populations. For yield improvement under their environmental and agronomic conditions, farmers make indiscriminate use of fertilizers, herbicides, and pesticides. These chemical stressors impart toxicity to wildlife and birds. Many desert reptiles, foxes, and birds are killed each year by widespread use of pesticides. In some cases paraquat, a deadly herbicide banned in the Western world, is in common use by farmers in this region.

The Middle East suffers a great deal of environmental pressure from urban development and ever-increasing fossil fuel demands of the energy-hungry world. Human activities in oil fields include exploration, development, production, and transportation. In the Middle East, wildlife has been lost as a result of development activities, land degradation (especially overgrazing and deforestation, leading to loss of plant cover), marine pollution, overfishing, hunting, and the overuse of freshwater, which affects plants and animals that rely on oases and wetlands. Because the region has such limited rainfall, the wetlands of the Middle East may be among the ecosystems most sensitive to changes in the amount and seasonality of rainfall and evaporation. Such changes may lead to local extinctions of some wildlife populations.

Middle East deserts are rich in many species. Approximately 15,500 flowering plants, 250 mammals, 950 birds, 350 reptiles and 15 amphibian species are endemic to these deserts. Wildlife indigenous to these deserts includes small herbivores, carnivores, rodents, birds, reptiles, arthropods, and gastropods. Populations are declining precipitously as a result of environmental pressures and human activities. Domestic animals such as camels, horses, donkeys, goats, and sheep also have an impact on wildlife.

The fragile coastal and marine ecosystems of the Middle East are impacted by the expansive petroleum industry. Many of the oil wells in the Middle East are located in the deep seas of the region, and oil spills are common. Kuwait alone produces over 2 million barrels of crude per day. Oil spills occurring in the Gulf usually come from deepwater oil wells, seepage from underwater, and from barge water. Thus, habitat degradation, oil spills, indiscriminate use of pesticides and herbicides, and outbreak of infectious diseases are important contributing factors to the ever-deteriorating status of wildlife in the region. Dead oiled birds, turtles, and fish are frequently seen floating in the Gulf waters. The Middle East is an important route and wintering area for numerous migrating birds, and these birds are very often killed by the oil slicks floating in Gulf waters.

Kuwait and many surrounding countries suffered an environmental catastrophe during the Gulf War when in February 1991, some 750 out of a total of 943 oil wells in eight oil fields were set ablaze by a retreating Iraqi army. The last oil well was capped in November 1991 (Tawfig 1991). Thus, in a span of 9 months, 4 to 6 million barrels of crude oil and 70 to 100 million square meters of natural gas were released each single day into the environment. Burning oil wells created a huge plume covering many countries in the Middle East, which spread to Asia and the Soviet Union, degrading air quality in the region (Draxler et al. 1994; El Desouky and Abdulraheem 1991). The plumes released sulfur dioxide, hydrogen sulfide, carbon monoxide, carbon dioxide, nitrogen oxides, and soot, carrying hydrocarbons and metals over a very wide area, an event unprecedented and unmatched in human history. The damaged oil wells also created over 300 oil lakes that contaminated 40 million tons of sand and earth. These oil lakes acted as traps, killing untold numbers of insects, wildlife, and birds. Carbon fallout was also devastating to human health and wildlife in the whole of the Middle East in general, and the state of Kuwait in particular. The state of the environment can be imagined from the fact that following the Gulf War, the sun was not visible, causing an unnatural night for many weeks. Two decades after the Gulf War, the Middle Eastern environment has still not recovered from hydrocarbon and metal toxicity. Wildlife that once was common to the region prior to the Gulf War has not returned to the region, either because some of the species became extinct and or because their habitat is completely destroyed.

Examples of species once so common that have been extirpated from the Middle East include the ostrich and two species of gazelles, the Arabian gazelle (*Gazella arabica*) and Queen Sheba's gazelle (*G. arabica bilkis*). In Yemen, all Arabian oryx (*Oryx leucoryx*), lions, leopards, and Nubian ibex (*Capra ibex nubiana*) have diminished to the point of near extinction. Leopards have disappeared from Jordan, Syria, Lebanon, and Egypt, with very few left in other countries of the Arab world, including Yemen, Oman, Saudi Arabia, and Palestine. Recently oryx have been reintroduced into Jordan's Wadi Rum. This decline of wildlife is due mainly to ever-increasing urbanization of the rural areas and over-hunting of ibex, hyrax, partridge, and other prey animals.

In conclusion, the major threats to wildlife populations in the Middle East are human activities associated with the oil boom and the Gulf War, which have resulted in an unprecedented environmental degradation in the region. Collective environmental remediation efforts from all governments of the region require immediate attention to repair and prevent further damage.

8.4 ASIA/CHINA

The People's Republic of China is the world's most populous country. With over a billion people at its disposal, China's economy, like those of other countries in Asia, has rapidly expanded over the past quarter century as a result of this extensive work

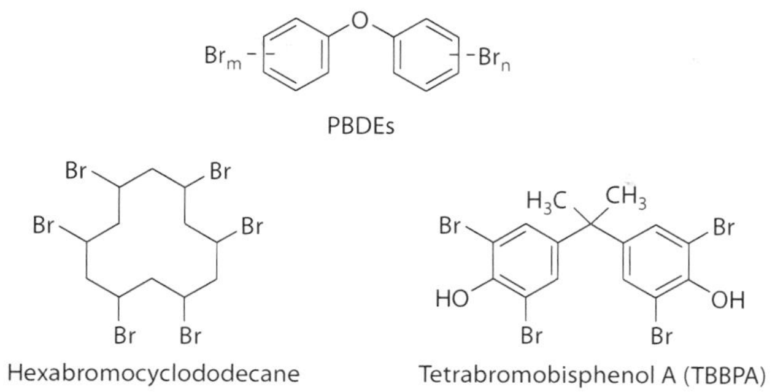

Br$_m$– –Br$_n$
PBDEs
Hexabromocyclododecane
Tetrabromobisphenol A (TBBPA)

FIGURE 8.2 Chemical structures of three brominated flame retardants (BFRs): polybrominated diphenyl ethers (PBDEs), hexabromocyclododecane (HBCD), and tetrabromobisphenol A (TBBPA).

force. An infinite array of consumer and electronic products are produced in this region and shipped to all parts of the globe. As a result, brominated flame retardants (BFRs; Figure 8.2) are of particular interest and concern in Asia. These compounds are used extensively here in the preparation of printed circuit boards for electronic equipment prior to their reactive incorporation in the products, and so the possibility of environmental release is significant (Law 2009). Growth in e-waste recycling in China has greatly increased the potential for release of PBDEs into the environment. To date, there have been fewer studies involving BFRs in Asia than in Europe and North America. Also, these have mostly been in Japan or mounted by Japanese institutes across a wider area. Now, though, studies from other Asian countries (such as China, Korea, and Taiwan) are beginning to emerge, and, while there is a long way to go, sufficient environmental data are becoming available for an assessment of the risks of the continuing use of BFRs, as has been the case in the European Union for the PBDEs, tetrabromobisphenol A (TBBPA), and hexabromocyclododecane (HBCD).

Environmental exposure and consequent contamination levels in the Asia-Pacific area have been reviewed by Tanabe et al. (2008), who also assessed spatial and temporal trends in concentrations of brominated diphenyl ethers (BDEs) and HBCD. BFRs have been found to be ubiquitous in the Asia-Pacific region. Examination of spatial trends showed that BFRs were relatively high in samples from Japan, Korea, and southern China. Evaluation of temporal trends in marine mammals (top predator species that can bioaccumulate lipophilic contaminants such as BFRs to high concentrations in blubber) showed drastic increases in concentrations of BFRs in the coastal waters of both China and Japan over the past 30 years. These changes were in line with trends in the production and use of the commercial formulations. In three sediment cores taken in Tokyo Bay in 2002, both BDEs and HBCD were detected (Minh et al. 2007). Concentrations of HBCD were lower than those of ΣBDEs and BDE-209 (representing the penta- and octa-mix, and deca-mix, of PBDE formulations, respectively). ΣBDE concentrations (penta- and octa-mixes) increased until the mid-1990s and decreased thereafter, while BDE-209 concentrations have continued to increase to the present day, reflecting usage patterns. HBCD first appeared in the mid-1970s, and concentrations have increased since then. In a similar study in the Pearl River delta in South China, Chen et al. (2007) reported that ΣBDE concentrations increased gradually from the mid-1970s to the early 1990s, following which trends varied by location, reflecting local inputs. Concentrations of BDE-209 remained constant until 1990, and then increased exponentially to the present in line with the increasing demand for the deca-mix PBDE formulation in China. Similar data had been reported earlier by Mai et al. (2005), who indicated that BDE-209 concentrations in areas affected by urban and industrial growth were toward the high end of those reported worldwide at that time.

In terms of wildlife exposure, a number of studies have addressed this aspect. The toxicology and likely effects are more difficult to elucidate. Generally, industry is one step ahead of the environmental chemists when introducing new products, and ecotoxicologists follow behind both. In the broadest terms, results from animal studies suggest reproductive and developmental effects, neurotoxicity, and endocrine disruption (Tanabe et al. 2008). Levels of both PBDEs and polybrominated biphenyls (PBBs) were determined in liver and eggs of cormorants (*Phalacrocorax carbo*) from Japan sampled in 2000. Both types of BFRs were detected in all samples, with ΣBDE concentrations from 0.3 to 2.6 μg/kg lipid weight in liver and from 0.6 to 3.3 μg/kg lipid weight in eggs. PBB concentrations were much lower, reflecting the lesser usage of these products, ranging from 0.003 to 0.3 μg/kg lipid weight in liver and from 0.003 to 0.8 μg/kg lipid weight in eggs. Lam et al. (2007) determined concentrations of BDEs, including BDE-209, in the eggs of waterbirds from South China. Eggs of little egrets (*Egretta garzetta*) and black-crowned night herons (*Nycticorax nycticorax*) were collected from Hong Kong, Xiamen, and Quanzhou in 2004. Eggs of cattle egrets (*Bubulcus ibis*) and Chinese pond herons (*Ardeola bacchus*) were also taken from Xiamen alone. Xiamen is an area of rapid and diverse industrial development, and Quanzhou is also an area under pressure from development of various sorts. Concentrations of ΣBDEs ranged from 30 to 1,000 μg/kg lipid weight, with marked concentrations of higher brominated congeners (BDE-183, BDE-196, BDE-197, BDE-206, BDE-207, and BDE-209) indicating the presence of the octa- and deca-mix PBDE formulations in these top predator species, as well as possible debromination of BDE-209 in the environment. In birds of prey from northern China—including the kestrel (*Falco tinnunculus*), sparrowhawk (*Accipiter nisus*), Japanese sparrowhawk (*Accipiter gularis*), scops owl (*Otus sunia*), little owl (*Athene noctua*), long-eared owl (*Asio otus*), common buzzard (*Buteo buteo*), and upland buzzard (*B. hemilasius*)—concentrations were high, particularly in kestrels, with the ΣBDE maxima at 31.7 mg/kg lipid weight in muscle and 40.9 mg/kg lipid weight in liver of one individual (Chen et al. 2007). Congener profiles were generally dominated by higher-brominated congeners (BDE-153, BDE-183, BDE-207, and BDE-209), again indicating presence of the octa- and deca-mix PBDE formulations in these terrestrial food chains, as well as possible debromination of BDE-209.

Raccoon dogs (*Nyctereutes procyonoides*) were collected in three areas of Japan from 2001 to 2006 (Kunisue et al. 2008). Relatively high concentrations of BDEs were observed in liver tissue; HBCD concentrations were much lower. BDE-209 was predominant in all animals, indicating that pollution from the deca-mix PBDE formulation is extensive across Japan.

Mussels are excellent sentinel organisms for pollution monitoring, as they are sedentary and effectively sample particulate-bound contaminants by filter-feeding, thereby reflecting local levels of contamination. Wang et al. (2009) determined BDE concentrations in mussels collected in 2006 from an inland sea in northeastern China. Median concentrations of ΣBDE (without BDE-209) and BDE-209 were 0.7 and 2.4 μg/kg dry weight, respectively. The deca-mix PBDE formulation was the primary source. In Korea, ΣBDE concentrations (including BDE-209) in bivalve molluscs (various species of mussels and oysters taken in 2004) ranged from 0.4 to 9.2 μg/kg dry weight (Moon et al. 2007). Concentrations of BDE-209 ranged from 0.15 to 8.6 μg/kg dry weight, hence it predominated in most samples (>60% of the total). In Hong Kong waters, high concentrations of PBDEs (on a worldwide comparative basis) were observed in green-lipped mussels (*Perna viridis*) affected by local sources of pollution from the Pearl River delta in southern China and local sources (Liu et al. 2005). In mussels, ΣBDE concentrations (including BDE-209) ranged from 27 to 84 μg/kg dry weight. In a wider study, mussels were taken for PBDE analysis from the coastal waters of Cambodia, China, Hong Kong, India, Indonesia, Japan, Korea, Malaysia, the Philippines, and Vietnam. PBDEs were detected in all the samples analyzed (Ramu et al. 2007) at ΣBDE concentrations (including BDE-209) ranging from 0.7 to 440 μg/kg lipid weight. This suggests that significant sources of PBDE contamination occur in industrialized areas across the Asia region.

PBDEs have been determined in fish from a number of Asian locations. In fish from six rivers and three estuaries in Taiwan, ΣBDE concentrations ranged from 2.9 to 1,240 μg/kg lipid weight (Peng et al. 2007). BDE-47 was the dominant congener in most samples; however, BDE-154 and BDE-183 dominated in some species. This may reflect the extensive use of the octa-mix PBDE formulation in Taiwan. Both PBDEs and HBCD were determined in freshwater fish from the Yangtze River in China taken in 2006 (Xian et al. 2008). Muscle concentrations ranged from 18 to 1,100 and 12 to 330 μg/kg lipid weight for ΣBDEs and HBCD, respectively. Relative to other areas of the world, PBDE concentrations were moderate and those of HBCD high. PBDEs were also detected in a species of catfish (*Pangasianodon hypophthalmus*) taken from the Mekong River delta in Vietnam in 2004, although levels were relatively low (Minh et al. 2006). Fish and shrimp of various species taken from the Pearl River estuary in South China in 2004 were analyzed for BDEs, yielding ΣBDE concentrations from 34 to 445 μg/kg lipid weight (Xiang et al. 2007). In some of the biota, as in sediments, BDE209 was the major congener. Ramu et al. (2006) studied BDEs in deep-sea fish from the Sulu Sea, located between the Philippines and Borneo. Despite the remoteness of this deep-sea area, ΣBDEs were detected at concentrations of 0.9 to 21 μg/kg lipid weight. Ueno et al. (2006) investigated the geographical distribution of HBCD across the Asia-Pacific region using skipjack tuna (*Katsuwonus pelamis*) as a bioindicator. HBCD was detected in all samples analyzed (even in the mid-Pacific region) except off the Seychelles. ΣHBCD concentrations ranged from <0.1 to 45 μg/kg lipid weight—rather lower than those reported for other studies in, for example, Spain and the North Sea (Law et al. 2008).

PBDEs have also been studied in a range of small cetacean species from Asian waters. In finless porpoises (*Neophocaena phocaenoides*) from the South China Sea, ΣBDE concentrations were up to 980 μg/kg lipid weight in 2001, and derived from the penta-mix PBDE formulation (Ramu, Kajiwara, Lam et al. 2006). BDE-209 was not detected. In a wider study, ΣBDE concentrations ranged from 6.0 μg/kg lipid weight in spinner dolphins (*Stenella longirostris*) from India to 6,000 μg/kg lipid weight in an Indo-Pacific humpback dolphin (*Sousa chinensis*) from Honk Kong (Kajiwara et al. 2006). The highest concentrations were found in animals from Hong Kong, followed by Japan, with much lower levels in animals from India and the Philippines. All these developing nations have PBDE sources, as reflected in the contamination observed in small cetaceans from these regions. An earlier study in Hong Kong waters had also shown elevated PBDE levels (Ramu et al. 2005). In finless porpoises and Indo-Pacific humpback dolphins, ΣBDE concentrations ranged from 9.9 to 470 μg/kg lipid weight. The authors suggested that their study indicated that PBDEs should be regarded as priority pollutants in Asia.

In summary, BFRs have been detected across the whole of the Asia-Pacific area in all environmental matrices studied. Unlike the situation within the European Union, no risk assessments of continued production and use of these compounds in this wide geographic area encompassing numerous countries have been conducted. For the purpose of wildlife protection, this is an emerging need.

8.5 AUSTRALIA

Australia, an island continent, is the sixth largest country in the world, covering a land area of approximately 7.7 million square kilometers, with 39% of its landmass in the tropical zone and 61% in the temperate zone. With a population of just over 20 million, Australia is one of the least densely populated countries in the world; however, over 70% of this population is concentrated along the eastern coast of the mainland (ABS 2008). Because of the diverse range of habitats, variable climate, and geographical isolation dating back almost 200 million years, Australia's biota is highly endemic and evolutionarily old, which has resulted in its listing as one of 17 mega-diverse countries in the world (Williams et al. 2001). The mainland of Australia is the driest in the world, and, although rainfall varies greatly, half of the continent receives an average annual rainfall of less than 300 millimeters (Pink 2008). As a result, arid and semi-arid regions occupy approximately 70% of the continental landmass, with a large proportion of these areas subjected to pressure from agricultural practices. This spread of agriculture has also unfortunately been accompanied by escalations in orthopteran populations, most notably locusts.

In Australia there are a number of locust species that can build to plague proportions, causing widespread crop damage. Of particular concern is the Australian plague locust (*Chortoicetes terminifera*), a species endemic to Australia (Szabo et al. 2003). In 1984 an Australian plague locust outbreak resulted in crop damage across two neighboring states (New South Wales and Victoria) estimated at over AU$ 5 million (Love and Riwoe 2005). Without control, potential losses would have been considerably higher (Abdalla 2007).

Two pesticides are currently used in Australia to control locust populations: fenitrothion, an OP, and fipronil, a phenyl pyrazole (Figure 8.3). These pesticides are used to combat locust bands and swarms through aerial application. Importantly, the climatic conditions that give rise to locust aggregations also stimulate the breeding

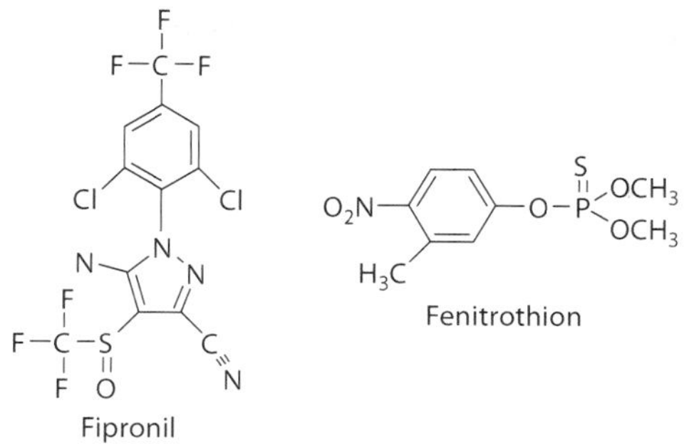

FIGURE 8.3 Chemical structures of fipronil and fenitrothion.

of many native vertebrates, including birds, mammals, and lizards. A risk assessment of locust control impacts on Australian arid-zone birds identified over 100 species that occur in areas common to locust-control spraying events (Szabo 2005), including the plains wanderer (*Pedionomus torquatus*), one of the rarest birds in Australia and currently listed as endangered in the state of New South Wales (Story et al. 2007).

Although fipronil is commonly used worldwide, there is still little information regarding its toxicological effects in vertebrates. Insectivorous birds are particularly susceptible, as fipronil-contaminated locusts can remain alive for up to 7 days (U.S. EPA 2001), making these and other insects a major route of pesticide exposure. Granivorous birds are also at high risk of exposure, as fipronil and its metabolites are detectable in seeds days after fipronil spraying events (JMPR 2001; Szabo 2005). Available avian toxicity information demonstrates a high species-specific variability in fipronil sensitivity (U.S. EPA 1996). There is no research, however, explaining this variability, nor is there understanding of the sub-lethal effects among avian species. This makes it extremely difficult to predict the toxicity of fipronil on unstudied species at high risk of exposure in the wild.

Recent research aimed at determining fipronil sensitivity in Australian native birds at risk of exposure to fipronil has demonstrated differences in the metabolism and bioaccumulation of fipronil between fipronil-sensitive and non-sensitive species (Kitulagodage unpublished data). Following exposure to fipronil, levels of the sulfone metabolite of fipronil (fipronil-sulfone) were progressively higher and persisted longer in tissues of the fipronil-sensitive species (bobwhite quail (*Colinus virginianus*)) and increased over time after exposure to fipronil, in contrast to lower peak values and rapidly decreasing fipronil-sulfone levels in tissues of less-sensitive species (zebra finch (*Taeniopygia guttata*) and house finch (*Carpodacus mexicanus*)). Considering that fipronil-sulfone is potentially more toxic than fipronil to vertebrates (Hainzl et al. 1998), these studies suggest that differences in the rate of conversion of fipronil to the sulfone may account for the varied sensitivity to fipronil among species.

Although adult zebra finches are moderately sensitive to fipronil (Kitulagodage et al. 2008), studies examining the impact of fipronil on embryonic and offspring development have demonstrated that breeding female zebra finches exposed to very low doses of fipronil transfer both fipronil and fipronil-sulfone residue into eggs, with subsequent hatchability rates greatly reduced. Behavioral and developmental abnormalities were also observed in chicks after in ovo exposure to fipronil, with fipronil and fipronil-sulfone residues being detectable in brain and liver of all chicks receiving in ovo application of fipronil (Kitulagodage, unpublished data). Fipronil and its metabolites have also reportedly been detected in milk of lactating cows and goats after exposure to fipronil (JMPR 2001), indicating that maternal transfer of fipronil and fipronil-sulfone to offspring is also likely to occur in lactating mammals.

Recent research into the sublethal effects of the ultra-low volume (ULV) fenitrothion formulation used for locust control in Australia on a common small marsupial, the fat-tailed dunnart (*Sminthopsis crassicaudata*), has revealed that severe locomotor impairments last up to 5 days after the ingestion of ecologically realistic doses (Buttemer et al. 2008). No effects were seen on metabolic rate during these exercise measures or on cold-induced peak metabolic rates. Subsequent studies (Story unpublished data) have repeated these experiments on a second dunnart species, *S. macroura*, and show that fenitrothion affects their circadian rhythm, resulting in phasic shifts in diel body temperature variation, with consequent mismatch between body temperature and activity/inactivity cycles. This is likely to reduce the metabolic performance of animals when they forage, defend a territory, engage in reproduction, and evade predators. Consequently, the pesticide effects identified in the lab could have significant effects on field populations of this and related species, and field verification studies are planned. This research will enable modeling to assist in the risk assessment of pesticide impacts on rare and threatened species.

Because ecotoxicological research pertaining to native Australian vertebrates is in its infancy, little is known about the effects of chemical insecticides on most Australian endemic vertebrates, and even less about how they might affect wild populations or ecological systems. Because of this limited understanding, registration of chemical pesticides in Australia currently relies on studies performed overseas on species that are distantly related to Australian native fauna and are ecologically distinct. Our studies on the extent of pesticide exposure in native species and its consequences on behavior and health have highlighted the shortcomings of current registration procedures. We are hopeful that as more research is directed at understanding the effects that pesticides have on native animals in an ecological context, the criteria for pesticide registration will become more aligned with protecting biodiversity.

8.6 INDIA

India is the seventh largest country in the world and its population ranks second in the world. It is one of the 12 mega-biodiversity countries of the world. Surrounded by Bangladesh, Burma, Bhutan, China, Nepal, and Pakistan, India covers an area of 3,287,590 square kilometers. The Tropic of Cancer, 23° 30′ N, divides India into two nearly equal halves. The land frontier of the country is 15,200 kilometers, and the total length of the coastline is over 7,516 kilometers. India hosts six major climatic subtypes, ranging from desert in the west, to alpine tundra and glaciers in the north, to humid tropical regions supporting rain forests in the southwest and island territories. Many regions have starkly different microclimates.

India's unique geography and geology strongly influence its climate; this is particularly true of the Himalayas in the north and the Thar Desert in the northwest. Seven major rivers along with their numerous tributaries make up the river system of India. Most of the rivers empty their waters into the Bay of Bengal. Some rivers, with courses taking them through the western part of the country and toward the east of the state of Himachal Pradesh, empty into the Arabian Sea.

India is a major producer and consumer of a wide range of agricultural, horticultural, fish, and other products. India is the world's second largest producer of food, after China, and has the potential of becoming the largest with the food and agricultural sector contributing nearly a quarter of India's Gross Domestic Product. Total food production in India is likely to double in the next 10 years. This also would mean that the agricultural input in terms of fertilizers and pesticides would likely increase.

8.6.1 Major Anthropogenic Stressors

Following the success of DDT in World War II, several organophosphate insecticides (OPs) were discovered for the control of vector-born diseases of crops, and the use of pesticides in agriculture commenced in India around 1948–49 (Gupta 1986). It increased manifold with the green revolution. Today India is the second largest manufacturer of pesticides in Asia and third largest consumer in the world. Usage of pesticides in India increased from 2350 metric tons in 1955–1956 to around 100 thousand metric tons in 2003 (Singhal 1999). According to Abhilash and Singh (2009), the present annual consumption is 41,020 metric tons. About 16% of the pest control practices rely on OPs, of which 76% are insecticides, compared to 44% globally (Mathur et al. 1992; Saiyed et al. 1999). The use pattern clearly shows that pesticides have, good or bad, become a major tool in today's agriculture, although alternatives to pesticides are being considered.

Although India is in the process of phasing out persistent organochlorine pesticides (OCs), many are still being used in substantial quantity either for agriculture or civic health. For example, DDT, which has been forbidden for agriculture (Ministry of Agriculture, Government of India's notification, dated 26 May 1989), is still being used in public health programs, especially in malaria eradication. India has sought exemption under the Stockholm Convention for use of 10,000 tons of DDT for restricted use in the public health sector (Lallas 2001). Hence, DDT will continue to pose problems in the environment. At present, usage of chemical pesticides is the highest in Andhra Pradesh (33%), followed by Punjab (14%), Karnataka (11%), Tamil Nadu (9%), Maharashtra (7%), Haryana (6%), Gujarat (5%), Uttar Pradesh (5%), and the remaining states 9.5% (Singhal 1999). Nearly 70% of pesticides used in India are for cotton (45%) and rice (25%), and this proportion of pesticide has remained almost unchanged over the last five decades.

Levels of PCBs in the environment merit attention in view of the rapid growth of industrial activities in India. Studies have reported varying levels of polychlorinated dibenzo-p-dioxin, dibenzofurans, and PCBs in human tissue, meat, fish, and wildlife samples (Senthilkumar et al. 1999), human adipose tissue (Rao and Banerji 1988), breast milk (Tanabe et al. 1990), and foodstuffs (Kannan et al. 1992) from India. Residues of PCBs were 120–1000 ng/g in strict resident birds, 190–890 ng/g in local migrants, 80–1900 ng/g in short-distance migrants, and 100–2000 ng/g in long-distance migrants.

Although there is some information available on the impact of pesticides on the environment, very little effort has gone into documenting the ill effects of PCBs and polycyclic aromatic hydrocarbons (PAHs) in India on a national scale. Efforts have been made to compile available information on levels of environmental contamination, pesticides, PCBs, and PAHs in wildlife, particularly among birds in India. Information presented is based on the published literature and data generated by the authors.

8.6.2 Highest Priority Emerging Contaminant-Related Threat(s) to Birds

During the last few decades, contamination by POPs such as PCBs, pesticides, and other organic pollutants has spread, as evidenced by their detection in a wide range of environmental media and biota (Tanabe 2001). Kaphalia et al. (1981) reported high levels of DDT in depot fat of crows, kites, and vultures collected from Lucknow. Misra (1994) studied the levels of organochlorine in a few biological and non-biological compounds, including six species of birds in the Mehala water reservoir in Rajasthan.

The residue levels of persistent OCs, such as hexachlorocyclohexane (HCH) isomers, DDT compounds, PCBs, and hexachlorobenzene (HCB) (Figure 8.4), were measured in wildlife. On the basis of overall concentrations, HCH ranked first, followed by DDT, PCBs, and HCB, reflecting the increasing usage of HCH in recent years in India. The residue levels of organochlorines in birds varied according to their feeding habits and showed the following pattern: inland piscivores and scavengers > coastal piscivores > insectivores > omnivores > granivores. High levels of HCH and DDT residues were recorded in pond herons (*Ardeola grayii*) and cattle egrets, which feed in the agricultural fields (Ramesh et al. 1992).

During the last two decades, impacts of pesticides on many species of birds have been abundantly obvious. Between 1987 and 1990, 18 sarus cranes (*Grus antigone*) and more than 50 collared doves (*Streptopelia decaocto*) fell victim to the pesticide aldrin in Keoladeo National Park, Bharatpur (Muralidharan 1993). Unsuccessful breeding of Himalayan gray-headed fish eagles (*Ichthyophaga ichthyaetus*) at Corbett National Park in Utter Pradesh was reported by Naoroji (1997). Higher concentrations of organochlorine in resident and migratory birds from south India were documented by Tanabe et al. (1998).

Senthilkumar et al. (1999) reported congener specific patterns of accumulation of PCBs in resident and wintering migrant birds collected from south India.

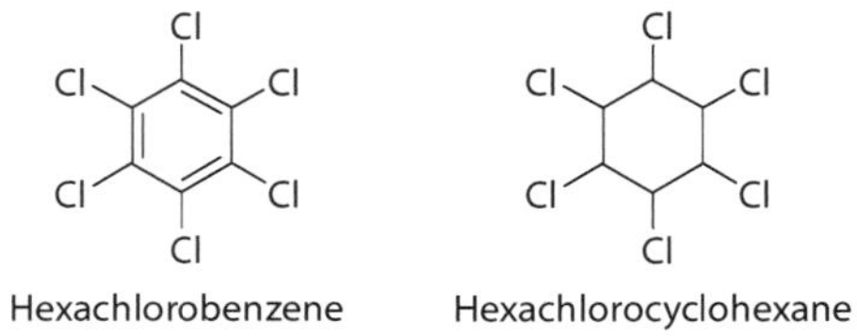

Cl Cl Cl Cl Cl Cl
Hexachlorobenzene
Cl Cl Cl Cl Cl Cl
Hexachlorocyclohexane

FIGURE 8.4 Chemical structures of hexachlorobenzene (HCB) and hexachlorocyclohexane (HCH).

Residue concentrations of PCBs were 120–1000 ng/g in resident birds, 190–890 ng/g in local migrants, 80–1900 ng/g in short-distance migrants, and 100–2000 ng/g in long-distance migrants.

Muralidharan (2002) reviewed the use of pesticides in India and subsequent impacts on birds. Reports on the impact of agricultural chemicals on agro-ecosystems and avifauna are also available (Muralidharan et al. 2004). Even after the ban on aldrin, the death of 15 sarus cranes due to monocrotophos poisoning was reported (Pain et al. 2004). There is also information available on residue levels of OCs in eggs of a few species of birds (Ranapratap 2004; Regupathy and Kuttalam 1990).

Mass mortality of peafowl (*Pavo cristatus*) due to endosulfan in Madhya Pradesh and some parts of western India during 1999 was reported by Muralidharan (2002). Although peafowl are also killed for illegal trade in feathers in many parts of the country, impact of pesticides is quite evident. From the print media (2004), it is understood that approximately 15 peafowl died from consuming pesticide-treated seeds in Gandhi Nagar, Gujarat, during 2004. During the last 3 years we have records of more than 32 peafowl deaths in the country. For many reasons, pesticide poisoning has become a cause for concern.

Between 1999 and 2009 around 1100 individuals from 105 species of birds have been received dead from various places in the country at Salim Ali Centre for Ornithology and Natural History. Dead birds were categorized into five groups based on their feeding habits: insectivorous (36.8%), omnivorous (31.6%), piscivorous (12.3%), carnivorous (8.8%), and herbivorous (10.5%). All species of birds studied had varying levels of many contaminants, namely DDT and its metabolites, HCH and its isomers, endosulfan and metabolites, heptachlor epoxide, and dieldrin (Muralidharan and Dhananjayan, unpublished data). Not all birds received had died from pesticide poisoning; yet except for a few of them, the presence of residues clearly indicate that they were exposed to pesticides. Continuous exposure to some of these chemicals result in toxicity and become manifest at the population level. Although available information on DDT does not indicate major threat to birds and other wildlife, DDT's presence at varying levels in the tissues and eggs of many species of birds in India (Kaphalia et al. 1981; Muralidharan 1993; Muralidharan et al. 2008; Muralidharan and Murugavel 2001; Ranapratap 2004) clearly indicates that the chemical is still available in the environment. This is because of the chemical's persistence and continued usage for disease control. Unless there is a complete ban on its usage, the environmental residue levels cannot be eliminated. This is applicable not only to DDT, but also to other persistent organic pollutants.

The frequency of occurrence of HCH, heptachlor epoxide, DDT, endosulfan, and dieldrin in the tissues of various species birds studied was 100, 98, 92, 85, and 45%, respectively. High levels of dieldrin in shikra and DDE in crows were indicative of poisoning. Other species of birds that we studied had varying levels of pesticide residues. Residues of chlorpyrifos, the most popular OP, have also been detected in the stomach contents of Asian koels (*Eudynamys scolopaceus*) collected from the agriculture field in Coimbatore. Reduced brain acetyl cholinesterase (AChE) activity supported by the presence of residues of methyl parathion confirmed that methyl parathion was responsible for the death of Indian bustards (*Ardeotis nigriceps*) received from Pune. Acetyl cholinesterase levels in plasma and brain tissues of 11 species of birds were documented (Dhananjayan 2009). Since the birds were victims of kite flying, the documented levels could serve as reference values for further research.

Nearly 50 individuals from the 55 species of birds collected between 2005 and 2007 in Ahmedabad and Coimbatore, India, (Dhananjayan 2009) exhibited widespread occurrence of OCs, PCBs, and PAHs in various tissues. About 12% of birds studied exceeded levels of OCs, PCBs, and PAHs reported to cause risk to birds or to wildlife. Overall, the OCs, PCBs, and PAHs ranged from below detectable limits to levels that could pose threat to birds or to animals that consume them. The highest concentrations of most of the contaminants were found in those species that are relatively higher in the food chain, such as vultures, kites, and crows. Birds received from industrial areas had higher levels of contaminants than birds of other areas. Additionally, among the majority of the birds included in the study, the pariah kite *(Milvus migrans)* and the blue rock pigeon (*Columba livia*) appeared to be the ideal species to be biosentinals, or indicator species, for continuous monitoring of residue levels in the environment (Dhananjayan 2009).

The abundance of white-backed vultures (*Gyps bengalensis*) declined up to 92% across their entire known distribution ranges in India (Pain et al. 2003). While Cunningham et al. (2003) reported that the most likely cause of vulture decline was a novel infectious disease, Pain et al. (2003) felt that there could be more than one factor responsible for such a mysterious situation, and that the role played by contaminants cannot be entirely discounted. Oaks et al. (2004) demonstrated that diclofenac, an anti-inflammatory drug used for treating cattle, was responsible for vulture mortalities. Although there is now a ban on the use of diclofenac for treating cattle in India, because diclofenac residues have been reported in vultures, including those that fell victim to kite flying in Ahmedabad, Gujarat, it is essential to monitor the residue levels in the environment so as to assess the effectiveness of the ban on diclofenac, particularly for veterinary use.

Birds and humans have a virtual commensal relationship, and therefore environmental factors that affect the welfare of natural bird populations could as well affect the welfare of people. Based on the information that we have, it can be inferred that pesticides have been creating problems for many species of birds in India. While we have proof beyond doubt for a few species of birds, there are indications in a few other cases. If the present trend continues unchecked, birds will certainly be one of the non-target organisms worst affected by pesticides. It is perhaps needless to say that agriculture occupies a place of prime importance in the economy of the country. However, large-scale and indiscriminate use of chemical pesticides in agriculture has lead to many problems, such as health hazards, pollution of environment, adverse effects on non-target organisms, insecticide resistance in target and non-target species, and unsustainable farming systems. This section largely focused at the contamination level only of persistent OCs in birds. However, there are a number of OPs and carbamate pesticides being used in considerable quantity in India. Although these chemicals degrade more quickly, they are also capable of creating ecological imbalance at various levels. Hence it is imperative to implement stricter environmental controls in the country in order to minimize potential risk to wildlife and humans.

8.7 WESTERN AND EASTERN EUROPE

Europe, one of the world's seven continents, represents the northwestern constituent of the larger landmass known as Eurasia. Europe covers about 10,180,000 square kilometers and is inhabited by some 731,000,000 people, with a population density of about 70/km^2. Europe's eastern frontier is now commonly delineated by the Ural Mountains in Russia. Europe is washed by the Arctic Ocean in the north, by the Atlantic Ocean in the west, by the Mediterranean Sea in the south, and by the Black Sea in the southeast. Politically, Europe comprises the member states of the European Union (EU), the few associated but non-member states (Norway, Iceland, and Switzerland), the European parts of the former USSR, and countries of the Balkan Peninsula, which covers a large part of the Eastern basin of the Mediterranean, and includes a part of Turkey: 50 countries in all. Often the word *Europe* is used, incorrectly, to refer only to the European Union, comprising 27 member states as of 2009. Based on the environment, culture, and economic similarities, Europe is divided into four regions: Eastern Europe, Western Europe, Northern Europe, and Southern Europe (Blij and Muller 2007). Eastern Europe (295 million people) is a term that applies to the geopolitical region encompassing the easternmost part of the European continent and is territorially the largest region. The southern part of Eastern Europe is referred to as the Balkans, or the Balkan Peninsula, named after a mountain range in Bulgaria. Western Europe refers to the countries in the westernmost half of Europe; it is the industrial heartland of Europe and the core of its economic power, with a total population of about 187 million. These definitions, however, separate these two areas mostly because of their political, religious, economic, cultural, and historical differences. Since the end of World War II, the term *Western Europe* has been commonly used to describe those countries allied with the United States through NATO.

From an ecological biogeography point of view, Europe, together with Northern Asia, is included in the Palearctic ecozone (Higgins et al. 2002), and it is divided into 11 biogeographical regions (terrestrial and freshwater) and seven seas. Biogeography strongly determines the community ecology of an area. In Western and Eastern Europe we can distinguish the Boreal, Continental, Atlantic, Mediterranean, Alpine, Pannonian, Steppic, and Black Sea regions.

8.7.1 Major Anthropogenic Stressors for Wildlife

Environmental pressure on biodiversity in Europe is intense. Over the past 100 years this region has been affected by numerous types of human interventions that, combined with natural climatic factors, have been having a negative impact on ecosystem quality, ecological status, and, consequently, on the health and the biodiversity of wildlife. These pressures and their impacts on wildlife biodiversity change between biogeographical regions, but, as a general rule, the following have been important:

- Destruction of habitats and the increasing uniformity of rural areas and physical disturbance of seas
- Habitat quality, in particular regarding the pollution of both air and water, and the fragmentation of areas

- Proximity and influence of human infrastructures on protected areas.
- Continuous changes in habitats, which generally become less rich in species and more uniform. Wetland decline appears to have halted in many countries, but permanent grasslands and old forests continue to decline.
- Competition with alien species (introduced through ballast water, aquaculture, active species introduction, and spread from crops, forestry, horticulture, fishing, hunting, or nature restoration).
- Continuous eutrophication of land and of fresh and salt waters.
- Climate changes that can be expected to induce changes to all levels of biodiversity of natural and cultivated species. The changes will affect all European habitat types and their species, in protected areas as well as outside of them.
- Contamination by "old" and emerging chemicals.

Moreover, Europe's wildlife is directly linked to the conditions of Europe's neighbor continents and seas through the exchange of migratory birds and fish, and other species groups such as marine mammals (seals, dolphins, and whales) and insects such as butterflies.

As an example, we can present the situation of the biodiversity status of the Continental biogeographical region that is the second largest in Europe, extending in a central east-west band over most of Europe. In this region occur at least 578 vertebrate species, not including fish; few are endemic to the region, and 124 are threatened species at the European level (Annex II of the Habitats Directive). The principal problems of this region are the infrastructures, dense in the western part, but also increasing in the east; changes in land use with a growing fragmentation of habitats and the creation of barriers, which isolate species populations; and loss of natural old forests and, in many areas, a poor condition of the forests affected by long-range air pollution.

In all European biogeographical regions, contaminants represent one of the many anthropogenic stressors. Contamination from and potential effects of the following compounds represent the principal causes of pollution in this area:

- Persistent organic pollutants (Figure 8.5)
    - Aldrin, chlordane, DDT, dieldrin, endrin, heptachlor, HCB, mirex, toxaphene, PCBs, polychlorinated dibenzo-p-dioxins (PCDDs), polychlorinated dibenzofurans (PCDFs)
- Polycyclic aromatic hydrocarbons
- Trace elements

Lethal, sublethal, or other detrimental effects of these contaminants on European wildlife are reported, in particular regarding the potential of several environmental chemicals to cause endocrine disruption at environmentally realistic exposure levels. In wildlife populations, associations have been reported between reproductive and developmental effects and endocrine-disrupting chemicals (EDCs). In the European environment, effects have been observed in mammals, birds, reptiles, fish, and mollusks (Vos et al. 2000). Observed abnormalities vary from subtle changes

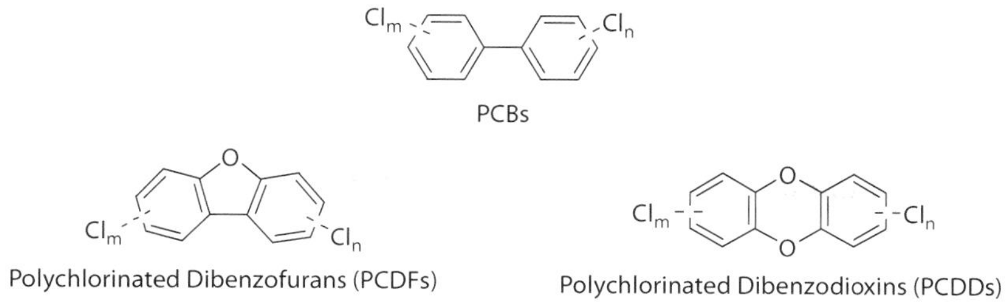

PCBs

Polychlorinated Dibenzofurans (PCDFs)

Polychlorinated Dibenzodioxins (PCDDs)

FIGURE 8.5 Chemical structures of polychlorinated biphenyls (PCBs), polychlorinated dibenzofurans (PCDFs), and polychlorinated dibenzodioxins (PCDDs).

to permanent alterations, including altered sex differentiation with feminized or masculinized sex organs, changed sexual behavior, and altered immune function. Intersex has been reported to varying degrees, in up to 100% of freshwater roach (*Rutilis rutilus*) at some locations in U.K. rivers (Jobling and Tyler 2003). Freshwater fish species in which abnormal intersex has been reported include the roach (*R. rutilus*), bream (*Abramis abramis*), chub (*Leuciscus cephalus*), gudgeon (*Gobio gobio*), barbel (*Barbus plebejus*), perch (*Perca fluviatilis*), stickleback (*Gasterosteus aculeatus*), and shovel-nosed sturgeon (*Scaphirhynchus platyorynchus*) (Jobling and Tyler 2003). However, estrogenic effects have been demonstrated not only in fish of freshwater systems, but also in estuaries and in coastal areas (Vos et al. 2000). Eggshell thinning, a notorious pollution-related effect on bird reproduction, is attributed to DDE exposure (Giesy et al. 2003). This has caused severe population declines in a number of raptor species in all Europe, but pollutant-related effects also include abnormal VTG production in male birds, deformities of the reproductive tract, embryonic mortality, reduced reproductive success, including egg-shell thinning, and poor parenting behavior. In 2007, Jiménez et al. (2007) found not only the presence, but also a rise, of the VTG levels in male peregrine falcons (*Falco peregrinus*), suggesting a potential ongoing threat to birds of prey. Mammal population declines have been correlated with OC pollution (Bernanke and Kohler 2009). Moreover, numerous studies have attributed reproductive and non-reproductive dysfunctions in mammals to EDC exposure. In rodents, for example, significantly reduced testes weights have been reported in male white-footed mice (*Peromyscus leucopus*) inhabiting PCB- and cadmium-contaminated land (Batty et al. 1990). The Eurasian otter (*Lutra lutra*) completely disappeared several decades ago in some U.K. and continental European rivers as a result of contaminant-induced reproductive problems, in particular PCBs (Mason and Macdonald 2004). But the best evidence in mammals comes from the field studies on Baltic gray (*Halichoerus grypus*) and ringed seals (*Pusa hispida*), and from the Dutch semi-field studies on harbor seals (*Phoca vitulina*), where both reproduction and immune function have been impaired by PCBs in the food chain (Reijnders 1986, 1990). Reproduction effects resulted in population declines, whereas impaired immune function has likely contributed to the mass mortalities due to *Morbillivirus* infections (Damstra et al. 2002; Härkönen et al. 2006), which occur in European cetacean populations as well (Di Guardo et al. 2005; Fossi and Marsili 2003).

8.7.2 Emerging Contaminant-Related Threats to Wildlife Populations

The list of chemical, physical, and biological stressors considered to be potentially dangerous to the environment has rapidly been growing, and the significant advancements in analytical technology have enabled chemists to identify an increasing number of emerging contaminants in the environment. Emerging contaminants include those substances in active or past production—including their metabolites and other transformation products and chemical by-products generated during production—that are presently unregulated or are not adequately regulated. Contaminants that have recently been discovered in the European environment include substances such as pharmaceuticals and personal care products (PPCPs), EDCs, nanoparticles, BFRs, and metabolites of pesticides and other POPs. One of the principal goals of scientists, managers, and policy makers is to identify and to assess public-health and environmental problems associated with these contaminants. These compounds, for example, have been listed as priority hazardous substances and are controlled under the EC Water Framework Directive 2006/60/EC. Importantly, wild populations are simultaneously exposed to several emerging contaminants, and, when evaluating the effects, we must consider both the synergism and the antagonism among these multiple stressors.

What follows is a detailed description of the highest priority emerging contaminant-related threat or threats to wildlife populations in Western and Eastern Europe. The Global Organic Contaminants include

- Flame retardants and their impurities that contain PBDEs, PBBs, polybrominated dibenzo-p-dioxins (PBDDs), polybrominated dibenzofurans (PBDFs), and HBCDs
- Perfluoronated compounds (PFCs; Figure 8.6) that contain perfluorooctane sulfonates (PFOS) and perfluoroctanoic acid (PFOA)

Occurrence of BFRs in biota was first reported by Anderson and Blomkvist (1981) in samples collected along the Visken River in Sweden. From that moment, their occurrence in biota has been demonstrated globally, even in remote habitats such as the Arctic. PBDEs have been measured in more than 50 species at different trophic levels in Asia, Europe, North America, and the Arctic. In most of these the lower-brominated congeners (BDE-47 and/or BDE-99) are dominant, although BDE-209 is also often observed (de Wit et al. 2006). Recently, more analysis has been done in the aquatic ecosystem, and levels were generally higher in marine organisms, although terrestrial food chains tended to accumulate more of the higher

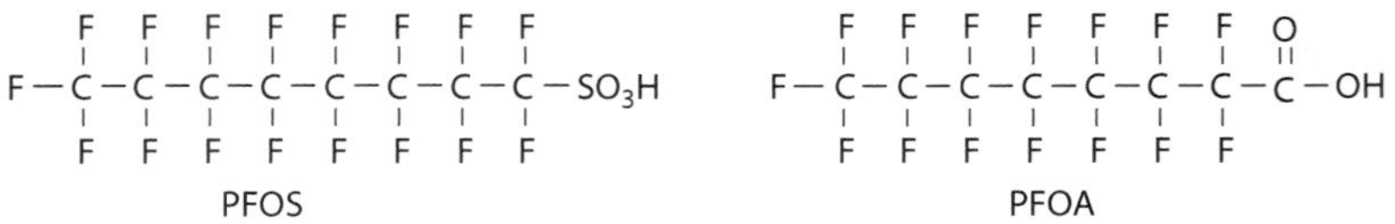

$$F-(CF_2)_8-SO_3H$$

PFOS

$$F-(CF_2)_7-C(=O)-OH$$

PFOA

FIGURE 8.6 Chemical structures of two perfluorinated compounds (PFCs): perfluorooctane sulfonates (PFOS) and perfluoroctanoic acid (PFOA).

brominated congeners (Christensen et al. 2005; Jaspers et al. 2006; Voorspoels et al. 2007). Bioaccumulation and biomagnification are of interest not only because of their importance in determining the fate of PBDEs in the environment, but also because correlations have been found between PBDE burdens in organisms and various indices of health or fitness, including immunosuppression and reproduction.

In fact, most toxicological studies have focused on the effects of PBDE exposure on the components of the endocrine system, a system that plays a critical role in the regulation of bodily processes, development of the brain and nervous tissues, sexual maturation and function, metabolism, and overall homeostasis. Many synthetic chemicals have the ability to mimic, block, or enhance the response to natural hormones. Such chemicals are generally labeled endocrine disruptors. All PBDE technical products, including the deca-BDE mixture, have been shown to have thyroid-disrupting properties (Birnbaum and Staskal 2004; Darnerud et al. 2001; Zhou et al. 2001). Competitive binding of PBDEs to thyroid hormone receptors is likely due to structural similarities between flame retardants and endogenous hormones. PBDEs can further interfere by altering liver function, leading to changes in thyroid hormones and vitamin A homeostasis, often resulting in over-elimination of the thyroid hormone T4 (Ellis-Hutchings et al. 2006; Zhou et al. 2001). Several studies (D'Silva et al. 2004; Herzke et al. 2005; Jaspers et al. 2006; Law et al. 2006; Lindberg et al. 2004; Sellstrom et al. 2003; Sinkkonen et al. 2004) indicate that penta-BDE is widespread in top predatory birds in Europe, such as the peregrine falcon, merlin (*Falco columbarius*), goshawk (*Accipiter gentiles*), golden eagle (*Aquila chrysaetos*), and buzzard. High levels were also detected in the eggs of top predatory birds such as the white-tailed sea eagle (*Haliaeetus albicilla*), peregrine falcon, osprey (*Pandion haliaetus*), and golden eagle (Herzke et al. 2005; Lindberg et al. 2004). High levels were also detected in the European harbor porpoise (*Phocoena phocoena*) (Covaci et al. 2002; Thron et al. 2004).

Perfluorinated compounds (PFCs), used as surface treatment chemicals, polymerization aids, and surfactants, are emerging environmental pollutants. They are ubiquitous and persistent, and they bioaccumulate in the environment. PFCs are being proposed as a new class of POPs, but their mechanism of toxicity is practically unknown. Preferentially, they are retained in the plasma and in the liver, and bind to proteins. In Europe, levels of PFCs were evaluated in Mediterranean fish, mammals, and birds (including eagles), and mammals of the Baltic Sea (Giesy and Kannan 2001; Kannan et al. 2002; Kannan, Franson et al. 2001; Kannan, Koistinen et al. 2001). Although tests have demonstrated numerous negative effects of PFCs in rodents, it is unclear whether exposure to PFCs may affect human health (Kovarova and Svobodova 2008).

8.7.2.1 Pharmaceuticals and Personal Care Products (PPCPs)

PPCPs include all prescriptions and over-the-counter drugs, diagnostic agents, dietary supplements, fragrances, soaps, conditioners, sunscreens, and cosmetics. This is the most diverse "category" of emerging substances of concern. Many are water soluble, and, in fact, the most common route into the environment is through wastewaters (municipal and septic drainage), application on land of sewage sludge and manure, and landfill leaking. The most common route into wildlife is through ingestion. Fent et al. (2006) critically reviewed the literature regarding environmental occurrence and fate of human pharmaceuticals in the aquatic environment to discuss potential mechanisms of action (based on present knowledge from mammalian studies) and to describe the acute and chronic ecotoxicological effects on aquatic organisms. Limited data are available on accumulation of PPCPs in organisms. In 2006, the Office of Science and Technology within the EPA's Office of Water initiated the EPA Pilot Study of PPCPs to investigate PPCP occurrence in fish tissues (Ramirez et al. 2009). Pharmaceutical compounds that occurred most frequently in any of the fillet or liver samples were diphenylhydramine, norfluoxetine (a fluoxetine metabolite), and sertraline.

8.7.2.2 Nanoparticles

Nanoparticles include fullerenes (a.k.a. buckyballs), nanotubes, quantum dots, nanopowders (metal oxides), and natural particles (e.g., soot). Nanoparticles are natural and synthetic structures in the 1-to-100-nanometer range and are used in nanotherapeutic pharmaceuticals, drug delivery, cosmetics, energy storage products, fabrics, lubricants, and even golf balls. Their environmental impact is largely unknown. Data on biological effects show that nanoparticles can be toxic to bacteria, algae, invertebrates, and fishes, as well as to mammals. However, much of the ecotoxicological data are limited to species used in regulatory testing and to freshwater organisms. Data on bacteria, terrestrial species, marine species, and higher plants are particularly lacking. Detailed investigations of absorption, distribution, metabolism, and excretion remain to be performed on all species but fish (Handy et al. 2008). Scown et al. (2009), in order to investigate tissue distribution and toxicity of titanium dioxide (TiO_2) nanoparticles in fish, have conducted a series of injection studies in rainbow trout (*Oncorhynchus mykiss*) and have concluded that upon a single high-dose exposure of TiO_2 nanoparticles via the bloodstream, TiO_2 accumulates in the kidneys but has only minimal effects on kidney function.

8.8 NORTH AMERICA

North America as a geospatial descriptor has various meanings depending upon context. For purposes of this discussion, North America will encompass only Canada and the United States, excluding Greenland, Bermuda, St. Pierre and Miquelon, and Mexico and other Central American countries and territories. Together, Canada and the United States support a population of nearly 350,000,000 people within an expansive landmass covering nearly 20,000,000 square kilometers. Both Canada and the United States are industrialized, technologically advanced, and rich in natural resources. In addition to industry and technology, both countries support extensive agricultural production in temperate regions. Thus North American ecosystems have been, and may continue to be, subjected to a wide variety of physical, biological, and chemical stressors. Physical and biological stressors include loss of habitat, habitat degradation and destruction associated with urban and suburban sprawl, increased utilization of marginal lands for agriculture, climate change, and introduction and expansion of invasive non-native species and diseases. Undoubtedly these factors are significant stressors on North American ecosystems and wildlife, and they often interact with one another and with chemical stressors in an additive or synergistic manner to produce adverse ecological effects. Yet the focus of this chapter is emerging global contaminant issues; therefore, the following discussion will focus on emerging chemical stressors with potential to impact North American wildlife populations into the future.

Historically, North American ecosystems and the wildlife they support have been subjected to stressors associated with mining of metal ores, coal, and other petroleum-related resources. These plentiful natural resources fueled rapid industrialization of North America in the late 1800s and early 1900s, which contributed to precipitous environmental degradation. A legacy of environmental issues from this era includes acid rain, mobilization of toxic metals, and atmospheric pollution. Rapid advancements in the field of synthetic chemistry in the mid-1900s led to widespread distribution of toxic and sometimes persistent chemicals over large expanses of North America. In response, government-sponsored environmental controls were implemented in the late 1900s by the U.S. EPA and Environment Canada. As a result, significant progress toward mitigation of historical environmental injuries has occurred, and procedures have been established to guard against future environmental insults.

Nonetheless, there are new and emerging chemical threats to North American ecosystems that will demand sound scientific assessment and careful regulatory control. The economies of North America rely increasingly on technological innovation, which carry with it novel stressors. There are several new classes of compounds that have potential for widespread environmental distribution and toxic effects among ecological receptors. These include PPCPs, which are used in increasing quantities by the aging population of North America. Additionally, interest in the development and use of nanomaterials has dramatically increased in North America. It is clear that PPCPs are entering environmental compartments in quantities that may cause significant harm to ecological receptors. Yet the extent to which nanomaterials may enter the environment and result in ecological exposures is less clear, though certainly of concern, considering the projected usage of these materials.

There is, however, one group of chemical compounds that has clearly emerged as a global threat to wildlife, and in particular to wildlife in North America. Brominated flame retardants are halogenated organic chemicals whose chemical properties retard and deter combustion reactions. There are a number of excellent reviews of BFRs, their chemical properties, and environmental occurrence in the literature that have evolved from considerable interest among European researchers (Birnbaum and Staskal 2004; Darnerud 2003; de Wit 2002; Hale et al. 2003; Law et al. 2003; Letcher and Behnisch 2003; Lorber 2008). BFRs can be combined with a wide variety of materials as either additives or reactants (Yogui and Sericano 2009a,b). BFRs as a group include PBDEs, TBBPA, HBCD, and to a lesser extent PBBs (de Wit 2002). Flame retardants are added to a wide variety of products including plastics, textiles, electronic equipment including televisions and computers, circuit boards, polyurethane foams, and many other products because of regulatory demands for non-flammable consumer goods (Lorber 2008). The demand for flame retardants in North America exceeds that of any other continent, with the majority of BFR usage occurring in the United States (de Wit 2002).

To date, PBDEs have been the most widely used group of BFRs in North America, followed by TBBPA, which is more heavily used than HBCD (Johnson-Restrepo et al. 2008). Voluntary and regulatory restrictions on use of certain PBDE formulations have given rise to increased usage of TBBPA and HBCD. To date, there is little information available on TBBPA exposure and effects in wildlife (Johnson-Restrepo et al. 2008). There are much more data available on environmental occurrence, exposure, and effects related to PBDEs than other BFRs; this discussion will therefore focus on PBDEs. This should not be construed as a statement regarding the potential impacts of TBBPA and HBCD on North American wildlife, but rather as an indication of currently available information. TBBPA and HBCD may indeed merit consideration as emerging contaminants as more data become available.

There are three main commercial PBDE formulations that have been used extensively or are currently in use, each of which differs in its composition of 209 possible brominated diphenyl ether congeners (de Wit 2002; Lorber 2008). Penta-BDE formulations are comprised primarily of penta-brominated congeners with accompanying tetra- and hexa-brominated congeners. Octa-BDE formulations consist primarily of hepta- and octa-brominated congeners with lesser proportions of hexa- and nona-brominated constituents. Deca-BDE formulations are predominantly deca-BDE, or BDE-209. The lower-brominated formulations (penta- and octa-) have been banned or voluntarily replaced with higher-brominated formulations since the early 2000s in many countries, including the United States (since 2004), because of their persistence and tendencies to biomagnify in humans and wildlife (Lorber 2008).

PBDEs and HBCD are included as additives in manufacturing but may leach from products over time because these chemicals are not fully integrated into product matrices. Once released into the environment, PBDEs are persistent and biomagnify. PBDEs and other BFRs are cause for significant concern because they are now ubiquitous pollutants in abiotic and biotic media (de Wit 2002; Letcher and Behnisch 2003; Lorber 2008). Though concentrations of these persistent organic pollutants are leveling off, or have leveled off in some regions of Europe, they are increasing rapidly in humans and wildlife of North America (Lorber 2008; She et al. 2007; Yogui and Sericano 2009b). PBDE concentrations in soil, dust, and air are higher in the United States than in other parts of the world (Lorber 2008). Exposure to PBDEs among humans in North America has been attributed to elevated concentrations of these substances emanating from household consumer products and associated dust (accounting for 82% of estimated human exposure; Lorber 2008). Human breast milk samples reflect the high degree of exposure among North Americans. She et al. (2007) documented PBDE concentrations in breast milk from 13 of 40 first-time mothers in the Pacific Northwest region of the United States that exceeded concentrations of PCBs by a factor of two. Moreover, breast milk samples from North American women are at least one order of magnitude greater than those measured in other parts of the world (Lorber 2008). PBDEs, especially the lower-brominated congeners, are rapidly surpassing PCBs as a major environmental concern in North America (She et al. 2007). Despite the fact that usage of BFRs in North America is among the highest in the world, and that extraordinarily high concentrations of PBDEs have been detected in humans and some compartments of the environment, research on environmental distribution of the compounds in North America lags behind Europe (Hale et al. 2003; Yogui and Sericano 2009b).

The disparity in PBDE-related data between Europe and North America is also quite evident when wildlife is considered (de Wit 2002), but there are little data for wildlife in general (Johnson-Restrepo et al. 2008). Nonetheless, trends observed in human PBDE exposure are mirrored in limited data on North American wildlife. Whereas concentrations of (some) PBDE congeners among a variety of ecological receptors in Europe are declining (Wolkers et al. 2008), concentrations of PBDEs are rising rapidly among wildlife of North America (de Wit 2002; She et al. 2007; Watanabe and Sakai 2003). PBDE concentrations in fish from high mountain lakes of the western United States were observed to be three times higher than concentrations in fish from similar lakes in Europe (Ackerman et al. 2008). Interestingly, these cold-water fishes had two to five times higher tissue concentrations than wild Pacific salmon, suggesting that inland transport mechanisms contributed to exposure among these fishes. PBDEs are also ubiquitous in lake trout (*Salvelinus namaycush*) in the Great Lakes region of North America (Luross et al. 2002). North American avian species, particularly piscivores, are also accumulating PBDEs to levels of concern. Nestling bald eagles (*Haliaeetus leucocephalus*) along the western coast of North America were found to contain PBDEs and metabolites approaching concentrations that may impact health and survival (McKinney et al. 2006). Elliott et al. (2005) documented exponentially increasing concentrations of PBDEs in great blue herons (*Ardea herodias*) and double-crested cormorants (*Phalacrocorax auritus*) in Canada. A doubling time for PBDE tissue concentrations was calculated to be approximately 5.7 in these birds from the late 1980s through the early 2000s. In contrast, tissue concentrations of PCBs and DDE in herons and cormorants remained stable over this period. Recent work by McKernan and colleagues (2009) indicates that the lowest-observed-effect level for PBDEs on pipping and hatching success in kestrels may be as low as 1.8 μg Σpenta-BDE/g egg, which, according to the authors, approaches levels found in free-ranging bird eggs.

PBDEs have been detected in polar bears in Canada (de Wit et al. 2006) and detected in brain tissues of wild-caught river otters (*Lutra canadensis*) from Canada (Basu et al. 2007). There has been considerably more research on BFR exposure among marine organisms than terrestrial organisms, and Yogui and Sericano (2009a,b) provide a thorough review of PBDEs in the marine environment of the United States.

The number of bromine molecules and bromination patterns affect metabolism and excretion of PBDEs (Staskal et al. 2006), but tissue distribution within an exposed organism is largely a reflection of the formulation to which the organism was exposed and not so dependent on degree of bromination (Huwe et al. 2007). In general, uptake appears to be low, with significant proportions of PBDEs eliminated via feces (de Wit 2002). Concentrations in birds and mammals tend to be higher than invertebrates and fishes (Yogui and Sericano 2009a). Birds occupying high trophic positions in food webs may biomagnify PBDEs to a greater extent than mammals, based on a comparison of three short food chains (Voorspoels et al. 2007). Further, terrestrial birds may be subject to greater exposure to higher brominated congeners than those utilizing marine ecosystems (Jaspers et al. 2006). BDE-47 (2,2′,4,4′-tetrabromodiphenyl ether) biomagnifies to the greatest extent, but all PBDEs have potential to biomagnify (de Wit 2002). PBDE congeners 47, 99, 100,

153, and 154 dominate congener profiles among both humans and wildlife (Staskal et al. 2006). Congeners associated with penta-BDE formulations have received the greatest attention from researchers because they are the main contributors to human and wildlife exposures. These lower brominated congeners were used extensively in North America prior to voluntary removal from the market. Additionally, higher brominated congeners may undergo selective debromination reactions, giving rise to lower brominated congeners (Lorber 2008; Staskal et al. 2006). Though many scientists believed that BDE-209 (2,2′,3,3′,4,4′,5,5′,6,6′-decabromodiphenyl ether) was too large to be bioavailable, recent data suggest that exposure to BDE-209 does occur among wildlife, and that it does biomagnify (Lindberg et al. 2004). There appears to be a trend toward higher PBDE body burdens among wildlife inhabiting areas near urban/heavily populated areas compared to rural regions (Van den Steen et al. 2009; Yogui and Sericano 2009a,b).

A wide variety of toxic responses have been observed in cell cultures, animal models, and wildlife following exposure to PBDEs. Effects attributed to PBDE exposure include changes in metabolism, development, neurological function/behavior, reproduction, and endocrine function. PBDE exposure among American kestrels (*Falco sparverius*) resulted in changes in thyroid hormone and vitamin A concentrations, glutathione metabolism, and oxidative stress (Fernie et al. 2005). Hydroxylated metabolites of BDE-47 disrupted oxidative phosphorylation in zebrafish (*Danio rerio*) embryonic fibroblast cells (PAC2) and mitochondria (Van Boxtel et al. 2008). PBDE exposure has been associated with altered development in amphibians (Balch et al. 2006), fish (Lema et al. 2007; Timme-Laragy et al. 2006), and mammals (Lorber 2008). Exposure to PBDEs during critical periods of neural development results in neurotoxicological effects that may become manifest as behavioral changes (de Wit 2002; Legler 2008; Lorber 2008; Timme-Laragy et al. 2006). PBDEs and their metabolites can be characterized as EDCs, with some congeners having potencies equal to, or greater than, well-known endocrine disruptors, drugs, and natural hormones (Hamers et al. 2006; Meerts et al. 2001). PBDEs also cause reproductive toxicity in fish (Timme-Laragy et al. 2006), birds (Van den Steen et al. 2009), and mammals (Stoker et al. 2005). Clearly, these compounds have significant potential to adversely affect the physiological status of exposed wildlife. Less clear are potential population-level responses of wildlife in response to global movement of PBDEs and other BFRs. As with PCBs and the American mink (*Mustela vison*), there likely exist wildlife species or groups of species in North America and elsewhere that are particularly sensitive to PBDEs.

The manner in which BFRs have emerged as global contaminants bares striking resemblance to the saga of PCBs as noted by several authors, who have opined that PBDEs are "the new PCB problem" (de Wit 2002; Santillo and Johnston 2003). Though BFRs are global contaminants, concentrations appear to be particularly rapidly increasing among environmental media and receptors in North America. Despite a transition from use of lower-brominated formulations (penta- and octa-BDE) to heavily brominated formulations (deca-BDE) in North America (and elsewhere) there does not appear to be an abeyance of this trend. Release of PBDEs to the environment can occur over the entire useful lifetime of a product, and beyond disposal (Petreas and Oros 2009). Limited information on worldwide BFR usage available in the scientific literature suggests that a disproportionate majority of BFRs produced over the past three decades were, or are currently, residing in North America. A recent study in California attempted to quantify PBDEs in waste streams. E-wastes (electronic wastes, including items such as computers, cell phones, printers, and radios) were the predominant repository for PBDEs in waste streams followed by autoshredder waste and then sewage sludge (Petreas and Oros 2009). However, only half of all PBDEs entering waste streams could be accounted for. The authors surmised that other components of waste streams, including household mixed wastes (old furniture, carpet, televisions, etc.), likely harbored a significant proportion of the total PBDE reservoir. Polyurethane foams used as padding in a variety of products, including furniture and automobiles, weathers and degrades upon exposure to sunlight, heat, and mechanical torsion and can contribute PBDEs to the environment. Laboratory-cultured amphibians accumulated high concentrations of PBDEs (~10 ppm), as did their food source (crickets: ~14 ppm) after polyurethane foam was added to tanks to serve as dry landing substrate (Hale et al. 2002). PBDEs have been identified in air near manufacturing facilities, and dust from televisions accounts for a significant proportion of exposure among humans (de Wit 2002; Lorber 2008; Takigami et al. 2008). Municipal waste facilities, if not managed properly, have significant potential to become sources of PBDE environmental contamination. Fish raised or collected near municipal dumping sites in Vietnam contained elevated concentrations of PBDEs and contributed to higher intake of PBDEs among people consuming fish raised near those areas (Minh et al. 2006). Disposal of sewage sludge via land farming poses another source of PBDE movement into the environment and exposure among North American wildlife (Swanson et al. 2004).

In summary, large repositories of previously used lower brominated PBDEs in North America and debromination of deca-BDE to lower brominated congeners pose risk to humans and wildlife of North America, but may also make North America a source of global PBDE contamination. Lower brominated congeners are transported via the vapor phase, whereas higher brominated congeners move to a lesser extent via particulate matter (Hale et al. 2006). It is clear that long-range transport of PBDEs is occurring; the presence of PBDEs in tissues of Antarctic organisms confirms global transport and distribution (Borghesi et al. 2008). When considered holistically, the chemical properties, persistence, bioavailability, biomagnification potential, and toxicity profiles of PBDEs position this class of BFRs as an emerging contaminant threat to North American wildlife.

8.9 LATIN AMERICA AND THE CARIBBEAN

The geographic region covered in this section includes Mexico, Central America, and South America. Central America includes Belize, Costa Rica, El Salvador, Guatemala, Honduras, Nicaragua, and Panama. South America includes Colombia, Venezuela, Guyana, Suriname, French Guyana, Argentina, Bolivia, Brazil, Chile, Ecuador, Paraguay, Peru, and Uruguay. The geographic region from Mexico to Tierra del Fuego in Chile has a terrestrial area of 21,069,501 km², almost 3.9% of the Earth's surface or 14.1% of its land surface area.

The region's human population has been estimated at more than 500 million and is one of the most diverse in the world. It is the most urbanized region in the developing world, with 77% of its people living in cities. Three of the megacities of the world are located in the region: Mexico City, Sao Paulo, and Buenos Aires. The region will continue urbanizing over the next two decades, when the proportion of the urban population will reach 85%.

The extremely heterogeneous relief and wide latitudinal range of the region give rise to a variety of landscapes, climates, and biomes. The climate of the region ranges from the hot and humid Amazon River basin to the dry and desert-like conditions of northern Mexico and southern Chile and Argentina. Forest is the predominant type of vegetation cover in the region, covering 52% of total area. The most extensive forest area, the Amazon rainforest, covers 29% of the region. The remaining forest is divided between various types of riparian or montane natural forest, deciduous forest with transition to savannah or semi-desert formations, and mosaics of degraded forest and secondary vegetation. The Argentine and Uruguayan pampas constitute the major grassland biome of nontropical Latin America.

Two of the most important freshwater ecosystems on Earth are located in the region, the Amazon River and the Rio de La Plata (Uruguay-Paraná Rivers), and the major rivers in the region carry more than 30% of the world's continental surface water. Its marine and coastal systems are among the most productive in the world. Several of the world's largest and most productive estuaries are found here, such as the Amazon and Plata Rivers on the Atlantic coast, and the Guayaquil and Fonseca on the Pacific.

The Neotropics are dominated by tropical forest and grassland biomes and contain 40% of all plant and animal species on the planet, being the most biodiverse of all the biogeographic realms (WWF 2008). Regarding vertebrate species, the Neotropics are also the most species-rich realm, both overall for terrestrial vertebrates and for each of the four taxa (amphibians, reptiles, birds, and mammals) (Millennium Ecosystem Assessment 2005). In relation to fish, the Neotropics have the highest number of freshwater fish families, doubling that of the Nearctic and Palearctic realms (Berra 2001). Also, the greatest diversity of amphibians in the world is found in the upper Amazon basin and in the Atlantic Forest of eastern Brazil (Young et al. 2004).

However, many vertebrate populations in the Neotropics are under threat from anthropogenic drivers of biodiversity loss, such as rapid climate change, land use changes, exploitation, pollution, pathogens, and the introduction of alien species. For example, between 2000 and 2005 the net loss of forests in South America was about 4.3 million hectares per year, exceeding that of all other regions (WWF 2008). As a result of the synergic interactions of many threats, many wildlife populations are known to be declining in the Neotropics. The best available estimate of global trends in populations is World Wildlife Fund's Living Planet Index. This index shows an average –76% decline from 1970 to 2004 (34 years) in 202 populations of 144 Neotropical species (WWF 2008). More than twice as many known populations of water birds are decreasing than increasing (Revenga and Kura 2003). It has been documented that 230 globally threatened birds—approximately 50% of threatened species that occur in the region—have become extinct across significant parts of their range (Millennium Ecosystem Assessment 2005). An assessment of the amphibians of the New World shows that 39%, or two out of every five amphibian species, are extinct or threatened (Young et al. 2004). Current data also indicate that seven of eight species of amphibians that become extinct in the New World are from the Neotropics (Young et al. 2004).

8.9.1 Major Anthropogenic Stressors for Wildlife

Pollutant stressors include the "legacy" compounds, such as DDT and PCBs. Though banned in most countries, they continue to occur in environmental samples. Mexico, Belize, and several South American countries continued DDT use for vector control until recent years. Aquatic ecosystems are especially at risk due to runoff from agricultural lands and from urban and industrial areas. Water and sediments from the most industrialized and urban areas, such as Río de La Plata (Argentina and Uruguay), Río de Janeiro (Brazil), and Biobio River (Chile), exhibit a wide range and high levels of persistent toxic substances (PTSs), in many cases with values exceeding international regulations (UNEP/GEFa 2002). Chlorinated pesticides, followed by PAHs, PCBs, and PCDD/Fs, are the most commonly reported PTSs in water and sediments from these countries. Furthermore, mining activities have expanded throughout the region, thus acid mine drainage, mercury and uranium releases, and the use of cyanide, mercury, and other chemicals associated with mining are major problems that must be dealt with.

8.9.1.1 Organochlorine Compounds in Wildlife

Agricultural use of DDT ceased in 1978, but was used for vector control in most tropical regions of Mexico until around 2000 (Mora 1997; Mora 2008). Organochlorine residues have been reported for migrant birds (Mora 1997; Mora 2008; Mora et al. 2008). Studies from north Mexico demonstrated elevated concentrations of DDE in aplomado falcons (*Falco femoralis*) from northern Chihuahua that could potentially negatively impact reproduction (Mora et al. 2008). Other studies indicate that DDE concentrations in birds from Mexico were similar to those from the southwestern United States during the same period (Mora 1997). Although there is concern that birds could be exposed to OCs during migration across Latin America, especially in Mexico because of its relatively recent use of DDT, a recent study indicates that there is not a significant buildup of persistent OCs in migrant and resident passerine birds in the central-pacific region of Mexico (Mora 2008).

A study of OC compounds in vertebrates in a conservation area in northwestern Costa Rica (Klemens et al. 2003) demonstrated widespread contamination both geographically and across taxonomic boundaries. The most frequently detected OC compounds in the area included DDE, BHC, heptachlor, and dieldrin. Turtles showed the highest frequency of contamination of the four groups analyzed, which included amphibians, rodents, and birds. Levels documented in turtles in this study have been linked to developmental abnormalities (Klemens et al. 2003), though these effects have not been documented in Costa Rica. Atmospheric long-distance transport of OCs was considered to be the major source of OC contamination, since the study area has been protected since the mid-1960s. Atmospheric transport of OCs has been documented in numerous studies in the Central American region (Daly, Lei, Teixeira, Muir, Castillo, Jantunen et al. 2007; Daly, Lei, Teixeira, Muir, Castillo, and

Wania 2007; Gouin et al. 2008). OC compounds have also been reported in crocodile species in Belize and Costa Rica (Pepper et al. 2004; Rainwater et al. 2007; Wu et al. 2000), and dolphins in Panamá (Borrell et al. 2004), and dolphins and porpoises collected along the Buenos Aires Atlantic coast (UNEP/GEF 2002).

8.9.1.2 Polycyclic Aromatic Hydrocarbon Compounds (PAHs)

Although PAHs are widespread pollutants, little information exists relative to most Latin American countries. Available information indicates occurrence of high concentrations in heavily populated and industrialized areas. In Brazil, emissions were estimated to be 467 to 6,607 t/yr, mostly due to wood combustion (UNEP/GEF 2002).

High concentrations of PAHs have been reported in coastal marine waters and sediments of the Argentine Patagonia and Gulf of Mexico impacted by crude oil extraction activities (Barra et al. 2007; Ponce Vélez and Botello 2005; Zapata-Perez et al. 2005). Muscle tissues of a detritivorous fish, Sabalo (*Prochilodus lineatus*), collected in the sewage-impacted Buenos Aires coastal area, contained extremely high concentrations of PAHs with a mean of approximately 1 mg/g. Concentrations reported are 1 to 3 orders of magnitude higher than those for Mediterranean fish (Colombo et al. 2007).

8.9.1.3 PCBs

PCBs predominate among all compounds monitored in the International Mussel Watch Program, which is the most comprehensive monitoring program for persistent chemicals in the South America coastal environment (UNEP/GEF 2002). PCB concentrations range from 4,000 to 13,000 μg/kg lipids in bivalves from the Río de la Plata (Argentina), Recife (Brazil), and Punta Arenas (Chile). Recent studies (Colombo et al. 2007) indicate extremely high concentrations of PCBs in fish in the Río de la Plata estuary. The high levels of PCBs reflect long-term chronic exposure from industrial-sewer discharges. Concentrations measured in these fish are comparable to those reported for the Great Lakes and Hudson River (Colombo et al. 2007). Concentrations of non-ortho PCBs in sparrows and ground doves from Baja California, Mexico, are among the highest concentrations reported for species of similar trophic status (Jimenez et al. 2005). Garcia Hernández et al. (2006) reported concentrations of non-ortho PCBs in eggs of doves and owls from the Colorado River delta in Mexico at concentrations that could affect hatchability.

8.9.1.4 Compounds of Emerging Concern

8.9.1.4.1 Pesticides

The use of pesticides in Latin America is high and Brazil and Mexico are considered as the main markets in the region (Mora 1997). Central America includes smaller countries, but they are among the most intensive pesticide consumers as compared with Europe (Heidorn 2002). Lindane and endosulfan are two potentially relevant PTS of emerging concern due to their widespread use in the region. Endosulfan and its degradation products are widespread contaminants in the environment (Daly et al. 2007b; Gouin et al. 2008; Hernández-Romero et al. 2004; UNEP/GEF 2002). Pentachlorophenol should also be considered even though its use and distribution is more limited within the region (UNEP/GEF 2002, 2002). Additionally, HCB, an impurity during the synthesis of several herbicides and pesticides, is a common environmental pollutant and has been documented in Mexico and Central America (Daly et al. 2007a; UNEP/GEF 2002).

Organophosphates and carbamates are agrochemicals commonly used globally and in Latin America, where they have caused mortalities in fish and birds, including migratory birds (Castillo et al. 2000; Goldstein et al. 1999). Fungicides such as mancozeb and benomyl are also of concern because of their extensive use and carcinogenic and teratogenic risks.

Studies carried out in the Atlantic zone of Costa Rica, where approximately 98% of bananas and 50% of pineapples are grown, have demonstrated extensive surface water contamination, the occurrence of pesticides in sediments, and biological impacts, especially near the banana and pineapple plantation areas. The frequency of occurrence and concentrations observed were highest near agricultural areas. In banana plantation areas, peak concentrations in surface waters were observed following ground and aerial application of pesticides. Fungicides used in packing plants (imazalil and thiabendazole), as well as several insecticide–nematicides (chlorpyrifos, diazinon, cadusafos, carbofuran, ethoprofos, and terbufos) were detected in concentrations that have the potential to adversely affect aquatic species (Castillo et al. 2000). Other fungicides frequently detected in surface waters include propiconazole and chlorothalonil and, more recently, difenoconazole. Several of the pesticides mentioned have been identified in recent years as the probable cause of fish mortalities in the area, specifically ethoprofos, terbufos, and chlorothalonil. These compounds have been detected in tissues of dead fish and in water collected at the time of the die-offs.

8.9.1.4.2 Metals

Metals reach ecosystems via natural geological processes, mining, and industrial activity (Lacher and Goldstein 1997). Consequently, the expansion of mining, industrialization, and other anthropogenic activities has led to metal contamination in many aquatic ecosystems of the Neotropics. Metals in freshwater ecosystems have been reported in different regions of the Neotropics. Analysis of scutes from Morelet's crocodiles (*Crocodylus moreletii*) from Belize and American crocodiles (*C. acutus*) from Costa Rica revealed the presence of mercury, cadmium, copper, lead, and zinc (Rainwater et al. 2007). In Lake Titicaca, levels of mercury in the pejerrey (*Basilichthyes bonariensis*) and the carachi (*Orestias* spp.) exceeded U.S. EPA fish tissue-based water quality criterion level of 0.30 μg/g (Gammons et al. 2006).

8.9.1.4.3 Other Compounds of Emerging Concern

With globalization, newer emerging pollutants are also a threat in the region arising from imports, industry, and mining activities. Surfactants and in particular linear alkylbenzene sulfonates (LASs) are widely employed in detergent formulations. Brazil is one of the main producers of synthetic detergents. It is estimated that Brazil is responsible for almost half the sales of cleaning products in Latin America. Eighty thousand tons per year of LAS and 220 thousand tons per year of linear alkylbenzenes (LAB), precursors to LASs, are produced, representing 10% of the world market (Penteado et al. 2006).

There are major data gaps regarding the presence, concentrations and effects of pollutants in most countries of the region. The effects on special and vulnerable ecosystems of the region (mountain forests, mangroves, coastal lagoons, and coral reefs) remain largely unknown. The interaction of the meteorological instability brought about by climate change and pollutants might be an additional stressor for wildlife and should be considered an area of priority study.

8.10 ARCTIC AND ANTARCTIC REGIONS

8.10.1 Arctic

According to the Arctic Council (http://arctic-council.org/), the Arctic covers an enormous area, over one-sixth of the surface of the Earth (>30 million square kilometers). The human population is relatively small at 4 million, but is culturally diverse, with over 30 indigenous peoples and dozens of languages. The Arctic is well known for its vast natural resources, and is often considered a "clean" environment, relatively speaking of course. The Arctic Monitoring and Assessment Programme (AMAP) is one of five working groups of the Arctic Council. The primary function of AMAP is to advise the governments of the eight Arctic countries (Canada, Denmark/Greenland, Finland, Iceland, Norway, Russia, Sweden, and the United States) on matters relating to threats to the Arctic region from pollution and on associated issues. Since this is the mission of AMAP, we will refer to its mission and products in support of our effort in this section to address the poles of the globe. We will supplement this information from other efforts, such as the Northern Contaminants Program in Canada.

AMAP implements parts of the Arctic Environmental Protection Strategy (AEPS) by "providing reliable and sufficient information on the status of, and threats to, the Arctic environment, and providing scientific advice on actions to be taken in order to support Arctic governments in their efforts to take remedial and preventive actions relating to contaminants."

Aside from the latitudinal definition of the Arctic (the "Arctic Circle"), defining the Arctic on a map is dynamic and contentious, and very subjective. In many ways the boundaries are in flux, and current investigations of climate change in the Arctic will undoubtedly alter those boundaries that are based on climate, vegetation, temperature, etc. The same features and definitions are applicable to the Antarctic as well.

Another Arctic Council group, known as CAFF (Conservation of Arctic Flora and Fauna), provides an excellent resource for understanding the biodiversity of the Arctic and some of the current management and research challenges (http://arcticportal.org/en/caff/aba). The Arctic Biodiversity Assessment (ABA) is an innovative and important exploration of Arctic biodiversity. Its purpose is to synthesize and assess the status and trends of biological diversity in the Arctic. The ABA is the Arctic Council's response to the United Nations UNEP/CBD 2010 global target to halt or significantly reduce biodiversity loss. It also responds to the Millennium Development Goal Nr. 4 to ensure environmental sustainability. It will serve as a baseline for use in assessments of Arctic biodiversity and to inform Arctic Council work. Much of this CAFF effort is based on concerns related to climate change as outlined in the rather large text (ACIA 2005).

8.10.2 Antarctic

The Antarctic involves international interests similar to those of the Arctic. The ice-dominated Antarctic is a single continent, unlike the Arctic, which is composed of parts of several continents. The Antarctic is currently only transiently occupied by humans, with no indigenous population, unlike the Arctic. The remote Southern Ocean encompasses Antarctica and appears affected by the Antarctic Convergence (cold Antarctic waters meeting warmer northern waters). The Antarctic Convergence can be considered a "biological barrier," and the associated marine system could be considered a closed system (though not in the absolute sense). However, the Antarctic has many commercial interests similar to the Arctic (e.g., fisheries, tourism, shipping, research, exploration, and development), and climate change may alter our view of these oceanographic "barriers" for both poles. The Committee for Environmental Protection (http://www.ats.aq/e/cep.htm) for the Antarctic was established in Article 11 of the Environment Protocol, and some participants have "observer status," as do the president of the Scientific Committee on Antarctic Research (SCAR) and the chairman of the Scientific Committee for the Conservation of Antarctic Marine Living Resources (SC-CAMLR). One can appreciate the parallels between the Arctic and Antarctic, as well as some of the differences.

The similarities can be so striking even at the acronym level, as yet another CAFF exists (adopted in 1964), this one for the Antarctic: Conservation of Antarctic Fauna and Flora. This CAFF considers the region relatively unmodified by human activities (this may be based on grossly observed disturbances and impacts, as chemical indicators may provide a different perspective). Protection of *native* fauna and flora has been a concern for the Consultative Parties to the Antarctic Treaty from the beginning. Considering human habitation and movement in the Arctic, there has been much more exchange of biota there compared with southern high latitudes. Some people have a sense of security that the "harsh" polar environments will keep the "exotic" species out. However, some of our biotoxin-producing and biotoxin-accumulating invertebrates and fish do not require major changes in oceanographic conditions to move to higher latitudes. Thus thermohaline barriers may merely be subtle obstacles easily altered that allow "invasive" species to move to much higher latitudes. These barriers will also alter xenobiotic inputs and dynamics via biotic and abiotic mechanisms. Many upper-trophic species have not adapted to the presence of these biotoxins (and certainly not to the xenobiotics) and thus are essentially naïve and vulnerable. Transport of inorganic and organic contaminants are not impeded by these "harsh" conditions either, and many unique polar mechanisms may actually exist that enhance polar transport and input (such as sea ice movement of sediments, cold condensation of volatiles (the "grasshopper effect"), polar sunrise deposition of Hg into snow via a complicated interaction between the sun and coastal environment, and the lipid-rich biota of the poles).

With respect to anthropogenic inputs, Article 3.2 of the Environment Protocol provides that activities to be undertaken in Antarctica shall be planned and conducted so as to avoid "detrimental changes in the distribution, abundance or productivity of species or populations of species of fauna and flora" and "further jeopardy to endangered or threatened species or populations of such species." Annex II to the Protocol sets out specific measures related to this. Thus several different policies are protecting Antarctic species, including the prohibition of introducing non-native species, except in accordance with a permit, and the designation of Specially Protected Species. Other conservation efforts include the Convention on the Conservation of Antarctic Marine Living Resources [http://www.ccamlr.org/pu/e/gen-intro.htm], formed as a result of concerns about increase in krill catches in the Southern Ocean, since populations of krill are important to marine life (birds, whales, seals, and fish). The Council of Managers of National Antarctic Programs (COMNAP) is the international association that brings together National Antarctic Programs from around the world to develop and promote best practices in managing the support of scientific research in Antarctica and includes the Antarctic Environmental Officers Network (AEON) to address the environmental management of Antarctic operations. In short, within these international management paradigms, the poles will be challenged by political (boundary disputes), logistical (remote and harsh conditions), exploration and development (resource extraction), and conservation issues, including the toxins and contaminants we discuss next.

Climate change and all variables associated with this phenomenon are likely the largest issue facing the poles, one that will alter the current state and management of these systems. As mentioned above, various "barriers" for the poles are changing or have already been easily breeched (e.g., by the transport of contaminants via air, water, and biota). This will open up habitable ranges for a variety of organisms, some of which represent toxin threats (e.g., harmful algal blooms). From the above discusion, the international nature of where the time zones thin and eventually merge should be well appreciated. Many of the physical and biological transports follow these same longitudinal trajectories and deposit transboundary agents upon these regions. These include biological and chemical agents. For both poles the loss of sea ice will alter many systems, including how contaminants are transported and enter food webs and the formation and distribution of biotoxins. Anthropogenic sources are not solely those emanating from industry, as they also come in the form of apparently innocent low-impact excursions (e.g., tourism). Touristic events have certainly proven to be disastrous, with ships running aground or sinking. The more insidious disposal of sewage and other cruise by-products should be of concern and very aggressively managed. The exploitation of charismatic megafauna in the Antarctic (e.g., scientific whaling) makes for great press and NGO fund raising, but many of these more insidious changes or impacts should be of concern (such as chronic whale watching). This may not be a popular notion, but the most obvious stressors may not be the most important. For example, changes in the pH of the ocean in the polar regions are receiving very little attention. Perhaps toxicologists should promote the hydronium ion as a marine toxicant since subtle changes in pH may have extremely large impacts on marine biochemistry. How will pH changes and increased nutrient inputs with a warming ocean alter the biochemistry of the polar marine environment?

8.10.3 Polar Issues

Open water ("ice free") in itself is a huge change for the poles, along with the drivers altering the ice; greater opportunities for fishing, shipping, exploration, development, habitation, research, and the military are going to change how we perceive the poles. Foresight is needed to appropriately take advantage of the opportunities presented by these ice-free waters. We need to be assured that the possibility of sustainability exists. Recent management decisions for the Chukchi Sea are encouraging, since fishing needs must be considered very deliberately and strategically. The North Pacific Fishery Management Council will not approve fishing north of the Bering Strait until we can better understand the potential impacts. Fishing is only one industry to consider; the Arctic may see more activity related to exploration for oil and gas and development, and increased shipping traffic (NW Passage/Trans-Siberian coastal shipment) in the highly commercialized and active Northern Hemisphere.

Harmful alga blooms are defined as harmful because they are detrimental to humans and other mammals. Yet many of these biotoxins are well tolerated by marine and aquatic invertebrates and fish, making the term *harmful* relative. There are no doubts these HABs are deadly to various biota and result in small- and large-scale morbidly and mortality events. Many have been well documented in regions with the investigative infrastructure and commercial interests (e.g., human foods), while many go uninvestigated or unnoticed in other areas. Because the Arctic and Antarctic are remote, it will be problematic to assess HABs unless they occur near human observers or are on a large scale. Changes in coastal and marine environments may favor larger-scale and more northerly HABs.

There are a number of emerging contaminants that have the potential to impact wildlife indigenous to the poles. One class is PBDEs, flame-retardant chemicals commonly used in plastics and foam products, described in detail in this and other chapters. During manufacturing and use in consumer products, PBDEs enter the air, water, and soil, and thus enter transport pathways, many of which reach the poles. The organohalogen class of contaminants has long been reported in studies conducted at the poles—for good reasons, since it is clear that the poles are not chemically isolated from the industrialized parts of the globe. Moreover, it is no surprise that chemicals with nearly identical properties behave in similar ways. Despite the obvious chemical and physical similarities between PCBs and PBDEs, management through predictive or prospective assessments of the use of these chemicals has been poorly implemented. Persistent halogenated organics are considered transportable to the poles and to have bioaccumulation and biomagnification potential. Their possible toxicity is still being debated, but not the mechanisms for polar transport and bioaccumulation and biomagnification in polar biota. Recent studies further document the presence of these compounds in various media, including biota.

PFOA is a component of Teflon®, a persistent breakdown product of chemicals used to coat food packaging and to make stain-resistant coatings for many items. PFOA is used as an essential processing aid in the production of fluoropolymers and fluoroelastomers, including polytetrafluoroethylene (PTFE) and polyvinylidine fluoride (PVDF). PTFE has hundreds of uses in many industrial and consumer products, including in soil-, stain-, grease-, and water-resistant coatings in textiles and carpets; in the automotive, mechanical, aerospace, chemical, electrical, medical, and construction industries; in personal care products; and in non-stick coatings on cookware. PVDF is used primarily in three major industrial sectors: electrical/electronics, building/construction, and chemical processing. These chemicals are persistent and of concern, but they do not accumulate in the same manner as other halogenated organics. Polar biota sampling designs must recognize this and consider the broad scope of species and tissues to sample for properly addressing the organohalogens.

8.11 OCEANS AND SEAS

The world ocean, or global ocean, is the continuous body of saline water that covers over 70% of the Earth's surface, an area of about 360 million square kilometers with a volume close to 1.3 billion cubic kilometers. It is conventionally divided into five oceans—in decreasing size, the Pacific, Atlantic, Indian, Southern, and Arctic—and into smaller seas. Ocean waters and resources (fish, oil, natural gas, ores, etc.) are critical to the global economy. In 1982, the United Nations Convention on the Law of the Sea was adopted by over 160 sovereign states, in a concerted effort to avoid mounting conflicts regarding the harvest of these resources. Since then, coastal states have had sole jurisdiction over a 12-mile zone along their shores and exclusive economic rights over a 200-mile zone, referred to as the Exclusive Economic Zone (EEZ). Close to 99% of the world fisheries are within EEZs and thus fall under national jurisdiction. The resources beyond the limits of EEZs are deemed the common heritage of humankind. Under the Convention, all participating states are obligated to protect and preserve the marine environment and to cooperate to facilitate this obligation. This obligation may require monitoring activities and anti-pollution measures for the six main sources of pollution specifically identified by the convention: "land-based and coastal activities; continental-shelf drilling; potential seabed mining; ocean dumping; vessel-source pollution; and pollution from or through the atmosphere" (United Nations 2007). In addition, states have to abide by the Convention on the Prevention of Marine Pollution by Dumping of Wastes and Other Matter, which was adopted in 1972, and amended into the 1996 Protocol, which came into force in 2006. This Protocol first introduced the precautionary approach into the control and prevention of marine pollution.

Beyond EEZs, the global ocean and all of its marine life belong to everyone, and yet to no one in particular, and thus the reality of ocean pollution monitoring and prevention is complex and challenging. Although it is feasible to regulate point source pollution, low-level, chronic diffuse pollution is more difficult to deal with. The deposition and transfer of pollutants and the movement and migration of marine life are not limited by artificial boundaries such as EEZs. Hence, marine pollution resulting from human activities is a global issue affecting all marine life and every marine ecosystem (Pew Oceans Commission 2003; Halpern et al. 2007). According to UNESCO's Global Ocean Observing System (2002) (UNESCO 2002), an enormous and increasing range of contaminants are of concern, including synthetic organics/POPs, pesticides, polycyclic aromatic and petroleum hydrocarbons, metals and organometals, radionucleides, pharmaceuticals, biotoxins, plastics, and most recently, nanomaterials. Marine pollution is recognized as a major factor in the current and global marine ecosystem crisis (United Nations 2007; United Nations Educational Scientific and Cultural Organization 2002). It affects ecosystem integrity, species survival, biodiversity, and seafood access, and presents an increasing threat to ocean and human health (Knap et al. 2002; Sala and Knowlton 2006; Schipper et al. 2008). Over a third of all marine mammal species are now considered threatened, with pollution ranked as the most prevalent threat after accidental mortality (Schipper et al. 2008). There is growing acknowledgment that marine ecosystem services are under threat, which will increase the economic burden on the already fragile global economy. Wildlife toxicology and ecotoxicology research is critically needed in order to evaluate the threats posed by environmental pollution to these and other marine wildlife species. It should also be noted that the context in which marine pollution studies will be pursued in the future is also changing. Pollutants are imposing pressures on marine wildlife at a time when they are increasingly challenged by climate change, ocean acidification, habitat loss, and human demographic changes. All of these stressors must be studied simultaneously. Emphasis should be placed on studies investigating the impact of pollution on a wide range of sentinel species from various taxa and positions within food webs, using suites of biomarkers rather than single endpoints, and relating findings within the framework of individual organism health as well as ecosystem health (Bowen and Depledge 2006; Depledge 1993; Depledge and Galloway 2005; Moore et al. 2004).

This section on oceans and seas will focus on marine species most generally considered wildlife as defined by Kendall (1994), that is, mammalian, avian, and reptilian species, with special focus on marine mammals, sea turtles, and seabirds. Because of their relatively high trophic position and long life spans, these species can offer a global perspective on emerging marine toxicology issues by providing integrated measures of ecosystem impact (Furness and Camphuysen 1997; Ross 2000). From a literature survey, we identified four emerging issues of concern related to the impact of pollution on marine wildlife health and biodiversity.

8.11.1 Harmful Algal Blooms

Over 200 species of microalgae and ciliata are known or suspected to produce biotoxins, including brevetoxins, ciguatoxins, saxitoxins, and domoic acids (Landsberg 2002). There is evidence that the frequency and distribution of related harmful algal blooms (HABs) are increasing (Hallegraeff 1993; Mos 2001; Smayda 1990; Van Dolah 2000). While the effects of these blooms on marine animal health are still poorly understood, morbidity and mortality events in marine mammals, seabirds, and sea turtles have been reported. Poisoning by paralytic toxins has been implicated in sudden mortality events in the highly endangered monk seal (*Monachus monachus*), Hawaiian monk seal (*M. schauinslandi*), California sea lion (*Zalophus californianus*), Florida manatee (*Trichechus manatus latirostris*), bottlenose dolphin (*Tursiops truncatus*), and humpback whale (*Megaptera novaeangeliae*), and has been linked to seizure, ataxia, and reproductive failure in California sea lions (Bossart et al. 1998; Brodie et al. 2006; Flewelling et al. 2005; Geraci et al. 1989; Gilmartin et al. 1980; Gulland et al. 2002; Hernandez et al. 1998; Scholin et al. 2000). In seabirds and sea turtles, observed toxic effects ranged from reduced feeding activity, impaired reproduction, ataxia, and death (Arthur et al. 2008; Beltran et al. 1997; Pierce and Henry 2008; Work et al. 1993). Predicted increases in water temperature, dust load through desertification, and the amount of nutrient-rich effluents released in the marine ecosystem are likely to increase the incidence of HABs (Harvell et al. 1999; Harvell et al. 2002; Mos 2001; Patz et al. 1996). Biomonitoring and toxicology programs focusing on biotoxins and their effects are therefore critically needed to further our understanding of HAB's impact on seafood safety, marine wildlife survival, and the overall integrity of marine food webs in the future.

8.11.2 Brominated Flame Retardants

BFRs and especially PBDEs have been detected at high concentrations in animals feeding at the top of marine food chains, such as Mediterranean swordfish (*Xiphias gladius*), European harbor seals, harbor porpoises (*Phocoena phocoena*), Southeast Atlantic bottlenose dolphins, and Indo-Pacific humpback dolphins (Boon et al. 2002; Corsolini et al. 2008; Fair et al. 2007; Ramu et al. 2005; Thron et al. 2004). In some marine mammal populations, contaminant burdens for these compounds have increased significantly over the last two decades and may surpass that of PCBs within the next 10 years (Ikonomou et al. 2002; Ross et al. 2009; She et al. 2002). Exposure to BFRs in laboratory animals is associated with various effects, including endocrine and reproductive toxicity, cancer, and neurotoxicity (Birnbaum and Staskal 2004; Hallgren and Darnerud 2002), but little is known in marine mammals. There is evidence that PBDEs may act as endocrine disruptors in gray seals (Hall et al. 2003). Fossi and collaborators (2006) investigated PBDE biotransformation in Mediterranean cetaceans in vitro using primary skin fibroblasts and observed induction of cytochrome P450 2B. In vitro investigation is particularly well suited to toxicological studies in marine mammals for which in vivo experimentation is usually precluded (Fossi and Marsili 2003; Fossi et al. 1992; Godard et al. 2004; Godard et al. 2006; Marsili 2000; White et al. 2000). PBDEs have been detected in many species of coastal birds and seabirds, including great blue herons (*Ardea herodias*), double-crested cormorants (*Phalacrocorax auritus*), storm petrels (*Ocaenodroma leucorhoa*), chinstrap and gentoo penguins (*Pygoscelis antarcticus* and *P. papua*), and south polar skuas (*Stercorarius maccormicki*) (Elliott et al. 2005; Yogui and Sericano 2009a). For some of these birds, egg and tissue burdens have significantly increased over the last two decades, approaching potentially toxic levels (Elliott et al. 2005). Additional studies, including in vitro, field, and epidemiology studies, that investigate the effects of BFRs in marine wildlife are deeply needed in order to better characterize their metabolism and toxicity.

8.11.3 Plastics

Plastics are synthetic organic polymers whose production and use has expanded at a very rapid pace in the last three decades (Derraik 2002). Most plastic products degrade very slowly, and associated debris and litter now represent an increasingly significant threat to marine mammals, sea turtles, and seabirds (Derraik 2002; Laist 1987; Moore 2008). Over 260 species of marine animals are affected by the presence of plastics in the oceans, including 43%, 44%, and 86% of all marine mammal, seabird, and sea turtle species, respectively (Laist 1997). Mechanical threats from entanglement in and ingestion of plastic litter appear to cause the most harm. Entanglement in packaging bands, synthetic ropes and lines, drift nets, and other plastic products can greatly reduce fitness by increasing energetic costs as a result of added weight, increased drag, and the physiological consequences of physical injury (Feldkamp et al. 1989). It can also lead to reduced ability to catch food, avoid predators, and keep from drowning (Laist 1987; Wallace 1984). Ingestion of plastics can lead to reduced food uptake and overall fitness in most species of marine mammals, sea birds, and sea turtles (Azzarello and Vanvleet 1987; Baird and Hooker 2000; Carr 1987). Ingestion can also result in blockage of gastric enzyme secretion and the intestinal tract, reduced feeding stimulus and ability to lay down fat, delayed ovulation, impaired reproduction, internal injury, and death (Azzarello and Vanvleet 1987; Balazs 1984; Moore 2008; Ryan 1988; Spear et al. 1995). Additionally, chemicals present in plastic particles can leak into an animal's tissues following ingestion. Such exposure route for PCBs and other contaminants was observed in seabirds and sea turtles (Bjorndal et al. 1994; Ryan 1988), and more research is needed to determine the exact impact of this threat in marine wildlife.

8.11.4 Coastal Pollution

While marine pollution is of great concern worldwide, coastal ecosystems appear to be at increased risk due to proximity to land-based sources of exposure, such as direct industrial discharge and agricultural run-off, and aggravating factors such as coastal development and shipping activities (United Nations Educational Scientific and Cultural Organization 2002). As the capacity of coastal ecosystems to assimilate these various stressors is exceeded, wildlife becomes more vulnerable to factors such as habitat loss, decline of prey populations, and exposure to environmental pollutants. Halpern and colleagues (2007) determined that coastal development is one of the most frequent and functionally impactful anthropogenic threats to marine ecosystems. Estuarine and coastal degradation has dramatically increased in the last 200 years with human activities, resulting in depletion of 91% of all animal and plant species, including most mammals, birds, and reptiles (Lotze et al. 2006). Overexploitation is the major culprit in this drastic loss in biodiversity, but pollution plays an increasing role that needs to be carefully monitored and managed.

8.12 CONCLUSIONS

Here we have assimilated from environmental professionals in various regions of the world disparate perspectives on emerging contaminant issues likely to impact wildlife populations. Despite the vast differences among geographic settings and incumbent availability of natural resources and biodiversity across the globe, some

TABLE 8.1
Chemical and Other Stressors Identified as Emerging Threats to World Wildlife Populations

	Stressors	
	Chemical	**Other Physical or Biological**
Africa	Mixtures of contaminants in effluent	Water distribution
	Pesticides	
	Industry and mining wastes	
Middle East	Petroleum products	Gulf War
	Fertilizers, pesticides	Oil production
		Overgrazing, overhunting
		Soot
		Oil lakes
Asia/China	Brominated flame retardants	Overpopulation
	E-wastes	
Australia	Insecticides	
India	Pesticides	
Western and Eastern Europe	Persistent organic pollutants	
	Brominated flame retardants	
	Perfluorinated compounds	
	Polycyclic aromatic hydrocarbons	
	Pharmaceuticals and personal care products	
	Nanoparticles	
North America	Brominated flame retardants	Habitat destruction, loss
	E-wastes	Invasive species
Latin America and the Caribbean	Pesticides	Deforestation
	Metals	
Arctic and Antarctic	Changes in pH	Climate change
	Brominated flame retardants	Harmful algal blooms
	Perfluorinated compounds	Tourists
Oceans and Seas	Brominated flame retardants	Plastics
	Industrial discharge to coastal regions	Harmful algal blooms
	Agricultural run-off to coastal regions	

thematic similarities have become evident from this effort. Perspectives on major anthropogenic threats to wildlife populations included within this chapter are summarized in Table 8.1.

It is clear that major emerging threats to wildlife are each in some way associated with food, energy, or the manufacture of consumer products. According to our coauthors, pesticides (including herbicides and insecticides) are emerging rather than declining threats to wildlife in Africa, India, and Australia. Both Africa and India have burgeoning human populations, which necessitate ever-increasing agricultural production. Synthetic pesticides increase efficiency of production and reduce competition from unintended consumers (pests). Alternatively, Australia has not experienced the increasing demands for food production from within as have Africa and India. The uniqueness of Australia's wildlife and reliance on regulatory data generated on non-native species may support the assertion that pesticides continue to be an emerging threat to wildlife in that country. Worldwide demand for oil and other petroleum products has made petro-associated chemicals major emerging threats, not only in the Middle East but elsewhere in regions where petroleum production is possible or necessary for fueling advanced economies.

Other emerging threats have been identified herein, but either they are of limited geographical distribution or there are insufficient data currently available to fully ascertain the degree to which they represent future threats to wildlife. These include mixtures of contaminants in effluent discharge, PPCPs, nanoparticles, harmful algal blooms, and plastics. Ironically, each of the agents of the mixtures is associated with water, and all may, in time, be inexorably linked with one another.

Interestingly, persistent organic pollutants, and specifically the BFRs, were identified as major emerging threats to wildlife in Europe, North America, China/Asia, the Arctic and Antarctic, and the world's oceans and seas. These compounds are now considered global contaminants because of their propensity for movement via mechanical transport (through international trade of consumer goods and electronics) and long-range transport (atmospheric and oceanic). Coincidentally, there are few internationally agreed-upon mechanisms in place to limit environmental dissemination of these types of compounds. It would seem that it is to some degree human nature to be less concerned about the occurrence of contaminants in someone else's back yard. Yet perspectives on emerging threats in the world's oceans and at its poles demonstrate that concepts of "back yard" may need to be adjusted. It is now cliché to speak of global economies, international communities, and small worlds, but the identification of BFRs and other POPs as major emerging threats in numerous regions of the world provide tangible supporting evidence.

Perhaps the perspectives provided within this chapter are too limited, distorted in some way, or prejudiced by the prolonged (and perhaps narrow) focus of the environmental scientists who shared them. These notions cannot be discounted. Likewise, each perspective should be considered separately and as part of a group. Taken in this manner, they offer insight into future trends and threats to wildlife populations across the globe. Further, they may also provide a means for interpolation of future threats to human health.

REFERENCES

Abdalla, A. 2007. Benefits of locust control in eastern Australia: A supplementary analysis of potential second generation outbreaks. Canberra, Australia: Australia Bureau for Agricultural and Resource Economics (ABARE).

Abhilash, P. C., and N. Singh. 2009. Pesticide use and application: An Indian scenario. *Journal of Hazardous Materials* 165 (1–3): 1–12.

ABS. 2008. Australian demographic statistics 3101.0. Canberra, Australia: Australian Bureau of Statistics.

ACIA. 2005. *Arctic Climate Impact Assessment*. Edited by C. Symon, L. Arris, and B. Heal. New York: Cambridge University Press.

Ackerman, L. K., A. R. Schwindt, S. L. M. Simonich, D. C. Koch, T. F. Blett, C. B. Schreck, M. L. Kent, and D. H. Landers. 2008. Atmospherically deposited PBDEs, pesticides, PCBs, and PAHs in Western U.S. National Park fish: Concentrations and consumption guidelines. *Environmental Science & Technology* 42 (7): 2334–2341.

Andersson, O., and G. Blomkvist. 1981. Polybrominated aromatic pollutants found in fish in Sweden. *Chemosphere* 10 (9): 1051–1060.

Aneck-Hahn, N. H., G. W. Schulenburg, M. S. Bornman, P. Farias, and C. De Jager. 2007. Impaired semen quality associated with environmental DDT exposure in young men living in a malaria area in the Limpopo Province, South Africa. *Journal of Andrology* 28 (3): 423–434.

Arthur, K. E., C. J. Limpus, and J. M. Whittier. 2008. Baseline blood biochemistry of Australian green turtles (*Chelonia mydas*) and effects of exposure to the toxic cyanobacterium *Lyngbya majuscule*. *Australian Journal of Zoology*, http://www.publish.csiro.au/?act=view_file&file_id=ZO08055.pdf.

Azzarello, M. Y., and E. S. Vanvleet. 1987. Marine birds and plastic pollution. *Marine Ecology-Progress Series* 37 (2–3): 295–303.

Baird, R. W., and S. K. Hooker. 2000. Ingestion of plastic and unusual prey by a juvenile harbour porpoise. *Marine Pollution Bulletin* 40 (8): 719–720.

Balazs, G. H. 1984. Impact of ocean debris on marine turtles: Entanglement and ingestion. Paper presented at Workshop on the Fate and Impact of Marine Debris, November 1984, Honolulu, Hawaii.

Balch, G. C., L. A. Vélez-Espino, C. Sweet, M. Alaee, and C. D. Metcalfe. 2006. Inhibition of metamorphosis in tadpoles of *Xenopus laevis* exposed to polybrominated diphenyl ethers (PBDEs). *Chemosphere* 64 (2): 328–338.

Barnhoorn, I. E. J., M. S. Bornman, G. M. Pieterse, and J. H. J. van Vuren. 2004. Histological evidence of intersex in feral sharptooth catfish (*Clarias gariepinus*) from an estrogen-polluted water source in Gauteng, South Africa. *Environmental Toxicology* (6), http://www3.interscience.wiley.com/journal/109747065/abstract?CRETRY=1&SRETRY=0.

Barra, R., C. Castillo, and J. P. M. Torres. 2007. Polycyclic aromatic hydrocarbons in the South American environment. *Reviews of Environmental Contamination and Toxicology, Vol 191* 191:1-22.

Basu, N., A. M. Scheuhammer, and M. O'Brien. 2007. Polychlorinated biphenyls, organochlorinated pesticides, and polybrominated diphenyl ethers in the cerebral cortex of wild river otters (*Lontra canadensis*). *Environmental Pollution* 149 (1): 25–30.

Batterman, S., S. Chernyak, Y. Gouden, J. Hayes, T. Robins, and S. Chetty. 2009. PCBs in air, soil and milk in industrialized and urban areas of KwaZulu-Natal, South Africa. *Environmental Pollution* 157 (2): 654–663.

Batty, J., R. A. Leavitt, N. Biondo, and D. Polin. 1990. An ecotoxicological study of a population of the white-footed mouse (*Peromyscus leucopus*) inhabiting a polychlorinated-biphenyls-contaminated area. *Archives of Environmental Contamination and Toxicology* 19 (2): 283–290.

Beltran, A. S., M. Palafox-Uribe, J. Grajales-Montiel, A. Cruz-Villacorta, and J. L. Ochoa. 1997. Sea bird mortality at Cabo San Lucas, Mexico: Evidence that toxic diatom blooms are spreading. *Toxicon* 35 (3): 447–453.

Bernanke, J., and H. R. Kohler. 2009. The impact of environmental chemicals on wildlife vertebrates. *Reviews of Environmental Contamination and Toxicology* 198:1–47.

Berra, T. M. 2001. *Freshwater fish distribution*. San Diego, Calif.: Academic Press.

Birnbaum, L. S., and D. F. Staskal. 2004. Brominated flame retardants: Cause for concern? *Environmental Health Perspectives* 112 (1): 9–17.

Bjorndal, K. A., A. B. Bolten, and C. J. Lagueux. 1994. Ingestion of marine debris by juvenile sea turtles in coastal Florida habitats. *Marine Pollution Bulletin* 28 (3): 154–158.

Blij, H. J. and M. Muller. 2007. *The World Today: Concepts and Regions in Geography.* 3rd edition. New York: Wiley.

Boon, J. P., W. E. Lewis, M. R. Tjoen-A-Choy, C. R. Allchin, R. J. Law, J. de Boer, C. C. ten Hallers-Tjabbes, and B. N. Zegers. 2002. Levels of polybrominated diphenyl ether (PBDE) flame retardants in animals representing different trophic levels of the North Sea food web. *Environmental Science & Technology* 36 (19): 4025–4032.

Borghesi, N., S. Corsolini, and S. Focardi. 2008. Levels of polybrominated diphenyl ethers (PBDEs) and organochlorine pollutants in two species of Antarctic fish (*Chionodraco hamatus* and *Trematomus bernacchii*). *Chemosphere* 73 (2): 155–160.

Bornman, R., J. van Vuren, H. Bouwman, T. de Jager, B. Genthe, and I. Barnhoorn. 2007. Endocrine disruptive activity and the potential health risk in the Rietvlei Natural Reserve. In *The Use of Sentinel Species to Determine the Endocrine Disruptive Activity in an Urban Nature Reserve*. Pretoria, South Africa: Water Research Committee.

Borrell, A., G. Cantos, T. Pastor, and A. Aguilar. 2004. Levels of organochlorine compounds in spotted dolphins from the Coiba archipelago, Panama. *Chemosphere* 54 (5):669–677.

Bossart, G. D., D. G. Baden, R. Y. Ewing, B. Roberts, and S. D. Wright. 1998. Brevetoxicosis in manatees (*Trichechus manatus latirostris*) from the 1996 epizootic: Gross, histologic, and immunohistochemical features. *Toxicologic Pathology* 26 (2): 276–282.

Bouwman, H. 2004. South Africa and the Stockholm Convention on persistent organic pollutants. *South African Journal of Science* 100 (7–8): 323–328.

Bouwman, H., A. Coetzee, and C. H. J. Schutte. 1990. Environmental and health implications of DDT-contaminated fish from the Pongolo Flood Plain South Africa. *Journal of African Zoology* 104 (4): 275–286.

Bouwman, H., A. Polder, B. Venter, and J. U. Skaare. 2008. Organochlorine contaminants in cormorant, darter, egret, and ibis eggs from South Africa. *Chemosphere* 71 (2): 227–241.

Bouwman, H., B. Sereda, and H. M. Meinhardt. 2006. Simultaneous presence of DDT and pyrethroid residues in human breast milk from a malaria endemic area in South Africa. *Environmental Pollution* 144 (3): 902–917.

Bowen, R. E., and M. H. Depledge. 2006. Rapid assessment of marine pollution (RAMP). *Marine Pollution Bulletin* 53 (10–12): 631–639.

Brodie, E. C., F. M. D. Gulland, D. J. Greig, M. Hunter, J. Jaakola, J. St Leger, T. A. Leighfield, and F. M. Van Dolah. 2006. Domoic acid causes reproductive failure in California sea lions (*Zalophus californianus*). *Marine Mammal Science* 22 (3): 700–707.

Buttemer, W. A., P. G. Story, K. J. Fildes, R. V. Baudinette, and L. B. Astheimer. 2008. Fenitrothion, an organophosphate, affects running endurance but not aerobic capacity in fat-tailed dunnarts (*Sminthopsis crassicaudata*). *Chemosphere* 72 (9): 1315–1320.

Carr, A. 1987. Impact of nondegradable marine debris on the ecology and survival outlook of sea turtles. *Marine Pollution Bulletin* 18 (6B): 352–356.

Castillo, L. E., C. Ruepert, and E. Solis. 2000. Pesticide residues in the aquatic environment of banana plantation areas in the North Atlantic Zone of Costa Rica. *Environmental Toxicology and Chemistry* 19 (8):1942–1950.

Chen, D., B. X. Mai, J. Song, Q. H. Sun, Y. Luo, X. J. Luo, E. Y. Zeng, and R. C. Hale. 2007. Polybrominated diphenyl ethers in birds of prey from Northern China. *Environmental Science & Technology* 41 (6): 1828–1833.

Chen, S. J., X. J. Luo, Z. Lin, Y. Luo, K. C. Li, X. Z. Peng, B. X. Mai, Y. Ran, and E. Y. Zeng. 2007. Time trends of polybrominated diphenyl ethers in sediment cores from the Pearl River estuary, South China. *Environmental Science & Technology* 41 (16): 5595–5600.

Christensen, J. R., M. Macduffee, R. W. Macdonald, M. Whiticar, and P. S. Ross. 2005. Persistent organic pollutants in British Columbia grizzly bears: Consequence of divergent diets. *Environmental Science & Technology* 39 (18): 6952–6960.

Colombo, J. C., N. Cappelletti, M. C. Migoya, and E. Speranza. 2007. Bioaccumulation of anthropogenic contaminants by detritivorous fish in the Rio de la Plata estuary: 1-aliphatic hydrocarbons. *Chemosphere* 68 (11):2128–35.

Corsolini, S., C. Guerranti, G. Perra, and S. Focardi. 2008. Polybrominated diphenyl ethers, perfluorinated compounds and chlorinated pesticides in swordfish (*Xiphias gladius*) from the Mediterranean Sea. *Environmental Science & Technology* 42 (12): 4344–4349.

Covaci, A., K. Van de Vijver, W. DeCoen, K. Das, J. M. Bouquegneau, R. Blust, and P. Schepens. 2002. Determination of organohalogenated contaminants in liver of harbour porpoises (*Phocoena phocoena*) stranded on the Belgian North Sea coast. *Marine Pollution Bulletin* 44 (10): 1157–1165.

Cunningham, A. A., V. Prakash, D. Pain, G. R. Ghalsasi, G. A. H. Welis, G. N. Kolte, P. Nighot, M. S. Goudar, S. Kshirsagar, and A. Rahmani. 2003. Indian vultures: Victims of an infectious disease epidemic? *Animal Conservation* 6:189–197.

D'Silva, K., H. Thompson, A. Fernandes, and M. Duff. 2004. PBDEs in heron adipose tissue and eggs from the United Kingdom. In *The Third International Workshop on Brominated Flame Retardants*. Toronto, Canada: BFR 2004.

Daby, D. 2006. Coastal pollution and potential biomonitors of metals in Mauritius. *Water Air and Soil Pollution* 174 (1–4): 63–91.

Daka, P. S., V. C. Obuseng, N. Torto, and P. Huntsman-Mapila. 2006. Deltamethrin in sediment samples of the Okavango Delta, Botswana. *Water SA* 32 (4): 483–488.

———. 2007. Bioaccumulation of anthropogenic contaminants by detritivorous fish in the Rio de la Plata estuary: 2-Polychlorinated biphenyls. *Chemosphere* 69 (8):1253–60.

Daly, G. L., Y. D. Lei, C. Teixeira, D. C. Muir, L. E. Castillo, L. M. Jantunen, and F. Wania. 2007a. Organochlorine pesticides in the soils and atmosphere of Costa Rica. *Environ Sci Technol* 41 (4):1124–30.

Daly, G. L., Y. D. Lei, C. Teixeira, D. C. Muir, L. E. Castillo, and F. Wania. 2007b. Accumulation of current-use pesticides in neotropical montane forests. *Environ Sci Technol* 41 (4):1118–23.

Damstra, Terri, Sue Barlow, Aake Bergman, Robert Kavlock, and Glen Van Der Kraak, eds. 2002. *Global Assessment of the State-of-the-Science of Endocrine Disruptors*. Geneva, Switzerland: International Programme on Chemical Safety/World Health Organization.

Darnerud, P. O. 2003. Toxic effects of brominated flame retardants in man and in wildlife. *Environment International* 29 (6): 841–853.

Darnerud, P. O., G. S. Eriksen, T. Johannesson, P. B. Larsen, and M. Viluksela. 2001. Polybrominated diphenyl ethers: Occurrence, dietary exposure, and toxicology. *Environmental Health Perspectives* 109: 49–68.

de Boer, J., H. Leslie, S. P. J. van Leeuwen, J. W. Wegener, B. van Bavel, G. Lindstrom, N. Lahoutifard, and H. Fiedler. 2008. United Nations Environment Programme Capacity Building Pilot Project: Training and interlaboratory study on persistent organic pollutant analysis under the Stockholm Convention. *Analytica Chimica Acta* 617 (1–2): 208–215.

de Wit, C. A. 2002. An overview of brominated flame retardants in the environment. *Chemosphere* 46 (5): 583–624.

de Wit, C. A., M. Alaee, and D. C. G. Muir. 2006. Levels and trends of brominated flame retardants in the Arctic. *Chemosphere* 64 (2): 209–233.

Dekock, A. C., P. B. Best, V. Cockcroft, and C. Bosma. 1994. Persistent organochlorine residues in small cetaceans from the east and west coasts of southern Africa. *Science of the Total Environment* 154 (2–3): 153–162.

Depledge, M. H. 1993. Ecotoxicology: A science or a management tool. *Ambio* 22 (1): 51–52.

Depledge, M. H., and T. S. Galloway. 2005. Healthy animals, healthy ecosystems. *Frontiers in Ecology and the Environment* 3 (5): 251–258.

Derraik, J. G. B. 2002. The pollution of the marine environment by plastic debris: A review. *Marine Pollution Bulletin* 44 (9): 842–852.

Dhananjayan, V. 2009. Levels of organic contaminants and select biomarkers in birds of Gujarat and Tamil Nadu, India. Bharathiar University, Coimbatore, India.

Di Guardo, G., G. Marruchella, U. Agrimi, and S. Kennedy. 2005. Morbillivirus infections in aquatic mammals: A brief overview. *Journal of Veterinary Medicine Series A – Physiology Pathology Clinical Medicine* 52 (2): 88–93.

Draxler, R. R., J. T. McQueen, and B. J. B. Stunder. 1994. An evaluation of air pollutant exposures due to the 1991 Kuwait oil fires using a Lagrangian model. *Atmospheric Environment* 28 (13): 2197–2210.

El Desouky, M., and M. Y. Abdulraheem. 1991 October. Impact of oil well fires on the air quality in Kuwait. Paper presented at The Environmental and Health Impact of the Kuwait Oil Fires, preceedings of an international symposium at the University of Birmingham, U.K.

Elfvendahl, S., M. Mihale, M. A. Kishimba, and H. Kylin. 2004. Pesticide pollution remains severe after cleanup of a stockpile of obsolete pesticides at Vikuge, Tanzania. *Ambio* 33 (8): 503–508.

Elliott, J. E., L. K. Wilson, and B. Wakeford. 2005. Polyhrominated diphenyl ether trends in eggs of marine and freshwater birds from British Columbia, Canada, 1979–2002. *Environmental Science & Technology* 39 (15): 5584–5591.

Ellis-Hutchings, R. G., G. N. Cherr, L. A. Hanna, and C. L. Keen. 2006. Polybrominated diphenyl ether (PBDE)-induced alterations in vitamin A and thyroid hormone concentrations in the rat during lactation and early postnatal development. *Toxicology and Applied Pharmacology* 215 (2): 135–145.

EPA. 2001. Assessment of the impact of insecticide spraying of Australian plague locusts. Environment Protection Agency, Government of South Australia: Department for Environment and Heritage.

Evans, S. W., and H. Bouwman. 2000. The geographic variation and potential risk of DDT in the blood of pied kingfishers from northern KwaZulu-Natal, South Africa. *Ostrich* 71 (1–2): 351–354.

Everts, J. W. 1997. Ecotoxicology for risk assessment in arid zones: Some key issues. *Archives of Environmental Contamination and Toxicology* 32 (1): 1–10.

Fair, P. A., G. Mitchum, T. C. Hulsey, J. Adams, E. Zolman, W. McFee, E. Wirth, and G. D. Bossart. 2007. Polybrominated diphenyl ethers (pbdes) in blubber of free-ranging bottlenose dolphins (*Tursiops truncatus*) from two southeast Atlantic estuarine areas. *Archives of Environmental Contamination and Toxicology* (3), http://www.springerlink.com/content/a5u215220l7680g6/.

Feldkamp, S. D., D. P. Costa, and G. K. Dekrey. 1989. Energetic and behavioral-effects of net entanglement on juvenile northern fur seals (*Callorhinus ursinus*). *Fishery Bulletin* 87 (1): 85–94.

Fent, K., A. A. Weston, and D. Caminada. 2006. Ecotoxicology of human pharmaceuticals *Aquatic Toxicology* 76 (2) 122–159; 78 (2): 207.

Fernie, K. J., J. L. Shutt, G. Mayne, D. Hoffman, R. J. Letcher, K. G. Drouillard, and I. J. Ritchie. 2005. Exposure to polybrominated diphenyl ethers (PBDEs): Changes in thyroid, vitamin A, glutathione homeostasis, and oxidative stress in American kestrels (*Falco sparverius*). *Toxicological Sciences* 88 (2): 375–383.

Flewelling, L. J., J. P. Naar, J. P. Abbott, D. G. Baden, N. B. Barros, G. D. Bossart, M. Y. D. Bottein, et al. 2005. Red tides and marine mammal mortalities. *Nature* 435 (7043): 755–756.

Fossi, M. C., and L. Marsili. 2003a. Effects of endocrine disruptors in aquatic mammals. *Pure and Applied Chemistry* 75 (11–12): 2235–2247.

———. 2003b. Organochlorines with endocrine disrupting capacity in marine mammals: The case study of Mediterranean cetaceans. *Organohalogen Compounds* 62:244–248

Fossi, M. C., L. Marsili, C. Leonzio, G. N. Disciara, M. Zanardelli, and S. Focardi. 1992. The use of nondestructive biomarker in Mediterranean cetaceans: Preliminary data on MFO activity in skin biopsy. *Marine Pollution Bulletin* 24 (9): 459–461.

Fossi, M. C., L. Marsili, S. Casini, and D. Bucalossi. 2006. Development of new tools to investigate toxicological hazard due to endocrine disruptor organochlorines and emerging contaminants in Mediterranean cetaceans. *Marine Environmental Research* 62:S200–S204.

Furness, R. W., and C. J. Camphuysen. 1997. Seabirds as monitors of the marine environment. *ICES Journal of Marine Science* 54 (4): 726–737.

Gammons, C. H., D. G. Slotton, B. Gerbrandt, W. Weight, C. A. Young, R. L. McNearny, E. Camac, R. Calderon, and H. Tapia. 2006. Mercury concentrations of fish, river water, and sediment in the Rio Ramis-Lake Titicaca watershed, Peru. *Sci Total Environ* 368 (2-3):637–48.

García-Hernández, J., Y. V. Sapozhnikova, D. Schlenk, A. Z. Mason, O. Hinojosa-Huerta, J. J. Rivera-Díaz, N. A. Ramos-Delgado, and G. Sánchez-Bon. 2006. Concentration of contaminants in breeding bird eggs from the Colorado River Delta, Mexico. *Environmental Toxicology and Chemistry* 25 (6):1640–1647.

Geraci, J. R., D. M. Anderson, R. J. Timperi, D. J. Staubin, G. A. Early, J. H. Prescott, and C. A. Mayo. 1989. Humpback whales (*Megaptera novaeangliae*) fatally poisoned by dinoflagellate toxin. *Canadian Journal of Fisheries and Aquatic Sciences* 46 (11): 1895–1898.

Giesy, J. P., and K. Kannan. 2001. Global distribution of perfluorooctane sulfonate in wildlife. *Environmental Science & Technology* 35 (7): 1339–1342.

Giesy, J. P., L. A. Feyk, P. D. Jones, K. Kannan, and T. Sanderson. 2003. Review of the effects of endocrine-disrupting chemicals in birds. *Pure and Applied Chemistry* (11–12), http://media.iupac.org/publications/pac/2003/pdf/7511x2287.pdf.

Gilmartin, W. G., R. L. DeLong, A. W. Smith, L. A. Griner, and M. D. Dailey. 1980. An investigation into unusual mortality in the Hawaiian monk seal, *Monachus schauinslandi*. Paper presented at Symposium on Status of Resource Investigations in the Northwestern Hawaiian Islands, at University of Hawaii, Honolulu.

Godard, C. A. J., R. M. Smolowitz, J. Y. Wilson, R. S. Payne, and J. J. Stegeman. 2004. Induction of cetacean cytochrome P4501A1 by beta-naphthoflavone exposure of skin biopsy slices. *Toxicological Sciences* 80 (2): 268–275.

Godard, C. A. J., S. S. Wise, R. S. Kelly, B. Goodale, S. Kraus, T. Romano, T. O'Hara, and J. P. Wise. 2006. Benzo[a]pyrene cytotoxicity in right whale (*Eubalaena glacialis*) skin, testis and lung cell lines. *Marine Environmental Research* 62:S20–S24.

Goldstein, M. I., T. E. Lacher, B. Woodbridge, M. J. Bechard, S. B. Canavelli, M. E. Zaccagnini, G. P. Cobb, E. J. Scollon, R. Tribolet, and M. J. Hopper. 1999. Monocrotophos-induced mass mortality of Swainson's hawks in Argentina, 1995–96. *Ecotoxicology* 8 (3):201–214.

Gouin, T., F. Wania, C. Ruepert, and L. E. Castillo. 2008. Field testing passive air samplers for current use pesticides in a tropical environment. *Environ Sci Technol* 42 (17):6625–30.

Gulland, F. M. D., M. Haulena, D. Fauquier, G. Langlois, M. E. Lander, T. Zabka, and R. Duerr. 2002. Domoic acid toxicity in Californian sea lions (*Zalophus californianus*): Clinical signs, treatment and survival. *Veterinary Record* 150 (15): 475–480.

Gupta, P. K. 1986. *Pesticides in the Indian Environment*. Edited by B. Bhatia and C. K. Varshney. New Delhi, India: Interprint Publishers.

Hainzl, D., L. M. Cole, and J. E. Casida. 1998. Mechanisms for selective toxicity of fipronil insecticide and its sulfone metabolite and desulfinyl photoproduct. *Chemical Research in Toxicology* 11:1529–1535.

Hale, Robert C., Mehran Alaee, Jon B. Manchester-Neesvig, Heather M. Stapleton, and Michael G. Ikonomou. 2003. Polybrominated diphenyl ether flame retardants in the North American environment. *Environment International* 29 (6): 771–779.

Hale, R. C., M. J. La Guardia, E. Harvey, and T. M. Mainor. 2002. Potential role of fire retardant-treated polyurethane foam as a source of brominated diphenyl ethers to the US environment. *Chemosphere* 46 (5): 729–735.

Hale, R. C., M. J. La Guardia, E. Harvey, M. O. Gaylor, and T. M. Mainor. 2006. Brominated flame retardant concentrations and trends in abiotic media. *Chemosphere* 64 (2): 181–186.

Hall, A. J., O. I. Kalantzi, and G. O. Thomas. 2003. Polybrominated diphenyl ethers (PBDEs) in grey seals during their first year of life: Are they thyroid hormone endocrine disrupters? *Environmental Pollution* 126 (1): 29–37.

Hallegraeff, G. M. 1993. A review of harmful algal blooms and their apparent global increase. *Phycologia* 32 (2): 79–99.

Hallgren, S., and P. O. Darnerud. 2002. Polybrominated diphenyl ethers (PBDEs), polychlorinated biphenyls (PCBs) and chlorinated paraffins (CPs) in rats: Testing interactions and mechanisms for thyroid hormone effects. *Toxicology* 177 (2–3): 227–243.

Halpern, B. S., K. A. Selkoe, F. Micheli, and C. V. Kappel. 2007. Evaluating and ranking the vulnerability of global marine ecosystems to anthropogenic threats. *Conservation Biology* 21: 1301–1315.

Hamers, T., J. H. Kamstra, E. Sonneveld, A. J. Murk, M. H. A. Kester, P. L. Andersson, J. Legler, and A. Brouwer. 2006. In vitro profiling of the endocrine-disrupting potency of brominated flame retardants. *Toxicological Sciences* 92 (1): 157–173.

Handy, R. D., R. Owen, and E. Valsami-Jones. 2008. The ecotoxicology of nanoparticles and nanomaterials: current status, knowledge gaps, challenges, and future needs. *Ecotoxicology* 17 (5): 315–325.

Härkönen, T., T. Dietz, P. Reijnders, J. Teilmann, K. Harding, A. Hall, S. Brasseur, et al. 2006. A review of the 1988 and 2002 phocine distemper virus epidemics in European harbour seals. *Diseases of Aquatic Organisms* (2), http://www.nrm.se/download/18.382ec74311 3b0c5a2a180001228/Harkonen_2006.pdf.

Harvell, C. D., K. Kim, J. M. Burkholder, R. R. Colwell, P. R. Epstein, D. J. Grimes, E. E. Hofmann, et al. 1999. Review: Marine ecology—Emerging marine diseases—Climate links and anthropogenic factors. *Science* 285 (5433): 1505–1510.

Harvell, C. D., C. E. Mitchell, J. R. Ward, S. Altizer, A. P. Dobson, R. S. Ostfeld, and M. D. Samuel. 2002. Ecology: Climate warming and disease risks for terrestrial and marine biota. *Science* 296 (5576): 2158–2162.

Heidorn, C. 2002. *Use of Plant Protection Products in the European Union: Data 1992–1999*. Luxembourg: Eurostat, Office for Official Publications of the European Communities.

Hernandez, M., I. Robinson, A. Aguilar, L. M. Gonzalez, L. F. Lopez-Jurado, M. I. Reyero, E. Cacho, J. Franco, V. Lopez-Rodas, and E. Costas. 1998. Did algal toxins cause monk seal mortality? *Nature* 393 (6680): 28–29.

Hernández-Romero, A. H., C. Tovilla-Hernández, E.A. Malo, and R. Bello-Mendoza. 2004. Water quality and presence of pesticides in a tropical coastal wetland in southern Mexico. *Marine Pollution Bulletin* 48 (11-12):1130–1141.

Herzke, D., U. Berger, R. Kallenborn, T. Nygard, and W. Vetter. 2005. Brominated flame retardants and other organobromines in Norwegian predatory bird eggs. *Chemosphere* 61 (3): 441–449.

Higgins, D., C. Berkley, and M. B. Jones. 2002. Managing heterogeneous ecological data using Morpho. Proceedings 14th International Conference on Scientific and Statistical Database Management, 69–76.

Husain, T. 1995. *Kuwait Oil Fires: Regional Environmental Perspectives*. 1st ed. Oxford, U.K.: BPC Wheatons Ltd.

Huwe, J., H. Hakk, and M. Lorentzsen. 2007. Bioavailability and mass balance studies of a commercial pentabromodiphenyl ether mixture in male Sprague-Dawley rats. *Chemosphere* 66 (2): 259–266.

Ikonomou, M. G., S. Rayne, and R. F. Addison. 2002. Exponential increases of the brominated flame retardants, polybrominated diphenyl ethers, in the Canadian Arctic from 1981 to 2000. *Environmental Science & Technology* 36 (9): 1886–1892.

Jaspers, V. L. B., A. Covaci, S. Voorspoels, T. Dauwe, M. Eens, and P. Schepens. 2006. Brominated flame retardants and organochlorine pollutants in aquatic and terrestrial predatory birds of Belgium: Levels, patterns, tissue distribution and condition factors. *Environmental Pollution* 139 (2): 340–352.

Jiménez, B., R. Rodriguez-Estrella, R. Merino, G. Gomez, L. Rivera, M. J. Gonzalez, E. Abad, and J. Rivera. 2005. Results and evaluation of the first study of organochlorine contaminants (PCDDs, PCDFs, PCBs and DDTs), heavy metals and metalloids in birds from Baja California, Mexico. *Environ Pollut* 133 (1):139–46.

Jiménez, B., G. Mori, M. A. Concejero, R. Merino, S. Casini, and M. C. Fossi. 2007. Vitellogenin and zona radiata proteins as biomarkers of endocrine disruption in peregrine falcon (*Falco peregrinus*). *Chemosphere* 67 (9): S375–S378.

JMPR. 2001. Pesticide residues in food: Fipronil — 2001 evaluations. Part I, FAO Plant Production and Protection Paper 171. Paper presented at Joint Meeting on Pesticide Residues: FAO and WHO.

Jobling, S., and C. R. Tyler. 2003. Endocrine disruption in wild freshwater fish. *Pure and Applied Chemistry* (11–12), http://media.iupac.org/publications/pac/2003/pdf/7511x2219.pdf.

Johnson-Restrepo, B., D. H. Adams, and K. Kannan. 2008. Tetrabromobisphenol A (TBBPA) and hexabromocyclododecanes (HBCDs) in tissues of humans, dolphins, and sharks from the United States. *Chemosphere* 70 (11): 1935–1944.

Kajiwara, N., S. Kamikawa, K. Ramu, D. Ueno, T. K. Yamada, A. Subramanian, P. K. S. Lam, et al. 2006. Geographical distribution of polybrominated diphenyl ethers (PBDEs) and organochlorines in small cetaceans from Asian waters. *Chemosphere* 64 (2): 287–295.

Kannan, K., S. Tanabe, A. Ramesh, A. Subramanian, and R. Tatsukawa. 1992. Persistent organochlorine residues in foodstuffs from India and their implications on human dietary exposure. *Journal of Agricultural and Food Chemistry* 40 (3): 518–524.

Kannan, K., J. C. Franson, W. W. Bowerman, K. J. Hansen, J. D. Jones, and J. P. Giesy. 2001. Perfluorooctane sulfonate in fish-eating water birds including bald eagles and albatrosses. *Environmental Science & Technology* 35 (15): 3065–3070.

Kannan, K., J. Koistinen, K. Beckmen, T. Evans, J. F. Gorzelany, K. J. Hansen, P. D. Jones, E. Helle, M. Nyman, and J. P. Giesy. 2001. Accumulation of perfluorooctane sulfonate in marine mammals. *Environmental Science & Technology* 35 (8): 1593–1598.

Kannan, K., S. Corsolini, J. Falandysz, G. Oehme, S. Focardi, and J. P. Giesy. 2002. Perfluorooctanesulfonate and related fluorinated hydrocarbons in marine mammals, fishes, and birds from coasts of the Baltic and the Mediterranean Seas. *Environmental Science & Technology* 36 (15): 3210–3216.

Kaphalia, B. S., M. M. Husain, T. D. Seth, A. Kumar, and C. R. K. Murti. 1981. Organochlorine pesticide-residues in some Indian wild birds. *Pesticides Monitoring Journal* 15 (1): 9–13.

Karlsson, H., D. C. G. Muir, C. F. Teixiera, D. A. Burniston, W. M. J. Strachan, R. E. Hecky, J. Mwita, H. A. Bootsma, N. P. Grift, K. A. Kidd, and B. Rosenberg. 2000. Persistent chlorinated pesticides in air, water, and precipitation from the Lake Malawi area, southern Africa. *Environmental Science & Technology* 34 (21): 4490–4495.

Keith, J. O., J. G. Ngondi, R. L. Bruggers, B. A. Kimball, and C. C. H. Elliott. 1994. Environmental effects on wetlands of Queletox(r) applied to ploceid roosts in Kenya. *Environmental Toxicology & Chemistry* 13 (2): 333–341.

Kendall, R. J. 1994. Using information derived from wildlife toxicology to model ecological effects of agricultural pesticides and other environmental contaminants on wildlife populations. In *Wildlife Toxicology and Population Modeling: Integrated Studies of Agroecosystems,* edited by R. Kendall and T. Lacher; 1–11. Boca Raton: CRC Press.

Kishimba, M. A., L. Henry, H. Mwevura, A. J. Mmochi, M. Mihale, and H. Hellar. 2004. The status of pesticide pollution in Tanzania. *Talanta* 64 (1): 48–53.

Kitulagodage, M., L. B. Astheimer, and W. A. Buttemer. 2008. Diacetone alcohol, a dispersant solvent, contributes to acute toxicity of a fipronil-based insecticide in a passerine bird. *Ecotoxicology and Environmental Safety* 71 (2): 597–600.

Klemens, J. A., M. L. Wieland, V. J. Flanagin, J. A. Frick, and R. G. Harper. 2003. A cross-taxa survey of organochlorine pesticide contamination in a Costa Rican wildland. *Environ Pollut* 122 (2):245–51.

Knap, A., E. Dewailly, C. Furgal, J. Galvin, D. Baden, R. E. Bowen, M. Depledge, et al. 2002. Indicators of ocean health and human health: Developing a research and monitoring framework. *Environmental Health Perspectives* 110 (9): 839–845.

Kovarova, J., and Z. Svobodova. 2008. Perfluorinated compounds: Occurrence and risk profile. *Neuroendocrinology Letters* 29: 599–608.

Kunisue, T., N. Takayanagi, T. Isobe, S. Takahashi, S. Nakatsu, T. Tsubota, K. Okumoto, S. Bushisue, K. Shindo, and S. Tanabe. 2008. Regional trend and tissue distribution of brominated flame retardants and persistent organochlorines in raccoon dogs (*Nyctereutes procyonoides*) from Japan. *Environmental Science & Technology* 42 (3): 685–691.

Lacher, T. E., Jr., and M. I. Goldstein. 1997. Tropical ecotoxicology: Status and needs. *Environmental Toxicology and Chemistry* 16 (1):100–111.

Laist, D. W. 1987. Overview of the biological effects of lost and discarded plastic debris in the marine environment. *Marine Pollution Bulletin* 18 (6B): 319–326.

———. 1997. Impacts of marine debris: Entanglement of marine life in marine debris including a comprehensive list of species with entanglement and ingestion records. In *Marine Debris: Sources, Impacts, and Solutions*, edited by J. M. Coe and D. B. Rogers. New York: Springer.

Lallas, P. L. 2001. The Stockholm Convention on persistent organic pollutants. *American Journal of International Law* 95 (3): 692–708.

Lam, J. C. W., N. Kajiwara, K. Ramu, S. Tanabe, and P. K. S. Lam. 2007. Assessment of polybrominated diphenyl ethers in eggs of waterbirds from South China. *Environmental Pollution* 148 (1): 258–267.

Landsberg, J. H. 2002. The effects of harmful algal blooms on aquatic organisms. *Reviews in Fisheries Science* 10 (2): 113–390.

Law, R. J., M. Alaee, C. R. Allchin, J. P. Boon, M. Lebeuf, P. Lepom, and G. A. Stern. 2003. Levels and trends of polybrominated diphenylethers and other brominated flame retardants in wildlife. *Environment International* 29 (6): 757–770.

Law, R. J., D. Herzke, S. Harrad, S. Morris, P. Bersuder, and C. R. Allchin. 2008. Levels and trends of HBCD and BDEs in the European and Asian environments, with some information for other BFRs. *Chemosphere* 73 (2): 223–241.

Law, Robin J. 2009. Tetrabromobisphenol A: Investigating the worst-case scenario. *Marine Pollution Bulletin* 58 (4): 459–460.

Law, R. J., C. R. Allchin, J. de Boer, A. Covaci, D. Herzke, P. Lepom, S. Morris, J. Tronczynski, and C. A. de Wit. 2006. Levels and trends of brominated flame retardants in the European environment. *Chemosphere* 64 (2): 187–208.

Legler, J. 2008. New insights into the endocrine disrupting effects of brominated flame retardants. *Chemosphere* 73 (2): 216–222.

Lema, S. C., I. R. Schultz, N. L. Scholz, J. P. Incardona, and P. Swanson. 2007. Neural defects and cardiac arrhythmia in fish larvae following embryonic exposure to 2,2′,4,4′–tetrabromodiphenyl ether (PBDE 47). *Aquatic Toxicology* 82 (4): 296–307.

Letcher, R. J., and P. A. Behnisch. 2003. The state-of-the-science and trends of brominated flame retardants in the environment: Present knowledge and future directions. *Environment International* 29 (6): 663–664.

Lindberg, P., U. Sellstrom, L. Haggberg, and C. A. de Wit. 2004. Higher brominated diphenyl ethers and hexabromocyclododecane found in eggs of peregrine falcons (*Falco peregrinus*) breeding in Sweden. *Environmental Science & Technology* 38 (1): 93–96.

Liu, Y., G. J. Zheng, H. X. Yu, M. Martin, B. J. Richardson, M. H. W. Lam, and P. K. S. Lam. 2005. Polybrominated diphenyl ethers (PBDEs) in sediments and mussel tissues from Hong Kong marine waters. *Marine Pollution Bulletin* 50 (11): 1173–1184.

Lorber, M. 2008. Exposure of Americans to polybrominated diphenyl ethers. *Journal of Exposure Science and Environmental Epidemiology* 18 (1): 2–19.

Lotze, H. K., H. S. Lenihan, B. J. Bourque, R. H. Bradbury, R. G. Cooke, M. C. Kay, S. M. Kidwell, M. X. Kirby, C. H. Peterson, and J. B. C. Jackson. 2006. Depletion, degradation, and recovery potential of estuaries and coastal seas. *Science* 312 (5781): 1806–1809.

Love, G., and D. Riwoe. 2005. Economic costs and benefits of locust control in Eastern Australia. Canberra, Australia: Australian Bureau for Agricultural and Resource Economics (ABARE) eReport 05.14.

Luross, J. M., M. Alaee, D. B. Sergeant, C. M. Cannon, D. M. Whittle, K. R. Solomon, and D. C. G. Muir. 2002. Spatial distribution of polybrominated diphenyl ethers and polybrominated biphenyls in lake trout from the Laurentian Great Lakes. *Chemosphere* 46 (5): 665–672.

Mai, B. X., S. J. Chen, X. J. Luo, L. G. Chen, Q. S. Yang, G. Y. Sheng, P. G. Peng, J. M. Fu, and E. Y. Zeng. 2005. Distribution of polybrominated diphenyl ethers in sediments of the Pearl River Delta and adjacent South China Sea. *Environmental Science & Technology* 39 (10): 3521–3527.

Marshall, D. J., and A. Rajkumar. 2003. Imposex in the indigenous *Nassarius kraussianus* (Mollusca : Neogastropoda) from South African harbours. *Marine Pollution Bulletin* 46 (9): 1150–1155.

Marsili, L. 2000. Lipophilic contaminants in marine mammals: Review of the results of ten years' work at the Department of Environmental Biology, Siena University (Italy). *International Journal of Environment and Pollution* 13 (1–6): 416–452.

Mason, C. F., and S. M. Macdonald. 2004. Growth in otter (*Lutra lutra*) populations in the UK as shown by long-term monitoring. *Ambio* 33 (3): 148–152.

Mathur, K. K., G. Harpalani, N. L. Kalra, G. G. K. Murthy, and M. V. V. L. Narasimham. 1992. Epidemic of malaria in Barmer district Thar Desert of Rajasthan during 1990. *Indian Journal of Malariology* 29 (1): 1–10.

Mbongwe, B., M. Legrand, J. M. Blais, L. E. Kimpe, J. J. Ridal, and D. R. S. Lean. 2003. Dichlorodiphenyltrichloroethane in the aquatic ecosystem of the Okavango Delta, Botswana, South Africa. *Environmental Toxicology & Chemistry* 22 (1): 7–19.

McKernan, M. A., B. A. Rattner, R. C. Hale, and M. A. Ottinger. 2009. Toxicity of polybrominated diphenyl ethers (DE-71) in chicken (*Gallus gallus*), mallard (*Anas platyrhynchos*), and American kestrel (*Falco sparverius*) embryos and hatchlings. *Environmental Toxicology & Chemistry* 28 (5): 1007–1017.

McKinney, M. A., L. S. Cesh, J. E. Elliott, T. D. Williams, D. K. Garcelon, and R. J. Letcher. 2006. Brominated flame retardants and halogenated phenolic compounds in North American west coast bald eaglet (*Haliaeetus leucocephalus*) plasma. *Environmental Science & Technology* 40 (20): 6275–6281.

Meerts, I. A. T. M., R. J. Letcher, S. Hoving, G. Marsh, A. Bergman, J. G. Lemmen, B. van der Burg, and A. Brouwer. 2001. In vitro estrogenicity of polybrominated diphenyl ethers, hydroxylated PBDEs, and polybrominated bisphenol A compounds. *Environmental Health Perspectives* 109 (4): 399–407.

Millennium Ecosystem Assessment. 2005. *Ecosystems and human well-being: current state and trends*. Washington, D.C., USA.

Minh, N. H., T. Isobe, D. Ueno, K. Matsumoto, M. Mine, N. Kajiwara, S. Takahashi, and S. Tanabe. 2007. Spatial distribution and vertical profile of polybrominated diphenyl ethers and hexabromocyclododecanes in sediment core from Tokyo Bay, Japan. *Environmental Pollution* 148 (2): 409–417.

Minh, N. H., T. B. Minh, N. Kajiwara, T. Kunisue, H. Iwata, P. H. Viet, N. P. C. Tu, B. C. Tuyen, and S. Tanabe. 2006. Contamination by polybrominated diphenyl ethers and persistent organochlorines in catfish and feed from Mekong River Delta, Vietnam. *Environmental Toxicology & Chemistry* 25 (10): 2700–2708.

Misra, V., and P. P. Bakre. 1994. Organochlorine contaminants and avifauna of Mahala–Water-Reservoir, Jaipur, India. *Science of the Total Environment* 144:145–151.

Moon, H. B., K. Kannan, S. J. Lee, and M. Choi. 2007. Polybrominated diphenyl ethers (PBDEs) in sediment and bivalves from Korean coastal waters. *Chemosphere* 66 (2): 243–251.

Moore, C. J. 2008. Synthetic polymers in the marine environment: A rapidly increasing, long-term threat. *Environmental Research* 108 (2): 131–139.

Moore, M. N., M. H. Depledge, J. W. Readman, and D. R. P. Leonard. 2004. An integrated biomarker-based strategy for ecotoxicological evaluation of risk in environmental management. *Mutation Research – Fundamental and Molecular Mechanisms of Mutagenesis* 552 (1–2): 247–268.

Mora, M. A. 2008. Organochlorine pollutants and stable isotopes in resident and migrant passerine birds from northwest Michoacan, Mexico. *Arch Environ Contam Toxicol* 55 (3):488–95.

Mora, M. A., A. B. Montoya, M. C. Lee, A. Macías-Duarte, R. Rodríguez-Salazar, P. W. Juergens, and A. Lafón-Terrazas. 2008. Persistent environmental pollutants in eggs of aplomado falcons from Northern Chihuahua, Mexico, and South Texas, USA. *Environment International* 34 (1):44–50.

Mora, M. A. 1997. Transboundary pollution: Persistent organochlorine pesticides in migrant birds of the Southwestern United States and Mexico. *Environmental Toxicology and Chemistry* 16 (1):3–11.

Mos, L. 2001. Domoic acid: A fascinating marine toxin. *Environmental Toxicology and Pharmacology* 9 (3): 79–85.

Mullie, W. C., A. O. Diallo, B. Gadji, and M. D. Ndiaye. 1999. Environmental hazards of mobile ground spraying with cyanophos and fenthion for Quelea control in Senegal. *Ecotoxicology and Environmental Safety* 43 (1): 1–10.

Mullie, W. C., and J. O. Keith. 1993. The effects of aerially applied fenitrothion and chlorpyrifos on birds in the savanna of northern Senegal. *Journal of Applied Ecology* 30 (3): 536–550.

Muralidharan, S. 1993. Aldrin poisoning of Sarus cranes (*Grus antigone*) and a few granivorous birds in Keoladeo National Park, Bharatpur, India. *Ecotoxicology* 2 (3): 196–202.

———. 2002. Pesticide contamination: A cause for population decline in Indian avifauna. Paper read at International conference on Life Cycle Assessment, at Bombay, India.

Muralidharan, S., and S. M. Murugavel. 2001. Pesticide contamination in select species of birds in Nilgiri District, Tamil Nadu. Paper read at International Seminar on Environment and Food Security, at New Delhi, India.

Muralidharan, S., V. Dhananjayan, and R. Sankardoss. 2004. Impact of agricultural chemicals on the Indian avifauna (1999–2003) – An overview. Paper read at National Workshop on Recent Trends in Pollution Control and Environmental Conservation, at Sathyamangalam, Tamil Nadu, India.

Muralidharan, S., V. Dhananjayan, R. Risebrough, V. Prakash, R. Jayakumar, and P. Bloom. 2008. Persistent organochlorine pesticide residues in tissues and eggs of white-backed vulture, *Gyps bengalensis*, from different locations in India. *Bulletin of Environmental Contamination and Toxicology* 81 (6): 561–565.

Naoroji, R. 1997. Contamination in egg shells of Himalayan greyheaded fishing eagle Ichthyophaga nana plumbea in Corbett National Park, India. *Journal of the Bombay Natural History Society* 94 (2): 398–400.

Ngabe, B., and T. F. Bidleman. 2006. DDT concentrations in soils of Brazzaville, Congo. *Bulletin of Environmental Contamination and Toxicology* 76 (4): 697–704.

Ntow, W. J. 2001. Organochlorine pesticides in water, sediment, crops, and human fluids in a farming community in Ghana. *Archives of Environmental Contamination and Toxicology* 40 (4): 557–563.

Oaks, J. L., M. Gilbert, M. Z. Virani, R. T. Watson, C. U. Meteyer, B. A. Rideout, H. L. Shivaprasad, et al. 2004. Diclofenac residues as the cause of vulture population decline in Pakistan. *Nature* 427 (6975): 630–633.

Osibanjo, O., H. Bouwman, N. H. H. Bashir, J. Okond'Ahoka, R. Choong Kwet Yve, and H. A. Onyoyo. 2002. Regionally based assessment of persistent toxic substances: Sub-Saharan Africa regional report. *UNEP Chemicals*, http://www.chem.unep.ch/pts/regreports/ssafrica.pdf.

Pain, D. J., A. A. Cunningham, P. F. Donald, J. W. Duckworth, D. C. Houston, T. Katzner, J. Parry-Jones, et al. 2003. Causes and effects of temporospatial declines of *Gyps* vultures in Asia. *Conservation Biology* 17 (3): 661–671.

Pain, D. J., R. Gargi, A. A. Cunningham, A. Jones, and V. Prakash. 2004. Mortality of globally threatened Sarus cranes *Grus antigon* from monocrotophos poisoning in India. *Science of the Total Environment* 326 (1–3): 55–61.

Patz, J. A., Epstein P. R., T. A. Burke, and J. M. Balbus. 1996. Global climate change and emerging infectious diseases. *The Journal of the American Medical Association* (3), http://jama.ama-assn.org/cgi/content/abstract/275/3/217.

Peng, J. H., C. W. Huang, Y. M. Weng, and H. K. Yak. 2007. Determination of polybrominated diphenyl ethers (PBDEs) in fish samples from rivers and estuaries in Taiwan. *Chemosphere* 66 (10): 1990–1997.

Penteado, J. C. P., O. A. El Seoud, and L. R. F. Carvalho. 2006. Alquilbenzeno sulfonato linear: uma abordagem ambiental e analítica. *Química Nova* 29:1038–1046.

Pepper, C. B., T. R. Rainwater, S. G. Platt, J. A. Dever, T. A. Anderson, and S. T. McMurry. 2004. Organochlorine pesticides in chorioallantoic membranes of Morelet's crocodile eggs from belize. *J Wildl Dis* 40 (3):493–500.

Petreas, M., and D. Oros. 2009. Polybrominated diphenyl ethers in California wastestreams. *Chemosphere* 74 (7): 996–1001.

Pew Oceans Commission. 2003. *America's Living Oceans: Charting a Course for Sea Change*. Arlington, VA: Pew Oceans Commission.

Pierce, R. H., and M. S. Henry. 2008. Harmful algal toxins of the Florida red tide (*Karenia brevis*): Natural chemical stressors in South Florida coastal ecosystems. *Ecotoxicology* 17 (7): 623–631.

Pink. 2008. 2008 Yearbook Australia. Canberra, Australia: Australian Bureau of Statistics.

Ponce Vélez, G., and A.V. Botello. 2005. Niveles de hidrocarburos en el Golfo de México. In *Golfo de México Contaminación e Impacto Ambiental: Diagnóstico y Tendencias*, edited by A. V. Botello, J. Rendón von Osten, G. Gold Bouchot and C. Agraz Hernández: Univ. Auton. De Campeche, Univ. Nal. Auton. De México, Instituto Nacional de Ecología.

Rainwater, T. R., T. H. Wu, A. G. Finger, J. E. Canas, L. Yu, K. D. Reynolds, G. Coimbatore, B. Barr, S. G. Platt, G. P. Cobb, T. A. Anderson, and S. T. McMurry. 2007. Metals and organochlorine pesticides in caudal scutes of crocodiles from Belize and Costa Rica. *Sci Total Environ* 373 (1):146–56.

Ramesh, A., S. Tanabe, K. Kannan, A. N. Subramanian, P. L. Kumaran, and R. Tatsukawa. 1992. Characteristic trend of persistent organochlorine contamination in wildlife from a tropical agricultural watershed, South India. *Archives of Environmental Contamination and Toxicology* 23 (1): 26–36.

Ramirez, A. J., R. A. Brain, S. Usenko, M. A. Mottaleb, J. G. O'Donnel, L. L. Stahl, J. B. Wathen, et al. 2009. Occurrence of pharmaceuticals and personal care products (PPCPs) in fish: Results of a national pilot study in the U.S. *Environmental Toxicology and Chemistry*, http://www.setacjournals.org/archive/1552-8618/preprint/2009/pdf/10.1897_08-561.1.pdf.

Ramu, K., N. Kajiwara, P. K. S. Lam, T. A. Jefferson, K. Y. Zhou, and S. Tanabe. 2006. Temporal variation and biomagnification of organohalogen compounds in finless porpoises (*Neophocaena phocaenoides*) from the South China Sea. *Environmental Pollution* 144 (2): 516–523.

Ramu, K., N. Kajiwara, H. Mochizuki, H. Miyasaka, K. A. Asante, S. Takahashi, S. Ota, H. M. Yeh, S. Nishida, and S. Tanabe. 2006. Occurrence of organochlorine pesticides, polychlorinated biphenyls and polybrominated diphenyl ethers in deep-sea fishes from the Sulu Sea. *Marine Pollution Bulletin* 52 (12): 1827–1832.

Ramu, K., N. Kajiwara, S. Tanabe, P. K. S. Lam, and T. A. Jefferson. 2005. Polybrominated diphenyl ethers (PBDEs) and organochlorines in small cetaceans from Hong Kong waters: Levels, profiles and distribution. *Marine Pollution Bulletin* 51 (8–12): 669–676.

Ramu, K., N. Kajiwara, A. Sudaryanto, T. Isobe, S. Takahashi, A. Subramamian, D.Ueno, et al. 2007. Asian mussel watch program: Contamination status of polybrominated diphenyl ethers and organochlorines in coastal waters of Asian countries. *Environmental Science and Technology* (13), http://pubs.acs.org/doi/pdf/10.1021/es070380p?cookieSet=1.

Ranapratap, S. 2004. Levels of organochlorine residues in eggs of a few species of birds with reference to usage pattern in select locations in Tamil Nadu. Bharathiar University, Coimbatore, Tamil Nadu, India.

Rao, C. V., and S. A. Banerji. 1988. Polychlorinated biphenyls in the human adipose tissue and liver samples of Bombay. *Toxicological and Environmental Chemistry* 17 (4): 313–317.

Regupathy, A., and S. Kuttalam. 1990. Monitoring of HCH and DDT residues in a few avian eggs in Tamil Nadu. In *Seminar on Wetland Ecology and Management*. Keoladeo National Park, Bharatpur.

Reijnders, P. J. H. 1986. Reproductive failure in common seals feeding on fish from polluted coastal waters. *Nature* (6096), http://www.nature.com/nature/journal/v324/n6096/abs/324456a0.html.

Reijnders, P. J. H. 1990. Progesterone and estradiol-17-beta concentration profiles throughout the reproductive-cycle in harbor seals (*Phoca vitulina*). *Journal of Reproduction and Fertility* 90 (2): 403–409.

Revenga, C., and Y. Kura. 2003. Status and Trends of Biodiversity of Inland Water Ecosystems In *Technical Series No. 11* Montreal, Canada: Secretariat of the Convention on Biological Diversity.

Ross, P. S. 2000. Marine mammals as sentinels in ecological risk assessment. *Human and Ecological Risk Assessment* 6 (1): 29–46.

Ross, P. S., C. M. Couillard, M. G. Ikonomou, S. C. Johannessen, M. Lebeuf, R. W. Macdonald, and G. T. Tomy. 2009. Large and growing environmental reservoirs of Deca: BDE present an emerging health risk for fish and marine mammals. *Marine Pollution Bulletin* 58 (1): 7–10.

Ryan, P. G. 1988. Effects of ingested plastic on seabird feeding: Evidence from chickens. *Marine Pollution Bulletin* 19 (3): 125–128.

Saiyed, H. N., V. K. Bhatnagar, and Rekha Kashyap. 1999. Impact of pesticide use in India. *Asian-Pacific Newsletter on Occupational Health and Safety* (6), http://www.ttl.fi/Internet/English/Information/Electronic+journals/Asian-Pacific+Newsletter/1999-03/05.htm.

Sala, E., and N. Knowlton. 2006. Global marine biodiversity trends. *Annual Review of Environment and Resources* 31:93–122.

Santillo, David, and Paul Johnston. 2003. Playing with fire: The global threat presented by brominated flame retardants justifies urgent substitution. *Environment International* 29 (6): 725–734.

Schipper, J., J. S. Chanson, F. Chiozza, N. A. Cox, M. Hoffmann, V. Katariya, J. Lamoreux, et al. 2008. The status of the world's land and marine mammals: Diversity, threat, and knowledge. *Science*, http://www.sciencemag.org/cgi/reprint/322/5899/225.pdf.

Scholin, C. A., F. Gulland, G. J. Doucette, S. Benson, M. Busman, F. P. Chavez, J. Cordaro, et al. 2000. Mortality of sea lions along the central California coast linked to a toxic diatom bloom. *Nature* 403 (6765): 80–84.

Scown, T. M., R. van Aerle, B. D. Johnston, S. Cumberland, Lead J. R., R. Owen, and C. R. Tyler. 2009. High doses of intravenously administered titanium dioxide nanoparticles accumulate in the kidneys of rainbow trout but with no observable impairment of renal function. *Toxicological Sciences (Advance Access)* (2), http://toxsci.oxfordjournals.org/cgi/content/abstract/kfp064v1.

Sellstrom, U., A. Bignert, A. Kierkegaard, L. Haggberg, C. A. De Wit, M. Olsson, and B. Jansson. 2003. Temporal trend studies on tetra-and pentabrominated diphenyl ethers and hexabromocyclododecane in guillemot egg from the Baltic Sea. *Environmental Science & Technology* 37 (24): 5496–5501.

Senthilkumar, K., K. Kannan, R. K. Sinha, S. Tanabe, and J. P. Giesy. 1999. Bioaccumulation profiles of polychlorinated biphenyl congeners and organochlorine pesticides in Ganges River dolphins. *Environmental Toxicology & Chemistry* 18 (7): 1511–1520.

She, J. W., M. Petreas, J. Winkler, P. Visita, M. McKinney, and D. Kopec. 2002. PBDEs in the San Francisco Bay Area: Measurements in harbor seal blubber and human breast adipose tissue. *Chemosphere* 46 (5): 697–707.

She, J., A. Holden, M. Sharp, M. Tanner, C. Williams-Derry, and K. Hooper. 2007. Polybrominated diphenyl ethers (PBDEs) and polychlorinated biphenyls (PCBs) in breast milk from the Pacific Northwest. *Chemosphere* 67 (9): S307–S317.

Singhal, C. S. 1999. People's participation in watershed management: Case study of village Nada, Hyderabad. *Journal of Rural Development* 18 (4): 557–564.

Sinkkonen, S., A. L. Rantalainen, J. Paasivirta, and M. Lahtipera. 2004. Polybrominated methoxy diphenyl ethers (MeO-PBDEs) in fish and guillemot of Baltic, Atlantic and Arctic environments. *Chemosphere* 56 (8): 767–775.

Smayda, T. J. 1990. Novel and nuisance phytoplankton blooms in the sea: Evidence for a global epidemic. In *Toxic Marine Phytoplankton,* Proceedings of the Fourth International Conference on Toxic Marine Phytoplankton, held June 26–30, 1989, in Lund, Sweden, edited by E. Graneli, B. Sundstrom, L. Edler, and D. Anderson. New York: Elsevier Science Ltd.

Spear, L. B., D. G. Ainley, and C. A. Ribic. 1995. Incidence of plastic in seabirds from the tropical Pacific, 1984–1991: Relation with distribution of species, sex, age, season, year and body weight. *Marine Environmental Research* (2), http://www.sciencedirect.com/science.

Staskal, D. F., H. Hakk, D. Bauer, J. J. Diliberto, and L. S. Birnbaum. 2006. Toxicokinetics of polybrominated diphenyl ether congeners 47, 99, 100, and 153 in mice. *Toxicological Sciences* 94 (1): 28–37.

Stoker, T. E., R. L. Cooper, C. S. Lambright, V. S. Wilson, J. Furr, and L. E. Gray. 2005. In vivo and in vitro anti-androgenic effects of DE-71, a commercial polybrominated diphenyl ether (PBDE) mixture. *Toxicology and Applied Pharmacology* 207 (1): 78–88.

Story, P. G., D. O. Oliver, T. Deveson, L. McCulloch, J.G. Hamilton, and D. Baker-Gabb. 2007. Estimating and reducing the amount of plains-wanderer (*Pedionomus torquatus* Gould) habitat sprayed with pesticides for locust control in the New South Wales Riverina. *Emu – Austral Ornithology* 107 (4): 308–314.

Swanson, R. Lawrence, Marci L. Bortman, Thomas P. O'Connor, and Harold M. Stanford. 2004. Science, policy and the management of sewage materials: The New York City experience. *Marine Pollution Bulletin* 49:679–687.

Szabo, J., L. B. Astheimer, P. G Story, and W.A. Buttemer. 2003. Birds, locusts and pesticides: Managing an ephemeral feast: The risks of locust control pesticides to Australian birds. *Wingspan* 13 (3): 10–15.

Szabo, J. K. 2005. Avian-locust interactions in eastern Australia and the exposure of birds to locust control pesticides. Doctoral dissertation, Department of Environmental Toxicology, Texas Tech University, Lubbock, Texas.

Takigami, H., G. Suzuki, Y. Hirai, and S. Sakai. 2008. Transfer of brominated flame retardants from components into dust inside television cabinets. *Chemosphere* 73 (2): 161–169.

Tanabe, S. 2001. Global movement of a chlorinated organic compound. *Kikan Kagaku Sosetsu (Quarterly Review of Chemistry)*, http://www.chemistry.or.jp/index-e.html.

Tanabe, S., K. Senthilkumar, K. Kannan, and A. N. Subramanian. 1998. Accumulation features of polychlorinated biphenyls and organochlorine pesticides in resident and migratory birds from South India. *Archives of Environmental Contamination and Toxicology* 34 (4): 387–397.

Tanabe, S., F. Gondaira, A. Subramanian, A. Ramesh, D. Mohan, P. Kumaran, V. K. Venugopalan, and R. Tatsukawa. 1990. Specific pattern of persistent organochlorine residues in human breast-milk from south India. *Journal of Agricultural and Food Chemistry* 38 (3): 899–903.

Tanabe, S., K. Ramu, T. Isobe, and S. Takahashi. 2008. Brominated flame retardants in the environment of Asia-Pacific: An overview of spatial and temporal trends. *Journal of Environmental Monitoring* 10 (2): 188–197.

Tawfig, N. I. 1991 October. Response by Saudi Arabia to the environmental crisis caused by the Gulf War. Paper presented at The Environmental and Health Impact of the Kuwait Oil Fires, proceedings of an international symposium, University of Birmingham, UK.

Thron, K. U., R. Bruhn, and M. S. McLachlan. 2004. The influence of age, sex, body-condition, and region on the levels of PBDEs and toxaphene in harbour porpoises from European waters. *Fresenius Environmental Bulletin* 13 (2): 146–155.

Timme-Laragy, A. R., E. D. Levin, and R. T. Di Giulio. 2006. Developmental and behavioral effects of embryonic exposure to the polybrominated diphenylether mixture DE-71 in the killifish (*Fundulus heteroclitus*). *Chemosphere* 62 (7): 1097–1104.

Torto, N., L. C. Mmualefe, J. F. Mwatseteza, B. Nkoane, L. Chimuka, M. M. Nindi, and A. O. Ogunfowokan. 2007. Sample preparation for chromatography: An African perspective. *Journal of Chromatography A* 1153 (1–2): 1–13.

Ueno, D., M. Alaee, C. Marvin, D. C. G. Muir, G. Macinnis, E. Reiner, P. Crozier, et al. 2006. Distribution and transportability of hexabromocyclododecane (HBCD) in the Asia-Pacific region using skipjack tuna as a bioindicator. *Environmental Pollution* 144 (1): 238–247.

UNEP. 2008. Africa: Atlas of Our Changing Environment. Nairobi, Kenya: UNEP.

UNEP/GEF. 2002. Regionally Based Assessment of Persistent Toxic Substances: Central America and the Caribbean Regional Report. Global Environment Facility, United Nations Environmental Programme. Geneva, Switzerland.

———. 2002. Regionally Based Assessment of Persistent Toxic Substances: Eastern and Western South America Regional Report. Global Environment Facility, United Nations Environmental Programme. Geneva, Switzerland.

UNEP/GEFa. 2002. Regionally Based Assessment of Persistent Toxic Substances: North American Regional Report. Global Environment Facility, United Nations Environmental Programme. Geneva, Switzerland.

UNESCO. 2002. The final design plan for the Health of the Ocean module of Global Ocean Observing System. Paris: UNESCO.

United Nations. *The United Nations Convention on the Law of the Sea: A Historical Perspective.* United Nations 2007. Available from http://www.un.org/Depts/los/convention_agreements/convention_overview_convention.htm

United Nations Educational Scientific and Cultural Organization. 2002. The Final Design Plan for the Health of the Ocean (HOTO) Module of the Global Ocean Observing System (GOOS). Paris: GOOS.

U.S. EPA. 1996. New pesticide fact sheet: Fipronil. Office of Prevention Pesticides and Toxic Substances.

Van Boxtel, A. L., J. H. Kamstra, P. H. Cenijn, B. Pieterse, M. J. Wagner, M. Antink, K. Krab, et al. 2008. Microarray analysis reveals a mechanism of phenolic polybrominated diphenylether toxicity in zebrafish. *Environmental Science & Technology* 42 (5): 1773–1779.

Van den Steen, E., V. L. B. Jaspers, A. Covaci, H. Neeis, M. Eens, and R. Pinxten. 2009. Maternal transfer of organochlorines and brominated flame retardants in blue tits (*Cyanistes caeruleus*). *Environment International* 35 (1): 69–75.

Van Dolah, F. M. 2000. Marine algal toxins: Origins, health effects, and their increased occurrence. *Environmental Health Perspectives* 108:133–141.

vanderValk, H. C. H. G. 1997. Community structure and dynamics in desert ecosystems: Potential implications for insecticide risk assessment. *Archives of Environmental Contamination and Toxicology* 32 (1): 11–21.

Vetter, W., M. Weichbrodt, E. Scholz, B. Luckas, and H. Oelschlager. 1999. Levels of organochlorines (DDT, PCBs, toxaphene, chlordane, dieldrin, and HCHs) in blubber of South African fur seals (*Arctocephalus pusillus pusillus*) from Cape Cross/Namibia. *Marine Pollution Bulletin* 38 (9): 830–836.

Voorspoels, S., A. Covaci, V. L. B. Jaspers, H. Neels, and P. Schepens. 2007. Biomagnification of PBDEs in three small terrestrial food chains. *Environmental Science & Technology* 41 (2): 411–416.

Vos, J. G., E. Dybing, H. A. Greim, O. Ladefoged, C. Lambre, J. V. Tarazona, I. Brandt, and A. D. Vethaak. 2000. Health effects of endocrine-disrupting chemicals on wildlife, with special reference to the European situation. *Critical Reviews in Toxicology* 30 (1): 71–133.

Vosloo, R., and H. Bouman. 2005. *Survey of Certain Persistant Organic Pollutants in Major South African Waters (POPS).* Pretoria, South Africa: Water Research Commission.

Wallace, N. 1984. Debris entanglement in the marine environment: A review. Paper presented at Workshop on the Fate and Impact of Marine Debris, November 1984, Honolulu, Hawaii.

Wang, Z., X. D. Ma, Z. S. Lin, G. S. Na, and Z. W. Yao. 2009. Congener specific distributions of polybrominated diphenyl ethers (PBDEs) in sediment and mussel (*Mytilus edulis*) of the Bo Sea, China. *Chemosphere* 74 (7): 896–901.

Watanabe, I., and S. Sakai. 2003. Environmental release and behavior of brominated flame retardants. *Environment International* 29 (6): 665–682.

Wenig, M., N. Spichtinger, A. Stohl, G. Held, S. Beirle, T. Wagner, B. Jahne, and U. Platt. 2003. Intercontinental transport of nitrogen oxide pollution plumes. *Atmospheric Chemistry and Physics* 3:387–393.

White, R. D., D. Shea, J. J. Schlezinger, M. E. Hahn, and J. J. Stegeman. 2000. In vitro metabolism of polychlorinated biphenyl congeners by beluga whale (*Delphinapterus leucas*) and pilot whale (*Globicephala melas*) and relationship to cytochrome P450 expression. *Comparative Biochemistry and Physiology C – Pharmacology Toxicology & Endocrinology* 126 (3): 267–284.

Wiktelius, S., and C. A. Edwards. 1997. Organochlorine insecticide residues in African fauna: 1971–1995. In *Reviews of Environmental Contamination and Toxicology*, edited by G. W. Ware. New York: Springer.

Williams, J., C. Read, T. Norton, S. Dovers, M. Burgman, W. Proctor, and H. Anderson. 2001. Biodiversity. Australia State of the Environment Report 2001 (Theme Report). Canberra, Australia: CSIRO Publishing on behalf of the Department of the Environment and Heritage.

Wilson, Edward O. 1999. *The Diversity of Life*. 2nd ed. New York: W. W. Norton & Company, Inc.

Wolkers, H., B. A. Krafft, B. van Bavel, L. B. Helgason, C. Lydersen, and K. M. Kovacs. 2008. Biomarker responses and decreasing contaminant levels in ringed seals (*Pusa hispida*) from Svalbard, Norway. *Journal of Toxicology and Environmental Health–Part a–Current Issues* 71 (15): 1009–1018.

Work, T. M., B. Barr, A. M. Beale, L. Fritz, M. A. Quilliam, and J. L. C. Wright. 1993. Epidemiology of domoic acid poisoning in brown pelicans (*Pelecanus occidentalis*) and brandt cormorants (*Phalacrocorax penicillatus*) in California. *Journal of Zoo and Wildlife Medicine* 24 (1): 54–62.

Wu, T. H., T. R. Rainwater, S. G. Platt, S. T. McMurry, and T. A. Anderson. 2000. Organochlorine contaminants in Morelet's crocodile (Crocodylus moreletii) eggs from Belize. *Chemosphere* 40 (6):671–8.

WWF. 2008. Living Planet Report 2008. WWF-World Wide Fund for Nature (formerly World Wildlife Fund). Gland, Switzerland.

Xian, Q. M., K. Ramu, T. Isobe, A. Sudaryanto, X. H. Liu, Z. S. Gao, S. Takahashi, H. X. Yu, and S. Tanabe. 2008. Levels and body distribution of polybrominated diphenyl ethers (PBDEs) and hexabromocyclododecanes (HBCDs) in freshwater fishes from the Yangtze River, China. *Chemosphere* 71 (2): 268–276.

Xiang, C. H., X. J. Luo, S. J. Chen, M. Yu, B. X. Mai, and E. Y. Zeng. 2007. Polybrominated diphenyl ethers in biota and sediments of the Pearl River Estuary, South China. *Environmental Toxicology & Chemistry* 26 (4): 616–623.

Yogui, G. T., and J. L. Sericano. 2009a. Levels and pattern of polybrominated diphenyl ethers in eggs of Antarctic seabirds: Endemic versus migratory species. *Environmental Pollution* 157 (3): 975–980.

———. 2009b. Polybrominated diphenyl ether flame retardants in the U.S. marine environment: A review. *Environment International* 35 (3): 655–666.

Young, B.E., S.N. Stuart, Chanson. J.S., N.A. Cox, and T.M. Boucher. 2004. Disappearing Jewels: The Status of New World Amphibians. NatureServe, Arlington, VA, USA.

Zapata-Perez, O., M. Del-Rio, J. Dominguez, R. Chan, V. Ceja, and G. Gold-Bouchot. 2005. Preliminary studies of biochemical changes (ethoxycoumarin O-deethylase activities and vitellogenin induction) in two species of shrimp (Farfantepenaeus duorarum and Litopenaeus setiferus) from the Gulf of Mexico. *Ecotoxicology and Environmental Safety* 61 (1):98–104.

Zhou, T., D. G. Ross, M. J. DeVito, and K. M. Crofton. 2001. Effects of short-term in vivo exposure to polybrominated diphenyl ethers on thyroid hormones and hepatic enzyme activities in weanling rats. *Toxicological Sciences* 61 (1): 76–82.

9 Ecological Risk Assessment and Emerging Issues in Wildlife Toxicology

Christopher J. Salice

CONTENTS

9.1 INTRODUCTION

It is clear that there are many challenging issues that lie ahead with regard to protecting natural resources from the impacts of anthropogenic chemicals. In particular, it is obvious that with the global impacts of many anthropogenic activities, understanding and predicting the effects of contaminants on wildlife populations in a changing world will require new methods, approaches, and ways of thinking. Aside from identifying the stress factors that may be impacting particular wildlife populations, the challenge is how to assess the ecological consequences of these impacts. Currently, ecological risk assessment (ERA) is the primary method used to assess the impacts that stressors exert on wildlife. Historically, ERA has emphasized understanding and predicting risks to wildlife of single chemical stressors, and hence current frameworks for ERA basically are restricted to consideration of stressors in isolation. For example, in pesticide registration, the ERA of pesticides is limited mostly to considering pesticides individually, although it is well known that agricultural practices commonly involve the simultaneous application of multiple products. Furthermore, any effects associated with pesticides are not considered in the context of non-chemical or naturally occurring stressors. While there is substantial legal precedence for this assessment method, topics discussed in this text point to the importance of interacting natural and anthropogenic factors on the persistence of populations, the stability of communities, and the function of ecosystems.

Existing information describing the multitude of chemical and interactive stressors suggests that we must move beyond considering effects and exposure in isolation and adopt a more comprehensive, system-based approach to understand and predict risks of anthropogenic chemicals and other sources of stress acting on wildlife populations. By necessity, this shift in paradigm requires an adjustment to system-based thinking and a move toward more sophisticated statistical and mechanistic models that more accurately and fully capture complexities of biological and physical phenomena.

The purpose of this chapter is to provide an overview of current ERA modeling practices and to highlight advances needed to protect ecological systems in a complex and dynamic world. While there are many facets of ERA that will require a shift in approach (problem formulation, for example), the focus on mathematical models in this chapter provides the basis for thinking about data, receptors, and the protection of wildlife. Importantly, the chapter is not intended to be comprehensive and is most useful as an introduction to the state of the science with regard to ecological risk assessment and as a guide to references that provide more detail regarding specific ecological risk assessments or tools. In all likelihood, protecting wildlife populations and ecosystems will require the maturation of techniques and approaches and a shift in our way of thinking about environmental contaminants, other sources of stress, and the consequences of anthropogenic activities on a global scale. To achieve this will require cooperation and interaction between scientists, risk managers, and policy makers.

9.1.1 Background

9.1.1.1 Overview of Current ERA Practices

Ecological risk assessment (ERA) is the primary method used by regulatory agencies such as the U.S. Environmental Protection Agency (U.S. EPA) and the European Environment Agency for evaluating the environmental safety of particular chemical contaminants (Suter 2006). The premise of ERA is that risk (the likelihood of an adverse effect on wildlife, for example) is a function of exposure and the sensitivity of a given receptor to a particular chemical (toxicity). Given the wide variety of environmental conditions, habitats, species, and chemicals, it is nearly impossible to collect data on all factors that can affect both exposure and toxicity. Therefore, there is widespread reliance within ERA on the use of statistical and mathematical models to inform conclusions regarding the potential for adverse ecological effects associated with a particular chemical.

In its simplest form, ERA is basically aimed at estimating risk as a function of exposure to a stressor and the response (toxicity, for chemicals) to the stressor. This is frequently expressed in equation form as

$$\text{Risk} = \text{Exposure}/\text{Toxicity}$$

For ERAs focused on estimating risks to chemical exposure, this quotient is called the *risk* or *hazard quotient* (hereafter, hazard quotient) and is a unitless value that is compared to a pre-determined level of concern to reach a risk conclusion. If, for example, the level-of-concern is 1.0, a potential for adverse effects may be assumed if the quotient is greater than this. Intuitively, this equation indicates that for a given exposure level, as toxicity increases, so does risk because higher toxicity is actually associated with lower endpoint values (e.g., an LD_{50} of 0.1 mg/kg represents greater toxicity than an LD_{50} of 10 mg/kg). It is important to acknowledge that the use of a risk or hazard quotient is intended for the initial phases of a risk assessment and is part of a larger, weight-of-evidence approach for ERA. That said, hazard quotients have, in many cases, served as the basis for risk conclusions, particularly in the case of pesticide registration (U.S. EPA 2004).

Although the risk quotient approach is a mainstay of ecological risk assessment, it has fallen under considerable criticism (Tannenbaum 2005). Most notably, it is reasonable to argue that reducing the complexities of exposure and toxicity for wild species to a single, unitless number may not be as informative as decision makers would like. Physical and biological properties can vary considerably, and this variability can have significant impacts on risk to species in the wild.

As an example, a screening-level risk assessment of pesticide effects on birds would, at a minimum, require several studies on the effects of the pesticide on birds. By law, the Federal Insecticide Fungicide and Rodenticide Act (FIFRA) requires acute toxicity studies of pesticides on either bobwhite quail or on mallard ducks and a passerine species (songbird). These acute toxicity data (LD_{50}s), an estimate of the foliar residue half-life for the pesticide, and some basic information on use rates are all that are needed to conduct a screening-level ecological risk assessment. With these data, and some fairly simple models, an estimate of exposure is divided by the lowest estimated toxicity value to yield the risk quotient, which, as stated, forms the basis of the risk conclusion regarding the given pesticide. Importantly, a variety of assumptions are folded into the screening-level process to allow the risk assessment to move forward and also to ensure safety. The screening-level risk assessment is designed to indicate when a chemical is safe and, hence, errs on the side of conservatism for a single chemical. On the other hand, if risks are identified at the screening level, the conclusion is more on the order of a "potential for adverse effects" instead of a certainty of adverse effects. Some examples of assumptions that are built into the screening-level ecological risk assessment for birds are listed in Table 9.1.

TABLE 9.1
Assumptions in the Screening-Level ERA for Birds

Assumption	Effect on Risk Estimates
All birds spend 100% time on treated field.	Overestimation of exposure
All birds consume 100% of diet from treated field.	Overestimation of exposure
There is no variation in pesticide residues on food items of a given category.	Central tendency of exposure estimate (actual exposures may be higher and lower)
Dietary is predominant route of exposure.	Uncertain effect on exposure given other assumptions about presence on field
Data on two species represent range of responses in all birds.	Uncertain effect on exposure

Although it may be assumed in a screening-level risk assessment, it is highly unlikely that all birds in agro-ecosystems spend 100% of their foraging time in agricultural (pesticide-treated) fields. Alternatively, it may also be reasonable to say that it is likely that birds are exposed to pesticide via various routes, not just dietary (Mineau 2002; Vyas 1999). Hence, a major criticism of the screening-level approach and the associated uncertainties is that biological and physical systems are complex and that there is typically significant variability. It could be argued that, in many cases, the use of quotients for risk assessments do not explicitly incorporate either complexity or variability. In addition, this framework provides little flexibility for exploring more complicated interactions of stressors such as disease and climate change that may modify wildlife responses to toxicants (see Chapters 5 and 6, this volume).

For example, if it was known that amphibians are at risk of exposure to pesticides and a pathogenic fungus (chytrid), there is no robust means of (a) identifying a level of concern (point of departure) or (b) making an "exposure estimate" that reduces to one value the combined risk of pesticides and pathogenic infection. While this type of scenario may not fall under a particular legislative requirement (unless the amphibian is an endangered species), it nevertheless represents a real and potentially likely scenario, especially in light of the well-documented decline in many amphibian populations and the contribution of contaminants and disease toward that end (Davidson 2004; Berger et al. 1998).

Methods are available that can be used to address biological, ecological, and physical complexities such as the example above and, in addition, provide a quantitative estimate of the likelihood and magnitude of risk. Furthermore, there is a drive toward producing estimates of risks that are placed in the context of ecologically relevant endpoints such as those that may more directly relate to the sustainability and maintenance of populations, communities, and ecosystems; this is particularly important given the wide range of emerging environmental issues (detailed within this book). A key element of these approaches is the development of more sophisticated mathematical tools that are likely to yield insights useful to the risk management process but also require more and better data and information. These are discussed further in the following sections.

9.1.2 Ecological Risk Assessment: Improvement through Modeling

9.1.2.1 Moving beyond the Screen

With the advent of greater computing power and more sophisticated mathematical and statistical methods, models used for ERAs can explicitly address variability, uncertainty, and ecological complexity (Suter 2006; Burmaster and Anderson 1994; Solomon et al. 2000; Thompson and Graham 1996; Macintosh et al. 1994). In addition, the effects of xenobiotics on wildlife has gained the attention of ecologists with the realization that contaminants are, at this point, ubiquitously distributed in the environment with effects on wildlife occurring in otherwise pristine habitats (Davidson 2004). This has the effect of bringing long-standing expertise in sophisticated ecological models into the applied realm of ecological risk assessment and environmental management. The ultimate aim of these developing approaches is to better predict the probability and magnitude of a deleterious effect associated with a particular xenobiotic, or in some cases, the interactive effects of multiple, simultaneous stressors. Several different modeling approaches are discussed with respect to certain types of stressors or effect predictions. The intent here is to provide a brief overview and some directions on the future of ERA.

9.1.2.2 Probabilistic Exposure and Effects Models

Probabilistic risk assessment (PRA) has been in development for some time (Thompson and Graham 1996) and is, by many accounts, a marked improvement over the screening-level ERA (Suter 2006; Thompson and Graham 1996). The concept behind PRA is that there is considerable variability and/or uncertainty in the biological and physical world and that this variability can impact estimates of risk. In part, variability is important because we are not necessarily concerned about individuals, per se, but collections of individuals (populations) and groups of these collections (communities). By using a probabilistic approach to risk estimation, risk can be placed in the context of populations and communities. A key output of probabilistic risk assessments is a likelihood of the occurrence and magnitude of an effect. For example, an output might be an 80% chance of 50% mortality. This type of output provides a more tangible sense of potential effects and can be used to initiate a dialogue with decision makers related to what this effect might mean for a population.

Although probabilistic risk assessment has been in development some time, major developments and implementation seem to have occurred over the last decade. Probabilistic risk assessments and approaches are available in the primary literature (Zolezzi et al. 2005; Giddings et al. 2001; Hall et al. 2000), and there has been a marked increase in the number of publications on PRA. Aside from the primary literature, PRA methods and approaches have been developed by the U.S. EPA and are currently in the process of finalization and vetting (U.S. EPA 2004).

The primary method behind PRA is the Monte Carlo Simulation, which involves randomly sampling model parameters from a specified distribution. A model is run numerous times, each time with a different set of parameter inputs based on the specified distribution. The resulting output is then a distribution that can be analyzed to determine the probability of an outcome. Figure 9.1 shows an example of a flow diagram for an avian exposure model.

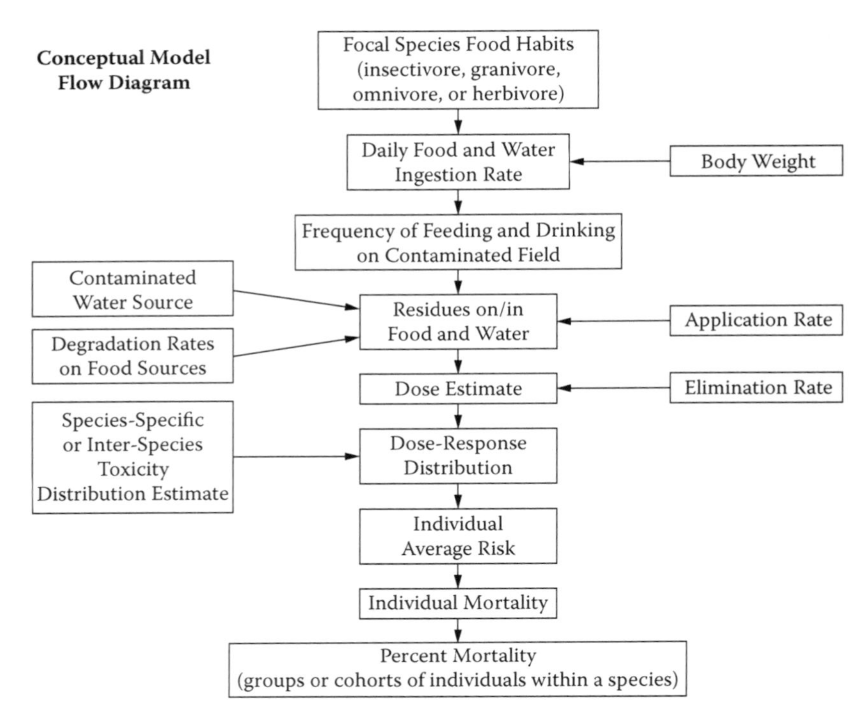

FIGURE 9.1 Flow diagram for probabilistic risk assessment for avian receptors in and around pesticide-treated fields. The flow diagram represents the modeling construct for an individual bird. For any given model run, many birds (10,000 commonly) are simulated and many of the inputs (e.g., body weight, frequency on treated field, residues on food and water) are described by distributions as opposed to point values.

As an example, the EPA's Terrestrial Investigation Model (U.S. EPA 2004) simulates exposure and mortality of birds on an agricultural field. The model follows or tracks exposure in many individual birds (frequently 10,000). Each bird is followed for a specified number of days, and exposure estimates are calculated every hour using allometric equations describing metabolic requirements for birds of a particular type (passerine or non-passerine) and size. Exposure estimates are compared at each hour to an assigned tolerance. Within the model, individual tolerance is assigned at the start of the model run and is based on a distribution of sensitivity derived from a number of toxicity studies. If exposure exceeds tolerance in any given hour, the bird is scored as "dead" and removed from further calculations. This modeling construct explicitly incorporates variability in parameters such as body size, feeding amount, foraging behavior, and pesticide residue distributions. Some model parameters and the distributions used to describe them are listed in Table 9.2, along with a comparison to the models used in screening-level ecological risk assessments.

As mentioned, the output from a probabilistic risk assessment is in the form of a probability. Probabilistic ecological risk assessment likely provides the most reasonable means by which to begin assessments on the effects of multiple, interacting

TABLE 9.2
PRA vs. Screening-Level ERA

Parameter or Calculation	Screening-Level ERA	PRA
Bird body weight	Varies according to species-specific data	Three body weights, irrespective of species
Diet	User-specified: can vary among 5 items for any given species (e.g. 50% insects, 25% seeds, 25% vegetation)	Does not vary and is not mixed within species (e.g., 100% seeds)
Time on field	User-specified: values determined by field studies	Does not vary: 100%
Pesticide residues	Varies: based on data from field studies (mean and std. dev.) and application rate	Does not vary: high-end estimate based on field data and application rate

stressors. A probabilistic construct as outlined above could be modified to include known effects of other types of stressors. An obvious example might be the interaction of a chemical contaminant and a pathogenic organism. These two stressors could be modeled in a variety of ways with specific constructs chosen based on what is known about how the chemical contaminant and the pathogen interact to impact the health of a receptor. Some possible interactions include (a) additive, independent effects of chemical and pathogen, (b) potentiating effect of chemical on the susceptibility to a pathogen, and (c) potentiating effect of a pathogen on chemical sensitivity. If data are available, a specific model strategy could be chosen. However, one very strong advantage of models is the ability to inexpensively explore the bounds of uncertainty and to conduct "virtual experiments." In this case, various scenarios could be modeled to gain an appreciation for the bounds on the expected responses of the receptor. If every modeled scenario results in widespread effects, this may suggest a potential for very real effects in the actual system. Alternatively, if only certain, specific scenarios generate adverse effects to the receptor, this may warrant further research focused on particular questions.

Probabilistic models provide a number of advantages for estimating risk to wildlife. By explicitly incorporating biological information and variability, PRA can produce estimates of risk that are inherently more meaningful than unitless quotients. These flexible models can be used to incorporate the effects of multiple stressors and can be programmed to produce outputs that are immediately meaningful. To date the limitation of PRA is related more to the endpoints or outcomes evaluated and not to the limitations of the approach. For example, PRAs in support of pesticide regulation are typically focused on generating an estimate of mortality associated with a defined exposure duration (for example, 28 days). The outcome is expressed as a likelihood of a certain percent mortality in birds in and around pesticide-treated fields. Where this falls short is in estimating effects to populations of receptors. In addition, PRAs are typically species or taxa specific, and an important advance in PRAs will be the development of various models reflecting the biology of a wide range of ecological receptors.

9.1.2.3 Population-Level Approaches

A long-standing criticism of ecological risk assessment and even ecotoxicology has been that observed effects or predicted risks are not placed in an ecologically meaningful context (Forbes, Calow, et al. 2001; Munns et al. 1997; Barnthouse and Sorenson 2007). As mentioned, the risk or hazard quotient provides little or no information regarding the magnitude of potential effects. Probabilistic approaches can provide estimates of the likelihood and magnitude of effects; however, most probabilistic ecological risk assessments have not gone the additional step to estimate the potential effects that may occur at higher levels of biological organization such as the population. As an example, a probabilistic risk assessment might generate an output such as a 50% probability of 20% adult mortality, but unless there is an established threshold of unacceptable mortality, it is difficult to know what this means from an ecological perspective. Complicating matters further is that effect levels are not created equally. For example, a 20% reduction in adult survival has different ecological implications for birds than it does for daphnia. The interaction between the observed effect and the life history of the organism will dictate what the ecological implications of the effects level might be (Calow et al. 1997; Raimondo et al. 2006). Here, PRA will benefit from the input and review of ecologists, toxicologists, and decision makers. Lastly, many chemicals have effects on more than one life stage or process in a given species, making ecological interpretation of effects more complicated. Given the complexities mentioned, how might an ecological risk assessor estimate the ecological significance of the effects of a given chemical?

There is a growing interest in using population models to estimate the effects of toxicants on particular organisms (Levin et al. 1996; Forbes, Calow, et al. 2001; Caswell 1996; Barnthouse et al. 2007; Akçakaya et al. 2008). Population models allow integration of effects on multiple components of an organism's life history (e.g., survival and reproduction) and can provide a quantitative estimate of population-level effects and incorporate species-specific life history information. For example, population modeling has been used to better understand the effects of PCBs and dioxin on fish (Munns et al. 1997), cadmium on snails (Salice et al. 2003, 2009), hydrocarbons on polychaetes (Levin et al. 1996), pesticides on mysid shrimp (Kuhn et al. 2000), and the effects of a number of contaminants on daphnia (Stark and Banks 2003). These are just a few of many examples using some measure of population modeling in the current literature.

Why the population level? Figure 9.2 provides some insight into why there has been a recent focus on developing methods to evaluate the population-level effects of xenobiotics. Within the hierarchy of biological organization, the general notion is that while the underlying mechanisms of a response can be understood at lower levels of biological organization, the significance is often understood at higher levels of organization. For example, the biochemical responses to a xenobiotic may provide an understanding of effects on individuals, but those effects have meaning only from an ecological perspective at the population level and higher (Landis 2000, 2002). In addition, there is a rich body of work from population ecology that can be adapted for use in ecotoxicology and risk assessment (Caswell 2001). Lastly, an improved understanding of ecotoxicology at the population level may provide a gateway to

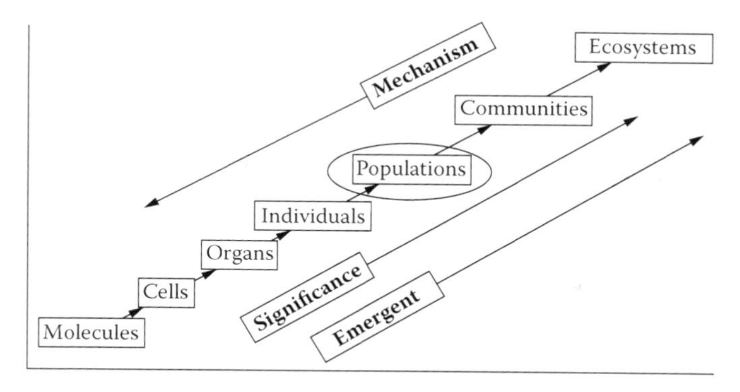

FIGURE 9.2 A representation of the hierarchy of biological organization. Populations occupy a key place between lower and higher levels of organization. Historically, ecotoxicology has primarily focused on individual and lower effects (e.g., growth, survival, biochemical responses). More recently, however, there is greater emphasis placed on understanding how individual- (and lower-) level effects translate to effects at the population level.

understanding effects that occur at the community and ecosystem level. The population level also has use from an environmental management perspective because (1) we are concerned about the sustainability of populations, and (2) populations are measurable and therefore management success can be evaluated. Hence, there is a reasonable justification for pursuing the incorporation of population-level ecotoxicology and ecological risk assessment (Barnthouse and Sorenson 2007).

In its simplest form, a population model relates the births and deaths of a given population to provide a sense of temporal dynamics. Ecotoxicologists and risk assessors are interested in how xenobiotics modify the rates of births and deaths and how these, in turn, affect population dynamics. Structured population models explicitly incorporate age, size, or otherwise defined stages (Caswell 2001). From an ecotoxicological perspective, these structured models are often the most useful because laboratory data tend to relate to the survival, growth, and reproduction of particular stages or ages of test organisms. Figure 9.3 is an example of a life-cycle graph of a snail species that is modeled to live 4 years. Each arrow represents a transition rate between ages and includes a reproductive rate as well.

The life-cycle graph can be easily translated into a matrix form that is simply called a matrix population model. Matrix models are often used in ecotoxicology, primarily because they have many useful and elegant mathematical properties (Caswell 2001; Akçakaya et al. 2008). In essence, a well-drawn life-cycle graph is translated directly into matrix notation (Figure 9.3). The simplest application of these models is to adjust known survival or reproductive rates (vital rates) based on observed toxicant-induced effects, with life table response experiments (Levin et al. 1996; Caswell 2001) being used to generate the data necessary to parameterize the models. Properties of the matrix such as eigenvalues and eigenvectors then provide a wealth of information relevant to understanding the dynamics of a given population (Caswell 2001). These include the growth rate of the population (is the population growing or declining), the age or stage structure, and the relative contribution of the different ages or stages

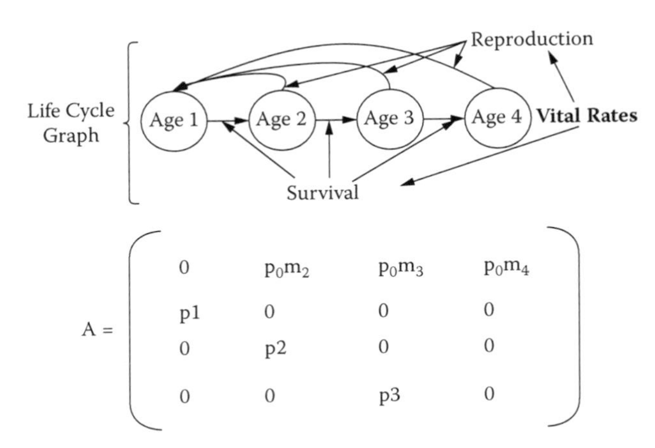

FIGURE 9.3 Example life cycle graph and corresponding matrix model for an aquatic snail that lives 4 years and reproduces in years 2 through 4. Matrix models are useful in ecotoxicology for placing individual-level effects in a population level context. Although many population models can be quite complex, the example here shows that, in actuality, matrix models are direct translations of graphical descriptions of organism life cycles and hence are very tractable.

to the overall growth rate. An important constraint on the use of population models, however, is that the data necessary to parameterize them often goes beyond what is normally available in standard toxicity tests. Nonetheless, careful parameterization of models and even more careful divulgence of model assumptions can be useful in providing insights into the population-level effects of xenobiotic exposures and how these might interact with other types of stressors.

An example of how to implement a population model to understand the impacts of multiple stressors can be found in Salice et al. (2009). The model is used to predict the extinction risk and population size of narrow-mouthed toads (*Gastrophryne carolinensis*) exposed to contaminants in an aquatic habitat, where development occurs, and terrestrial habitat loss, where toads spend most of their lives. The model incorporated density-dependent survival effects occurring in the aquatic and terrestrial phase with terrestrial density dependence used as a surrogate for habitat loss. Environmental stochasticity was incorporated by explicitly modeling reproductive failure associated with drought events, the frequency of which were specified from historical records. The results provide insights that would not have been obvious without explicit consideration of multiple stressors. For example, because of the strength of density dependence in the developmental stage, effects of aquatic contaminants actually had a beneficial effect on the population via a relief from density-dependent effects (Figure 9.4). Terrestrial habitat loss had significant impacts on the population, as expected, although there appeared to be a threshold response; once terrestrial habitat is reduced to a certain point, the population risk increases significantly (Figure 9.5). In a series of modeling experiments, simulations indicated that if there were carry-over effects associated with exposure to contaminants in the aquatic phase, the risks of population extinction increased

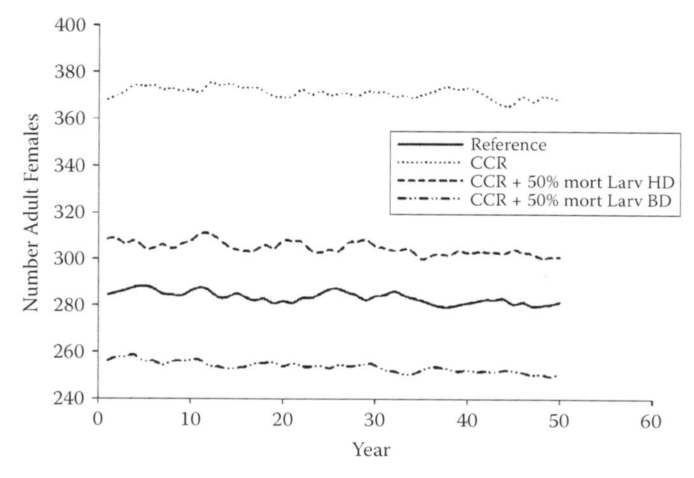

FIGURE 9.4 Total number of toads in contaminated (CCR) and reference habitats. CCR + 50% mortality represents a simulated additional effect of CCR on later-age larvae. Here, regardless of the effects on larval survival, the effects of contaminants reduce the impacts of density-dependent survival and the population actually has more female toads unless the additional late-age larval mortality applies to both high and low density larval conditions (CCR + 50% mort Larv BD). These results highlight the importance of considering ecological factors in understanding the risks of chemical contaminants. The significance of this type of modeling effort is the attention drawn to factors thought to be important for risk; in this case a clear understanding of the interactions between CCR-induced toxicity and density-dependent effects is critical.

significantly (also Figure 9.5). This last result has important consequences for amphibian ecotoxicology. The importance of carry-over effects at the population level suggests a change in research focus. There are few studies that have explicitly looked for carry-over effects of anthropogenic stressors, although the effect has been observed (Rohr and Palmer 2005; Rowe et al. 2001; Pahkala et al. 2001). Hence, this example illustrates the utility in using slightly more sophisticated ecological models to understand responses to multiple stressors. At a minimum, model results can frequently be used to identify research needs, which can be pursued to reduce uncertainty.

To date the vast majority of modeling in risk assessment has focused on probabilistic models of effects and population models to better understand the ecological effects of contaminant exposure. There are several examples, however, of higher-level models and approaches designed to understand effects at the meta-population or landscape level. These models essentially estimate effects and outcomes in a spatially explicit manner. Although these will only be mentioned here, the EPA's HexSim, or Life History Simulator for Terrestrial Wildlife Populations, is a working example of a spatially explicit, individual-based, multi-species simulation model aimed at understanding population dynamics and interactions in terrestrial wildlife populations. The model can be used to track multiple populations in a landscape and can integrate the effects of a variety of stressors. While sophisticated and useful in

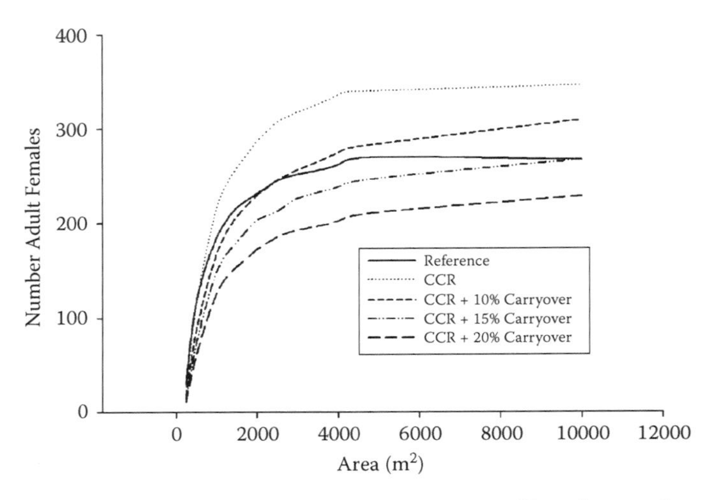

FIGURE 9.5 Total number of adult female toads in and around breeding pond contaminated with coal combustion residues (CCR) and surrounded by different amounts of terrestrial habitat (*x* axis). The "Carryover" simulations represent terrestrial phase mortality of toads that is "carried over" from exposures that had occurred during the larval aquatic stage. Again, as in the previous figure, CCR in the aquatic habitat results in some mortality of larvae, thereby generating some relief from density dependent effects and yielding a greater number of adult females. This is not the case when there is a simulated Carryover effect. Moreover, once terrestrial habitat is reduced, it tends to override other simulated effects of contaminants. The vast majority of amphibian ecotoxicology and ecology studies have focused on the aquatic phase but more research emphasis is being placed on the terrestrial stage, largely as a result of some significant population modeling efforts.

many contexts, it is important to keep in mind that the data requirements of such a model likely go beyond what is either required or typically available.

9.1.2.4 Other Population-Based Models

A key element in predicting the effects of anthropogenic activities on wildlife populations will be estimating the impacts of global climate change. Already, a number of analyses have described changes in species phenology (timing of events) and geographical ranges (Parmesan and Yohe 2003; Pearson and Dawson 2003). Species distribution models are used to predict future changes in the geographical location of species. SDMs are discussed in Chapter 6 and will be presented here only in relation to the potential application toward ecological risk assessment.

SDMs are statistical models that relate observed species occurrences to underlying environmental gradients or climatological data. They can be used to predict how species distribution will change when climatological patterns change. To date, there are no known ERAs that explicitly consider the interactive effects of contaminants and climate change (or temperature). In effect, the state of the science of risk assessment has only started to embrace the reality that wildlife populations are subject to multiple sources of stress, both anthropogenic and natural. SDMs may provide a means of linking different sources of stress to evaluate the combined impact on risk to wildlife populations.

Furthermore, the basic approach of SDMs may find utility in predicting the response of species to multiple stressors. Here, instead of a mechanistic model (such as a structured population model), inferences about a system are made based on statistical relationships between an endpoint (e.g., species abundance) and the concomitant occurrence of different habitat, physical, or biological characteristics. Many of the methods are in place and can be adapted from available SDMs.

While there are significant advantages and stimulating test cases of this type of statistical approach, it will be strengthened by a mechanistic understanding of the processes that regulate populations. Because of the multitude and complexity of the systems we are trying to protect (various communities, habitats, stressors, etc.), a mechanistic understanding of the processes that dictate population persistence of extinction is necessary. The combined approaches of inductive and deductive modeling will likely provide the most robust understanding and predictive ability related to the response of ecological receptors to anthropogenic stressors.

9.1.2.5 ERA of Communities and Ecosystems

As mentioned previously (Figure 9.2), the significance of effects on ecological systems manifests at higher levels of biological organization. Currently, considerable emphasis in wildlife toxicology is placed on understanding population-level responses to xenobiotics (Forbes and Calow 2002; Barnthouse and Sorenson 2007; Akçakaya et al. 2008; Forbes and Calow 1999). That said, there is also considerable effort made to understand how communities respond to environmental stressors (Clements and Rohr 2009; Clements 1994; Clements and Kiffney 1994; Rohr et al. 2006). Because communities are a collection of populations and because there is wide variation in community composition, generalities regarding how communities respond to anthropogenic stressors are elusive. However, hypotheses such as the community conditioning hypothesis (Landis et al. 2000; Matthews et al. 1996, 1996) and trait-mediated indirect effects (Relyea and Hoverman 2006) provide the means by which to address and understand community-level responses. These approaches have not, however, become commonplace within ERA. In part, this is likely because of the aforementioned complexity associated with community composition, which can have a direct and strong impact on response to anthropogenic stressors.

9.1.2.5.1 Community-Level Ecological Risk Assessment

One example of a community-based ERA approach is the re-registration ecological risk assessment for atrazine conducted by the U.S. EPA. As a condition of re-registration of atrazine, the U.S. EPA required the registrant to develop a monitoring program for atrazine in streams with the intent of identifying the extent of the exceedance of levels of concern (LOC) for atrazine in flowing surface waters for effects on community-level endpoints (U.S. EPA 2008). The atrazine registrant developed the Comprehensive Aquatic System Model (CASM) (Bartell et al. 1999) for use in a community-level ecological risk assessment. The model has both statistical and mechanistic components and relates certain physical and/or biological properties to community-level endpoints such as phytoplankton productivity. The model is based on a series of 33 microcosm and mesocosm studies of the effects of atrazine on aquatic systems. Importantly, the range of micro- and mesocosm studies creates a range of time-series, system differences, and design details that require the use of a Model Effects Index (MEI) to provide a general measure of effect. The CASM model, in this case, was not intended to simulate aquatic systems per se, but to relate results from atrazine monitoring efforts to potential community-level effects as determined by the results of the mesocosm studies. The methods and models are currently undergoing review and continued vetting within the EPA and have recently undergone a second review by the U.S. EPA Science Advisory Panel.

9.1.2.5.2 Ecosystem-Level Ecological Risk Assessment

Regional ecological risk assessments represent the most encompassing of the risk assessment approaches. The focus is set at the regional level to include ecological and human elements. A series of papers on the subject provides substantial background and specifics on the merits, uncertainties, and applications of regional risk assessments (Landis and Wiegers 1997, 2007; Graham et al. 1991; Wiegers et al. 1998; Landis 2004).

Perhaps a key component of regional ecological risk assessments is that multiple sources of stress must be considered. In this regard, typical approaches to risk assessment, such as the hazard quotient, are incapable of handling the complexity evident at larger spatial and temporal scales (Landis and Wiegers 2007). Furthermore, because no two systems are the same, particularly at regional scales, the concept of a reference site does not apply. This further limits the utility of more commonly used risk assessment approaches.

Importantly, for risk assessments at this scale, there is no single model that adequately captures the salient and driving properties of the system. By necessity, the regional level involves multiple endpoints, receptors, exposure pathways, and stressors (Landis 2004). One method for handling this type of complexity is to use the "relative risk model," or RRM (Wiegers et al. 1998), which circumvents the need for a reference site or a stress-free baseline. The premise of the RRM is a departure from the standard risk assessment framework that was intended for use with single stressors (chemical) affecting a single receptor. The RRM approach offers a way to integrate the spatial, temporal, and compositional complexity of regions by ranking sources of stressors and habitats by sub-regions within the assessment area. The ranks are obtained from quantitative determinations of the risks in the given sub-regions, which can then be used to facilitate decision-making. A number of successful regional risk assessments using RRM have been conducted, including a fjord in Port Valdez, Alaska (Wiegers et al. 1998); Mountain River, Tasmania (Walker et al 2001; Obery and Landis 2002); Codorus Creek, Pennsylvania (Obery and Landis 2002); and Cherry Point, Washington (Hayes and Landis 2004).

Because of the complex and expansive nature of many of the systems considered in regional-scale ecological risks assessments, sufficient data are often lacking to adequately understand and generate hypotheses regarding the multitude of interactions present at this scale (Landis and Wiegers 2007). It seems likely that if regional-scale risk assessment becomes more commonplace, data needs will drive data collection and more scale-appropriate methods will be employed to adequately characterize the system. The use of sophisticated statistical and mechanistic models will continue to find applicability in these types of risk assessment. In particular, geographic information systems (GIS) have been used to provide support with regard to spatial relationships between ecosystem and region components. Monte Carlo analysis also has been applied to address uncertainty and variability in a robust, quantitative manner (Hayes and Landis 2004).

9.2 ERA: CHALLENGES AND PATHS FORWARD

This section highlights some current approaches in ecological risk assessment and ecology for understanding the response of systems (populations, communities, and ecosystems) to anthropogenic disturbances that may include chemical contaminants, climate change, and habitat alteration. These are only some examples of the many models available in ecology, conservation biology, and toxicology that may be used to provide useful insights into how wildlife populations may respond to stress.

What emerges is a hierarchy of approaches with the optimal approach based on the assessment endpoint of interest (see Figure 9.6, showing ERA approaches and assessment scale). Minimally, it seems reasonable to expect that PRA will be used

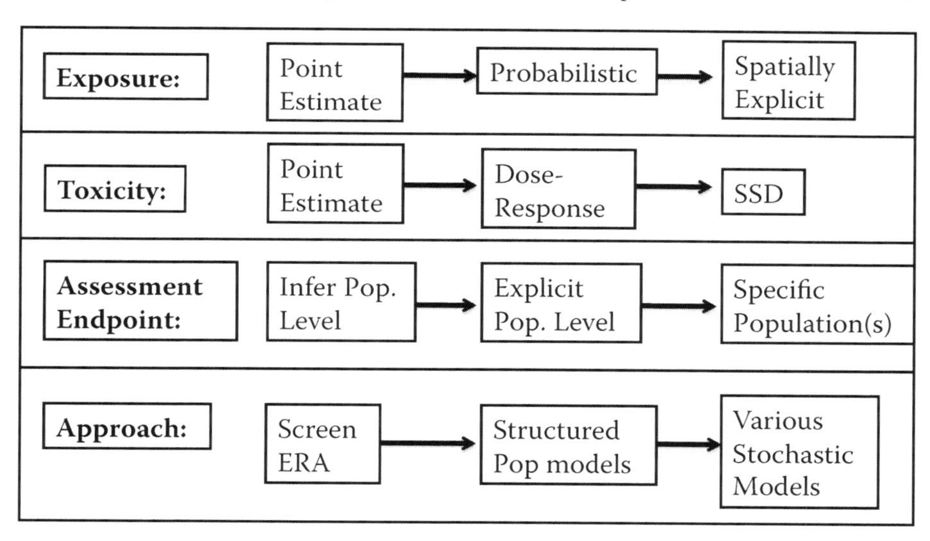

FIGURE 9.6. Hierarchy of assessment endpoints and risk assessment and modeling approaches that could be used. Note that as the scale of the assessment increases (from a generic or chemical-specific risk assessment to a site or landscape assessment), the data needs increase markedly. Also, at lower tiers of ERA we may infer ecological effects, while at higher tiers, there is more emphasis on predicting ecological effects for specific populations or habitats.

to understand the effects of a chemical on a population of receptors. More regular incorporation of population-level principles will likely improve the state of science with regard to estimating risk to wildlife (Barnthouse and Sorenson 2007; Landis 2002). It also seems likely that regional-scale and community-level approaches will see more widespread use. Importantly, with the acknowledgment that wildlife populations and communities are faced with a mixture of different stressors operating on sometimes disparate spatial and temporal scales, more sophisticated modeling and statistical and research methods are required. Without a doubt, future improvements will be dependent on the collaborative efforts of ecotoxicologists, conservation biologists, ecologists, and decision makers.

Specific recommendations include a more robust education of students with regard to the use of mathematical approaches in risk assessment and wildlife toxicology. To our knowledge, there are few examples of curriculum in environmental toxicology and applied ecology offered to students. Minimally, courses that provide the basics of ERA to include screening-level analyses, probabilistic methods, statistical inference, and population modeling would be of immediate utility to students who are the future scientists, regulators, and decision makers.

9.2.1 Non-Technical Challenges

9.2.1.1 Constraints Imposed by Environmental Regulation

The implementation of environmental laws has specified the type of research or scientific evidence needed to regulate and enforce contaminant and pesticides in the environment (CERCLA, FIFRA, CWA). As an example, FIFRA states specifically which toxicity studies must be completed in support of the registration of a given pesticide. Moreover, the details of study design are specific and non-flexible; studies must be conducted in a very specific way. This rigorous attention to design facilitates defense of scientific results and is useful in the regulatory context. However, it simultaneously serves to constrain the science and potentially limits overall progress in terms of effective environmental management. Hence, policy has driven the science, and not the opposite. The environmental consequences of this may be playing out presently.

Data and regulatory requirements are limited to single chemical stressors and are absent of any knowledge of other natural or anthropogenic stressors (including pesticides) that may be acting on wildlife populations. To what extent these legal requirements constrain the development of new approaches or methods is unknown. However, given that environmental regulations such as FIFRA have remained essentially unchanged since their original inception, it seems likely that the constraints imposed are real. Indeed, this bears out when reviewing current ecotoxicological literature and when attending environmental conferences; much of the research presented is descriptive in nature. While these data may prove useful, without a research agenda focused on devising generalities of stressor response, research will merely stay descriptive.

9.2.1.2 Constraints Imposed by Lack of Knowledge

Within the field of environmental toxicology, there has traditionally not been an emphasis on understanding ecological principles with respect to chemical stressors. Again, this is likely driven in part by the emphasis placed on conducting research in a regulatory context (see previous paragraph). As a result, exotoxicologists have just started realizing the utility and benefit of incorporating more sophisticated models (PRA) and in placing toxicant-induced effects in an ecological context (Akçakaya et al. 2008; Barnthouse and Sorenson 2007). Conversely, within ecology, effects of toxicants were generally considered to be "too applied" to be useful for developing and understanding general ecological principles. Granted, if toxicants were spatially limited, this view may have been reasonable. In light of the widespread distribution of contaminants and anthropogenic chemicals, this view is outdated. Even organisms in isolated areas are subject to contaminant exposure as a result of environmental fate and transport processes (Martin et al. 2004; McConnell et al. 1998).

9.2.1.3 Constraint Imposed by Historical Funding Priorities

Generally, governmental agencies that fund scientific research such as the National Science Foundation (NSF) and the National Institutes of Health (NIH) have not provided significant support toward understanding the effects of anthropogenic stressors (such as toxicants) on ecological receptors. As mentioned in the previous paragraph, research on this topic was likely considered "too applied" by NSF and "not relevant to human health" by NIH. In light of the complex and prevalent interactions among stressors (some of which were outlined in this text), new priorities are needed. Openness is needed to the view that toxicants are ubiquitously distributed and therefore relevant to all natural, ecological systems. In addition, the interrelatedness of wildlife and human populations deserves increased recognition. An excellent example of this latter case is how chemical stressors and climate change can influence the prevalence of vectors for human disease. Specifically, some climate change predictions show an increase in the range of snails that are vectors for the second most prevalent disease in human, schistosomiasis (Zhou et al. 2008). Here, an ecological understanding of how snails respond to environmental change (and management) is critical in limiting infection among humans. Hence, an acceptance by funding agencies that anthropogenic stressors are important to a variety of organisms and systems would serve to foster collaborations among ecologists, conservation biologists, and ecotoxicologists. Additionally, this would move the state of the science beyond the descriptive realm to one engaged in trying to understand generalities in response to multiple anthropogenic and natural stressors.

9.2.2 Moving Forward with ERA

In terms of identifying a path forward for improving ERA in light of the complexities faced by wildlife, it seems clear that new approaches to scientific research and risk communication are needed to effectively manage natural resources. Perhaps the most direct path to be pursued immediately would be fostering of collaborations between researchers from different fields. Most rapid and perhaps comprehensive advances can be made when a combination of approaches is applied to certain problems. This, generally, is in accord with a weight of evidence approach in ERA that is considered to provide the most robust and defensible risk conclusions. Now, however, the issue is not so much the risk of a single chemical but a complex mixture of stressors on a complex system. This complexity necessitates a more comprehensive and broad way of thinking about environmental problems that must incorporate the talent and experience of a range of environmental scientists. There are examples in the literature of mergers of ecological, toxicological, and conservation methods and approaches to environmental problems. These are not nearly frequent enough or as wide-reaching enough as may be needed, particularly in light of some of the issues highlighted in this text.

Beyond identification of the complexity of environmental issues, an acceptance by funding agencies, decision makers, and scientists regarding more flexible, sophisticated approaches to assessing risks is a necessary first step. This requires a move beyond the concept of single stressor vs. specified response to a more comprehensive concept of how species interact with multiple anthropogenic stressors in the context of natural stressors. Importantly, it must be acknowledged that anthropogenic chemicals are, arguably, omnipresent. Thus, they may be considered abiotic factors faced by almost all populations of wildlife across the globe.

Tied to the need for greater collaboration between scientists from different fields, policy makers, and decision makers is a need for harmonization among regulatory agencies with regard to ERA approaches, natural resource priorities, and environmental regulations. Because there are apparent global issues in environmental toxicology and anthropogenic stressors, and because many of the effects of these extend beyond national borders, international interactions will likely have important implications for the protection of wildlife in the future.

9.3 SUMMARY

Models are integral components of ERA and are used regularly. The first tier of an ecological risk assessment or screening level is characterized by the use of deterministic models that generate a point value: the unitless risk quotient or hazard quotient. The use of point values such as the risk quotient does not provide a quantitative sense of the probability and magnitude of potential effects and does not explicitly incorporate variability or uncertainty. Newer methods and models such as probabilistic risk assessment can address some of these shortcomings by including variability in biological and physical processes that are known to influence exposure and risk. Examples include variability in the distribution and uptake of xenobiotics or variability in the behavior of organisms that modify how they experience the environment (for example, foraging behavior). Population-level ecological risk assessment is seeing a surge in interest as risk assessor and risk managers attempt to better understand (and regulate) based on effects to populations of organisms. Here, population modeling is used to place individual-level effects (survival, growth, and reproduction) in a population-level context. This represents an area of active research and will likely become a permanent component of ecological risk assessment. As a natural jumping off from population-level assessments, models have been developed that address ecological risk at the landscape level. These models (PRA, population, landscape) represent step-wise refinements on ecological risk assessment. With each higher tier comes additional insight into the ecological effects of xenobiotics and greater flexibility in incorporating other sources of environmental stress. One drawback with higher tier, more sophisticated models is that with each additional layer of biological organization and spatial scale comes ever-increasing data needs. So while more complex models are appealing from the perspective of greater incorporation of ecological information, the enthusiasm must be tempered with the realization that there are significant data needs. Furthermore, with increases in scale and complexity, there are associated increases in uncertainty and greater consideration of outside factors that may interact with xenobiotics. For example, at the population-level a robust assessment would have to include "natural" stressors such as disease, density dependence, or predation, as these are known to interact significantly with xenobiotic stressors (Taylor et al. 1999; Coors et al. 2008; Maul et al. 2006; Forbes et al. 2001). Ultimately, the aim of these refinements is to facilitate better management of natural resources through improving information provided to decision makers. To that end, many significant challenges remain.

While it is apparent that there are many intellectual challenges related to the development and interpretation of sophisticated probabilistic, population, and landscape models, there are other challenges that are equally important but perhaps less obvious. In the light of the multitude of anthropogenic stressors and the interaction of these with naturally occurring stressors, there is greater need for collaboration among ecologists, conservation biologists, and ecotoxicologists. Moreover, to effectively manage wildlife populations will require a wider-scale acceptance of a new paradigm in assessing risks. Policy makers, decision makers, and scientific funding agencies will have to broader their perspectives on the factors that modify how natural systems respond to anthropogenic chemicals in light of the many other anthropogenic stressors faced by many populations and communities. A system-based thinking may be necessary to move into the next phase of wildlife toxicology and risk assessment. This will incorporate the development and implementation of inductive and deductive methods for understanding ecological responses to multiple stressors.

REFERENCES

Akçakaya, H.R., J.D. Stark, and T.S. Bridges. 2008. *Demographic Toxicity: Methods in Ecological Risk Assessment*. Oxford, UK: Oxford University Press.

Barnthouse, L.W., and M.T. Sorenson. 2007. *Population-Level Ecological Risk Assessment*: Boca Raton, FL: CRC Press.

Bartell, S.M., G. Lefebvre, G. Kaminski, M. Carreau, and K.R. Campbell. 1999. An ecosystem model for assessing ecological risks in Quebec rivers, lakes, and reservoirs. *Ecological Modelling* 124 (1): 43–67.

Berger, L, R Speare, P Daszak, DE Green, AA Cunningham, CL Goggin, R Slocombe, MA Ragan, AD Hyatt, and KR McDonald. 1998. Chytridiomycosis causes amphibian mortality associated with population declines in the rain forests of Australia and Central America. *Proceedings of the National Academy of Sciences* 95 (15):9031.

Burmaster, D.E., and P.D. Anderson. 1994. Principles of good practice for the use of Monte Carlo techniques in human health and ecological risk assessments. *Risk Analysis* 14 (4): 477–481.

Calow, Peter, Richard M. Sibly, and Valery E. Forbes. 1997. Risk assessment on the basis of simplified life-history scenarios. *Environmental Toxicology & Chemistry* 16 (9): 1983–1989.

Caswell, H. 2001. *Matrix Population Models*. 2nd ed. Sunderland, MA: Sinauer Associates.

Clements, W.H. 1994. Benthic invertebrate community responses to heavy metals in the Upper Arkansas River Basin, Colorado. *Journal of the North American Benthological Society* 13 (1): 30–44.

Clements, W.H., and PM Kiffney. 1994. Assessing contaminant effects at higher levels of biological organization. *Environmental Toxicology & Chemistry* 13 (3): 357–359.

Clements, W.H., and J.R. Rohr. 2009. Community responses to contaminants: Using basic ecological principles to predict ecotoxicological effects. *Environmental Toxicology & Chemistry/SETAC*:1.

Coors, A., E. Decaestecker, M. Jansen, and L. De Meester. 2008. Pesticide exposure strongly enhances parasite virulence in an invertebrate host model. *Oikos* 117 (12):1840.

Davidson, C. 2004. Declining downwind: amphibian population declines in California and historical pesticide use. *Ecological Applications* 14 (6):1892-1902.

Forbes, V. E., and P. Calow. 1999. Is the per capita rate of increase a good measure of population-level effects in ecotoxicology? *Environmental Toxicology & Chemistry* 18 (7): 1544–1556.

Forbes, V. E., P. Calow, and R. M. Sibly. 2001. Are current species extrapolation models a good basis for ecological risk assessment? *Environmental Toxicology & Chemistry* 20 (2): 442–447.

Forbes, V.E., R.M. Sibly, and P. Calow. 2001. Toxicant impacts on density-limited populations: A critical review of theory, practice, and results. *Ecological Applications* 11 (4): 1249–1257.

Forbes, V.E., and P. Calow. 2002. Population growth rate as a basis for ecological risk assessment of toxic chemicals. *Philosophical Transactions of the Royal Society–Series B–Biological Sciences* 357 (1425): 1299–1306.

Giddings, J.M., K.R. Solomon, and S.J. Maund. 2001. Probabilistic risk assessment of cotton pyrethroids: II. Aquatic mesocosm and field studies. *Environmental Toxicology & Chemistry* 20 (3): 660–668.

Graham, R.L., C.T. Hunsaker, R.V. O'Neill, and B.L. Jackson. 1991. Ecological risk assessment at the regional scale. *Ecological Applications* 1 (2): 196–206.

Hall, L.W., M.C. Scott, W.D. Killen, and M.A. Unger. 2000. A probabilistic ecological risk assessment of tributyltin in surface waters of the Chesapeake Bay watershed. *Human and Ecological Risk Assessment* 6 (1): 141–179.

Hayes, E.H., and W.G. Landis. 2004. Regional ecological risk assessment of a near shore marine environment: Cherry Point, WA. *Human and Ecological Risk Assessment* 10 (2): 299–325.

Kuhn, Anne, Waryne R. Munns, Jr., Sherry Poucher, Denise Champlin, and Suzanne Lussier. 2000. Prediction of population-level response from mysid toxicity test data using population modeling techniques. *Environmental Toxicology and Chemistry* 19 (9):2364–2371.

Landis, W. G. 2000. The pressing need for population-level risk assessment. *SETAC Globe* 1 (2): 44–45.

———. 2002. Population is the appropriate biological unit of interest for a species-specific risk assessment. *SETAC Globe* 3 (5): 31–32.

———. 2004. *Regional Scale Ecological Risk Assessment: Using the Relative Risk Model.* Boca Raton, FL: CRC Press.

Landis, W.G., and J.A. Wiegers. 1997. Design considerations and a suggested approach for regional and comparative ecological risk assessment. *Human and Ecological Risk Assessment: An International Journal* 3 (3): 287–297.

Landis, W.G., and J.K. Wiegers. 2007. Ten years of the relative risk model and regional scale ecological risk assessment. *Human and Ecological Risk Assessment* 13 (1): 25–38.

Landis, W. G., A. J. Markiewicz, R. A. Matthews, and G. B. Matthews. 2000. A test of the community conditioning hypothesis: Persistence of effects in model ecological structures dosed with jet fuel JP–8. *Environmental Toxicology & Chemistry* 19 (2): 327–336.

Levin, L., H. Caswell, T. Bridges, C. DiBacco, D. Cabrera, and G. Plaia. 1996. Demographic responses of estuarine polychaetes to pollutants: Life table response experiments. *Ecological Applications* 6 (4): 1295–1313.

Macintosh, D.L., G.W. Suter, and F.O. Hoffman. 1994. Uses of probabilistic exposure models in ecological risk assessments of contaminated sites. *Risk Analysis* 14 (4): 405–419.

Martin, J.W., M.M. Smithwick, B.M. Braune, P.F. Hoekstra, D.C.G. Muir, and S.A. Mabury. 2004. Identification of long-chain perfluorinated acids in biota from the Canadian Arctic. *Environmental Science Technology* 38 (2): 373–380.

Matthews, R.A., W.G. Landis, and G.B. Matthews. 1996. Community conditioning: An ecological approach to environmental toxicology. *Environmental Toxicology & Chemistry* 15:597–603.

———. 1996. The community conditioning hypothesis and its application to environmental toxicology. *Environmental Toxicology & Chemistry* 15 (4): 597–603.

Maul, J.D., J.L. Farris, and M.J. Lydy. 2006. Interaction of chemical cues from fish tissues and organophosphorous pesticides on *Ceriodaphnia dubia* survival. *Environmental Pollution* 141 (1): 90–97.

McConnell, L.L., J.S. LeNoir, S. Datta, and J.N. Seiber. 1998. Wet deposition of current-use pesticides in the Sierra Nevada mountain range, California, USA. *Environmental Toxicology & Chemistry* 17 (10): 1908–1916.

Mineau, P. 2002. Estimating the probability of bird mortality from pesticide sprays on the basis of the field study record. *Environmental Toxicology & Chemistry* 21 (7): 1497–1506.

Munns, W. R., Jr., D. E. Black, T. R. Gleason, K. Salomon, D. Bengtson, and R. Gutjahr-Gobell. 1997. Evaluation of the effects of dioxin and PCBs on *Fundulus heteroclitus* populations using a modeling approach. *Environmental Toxicology & Chemistry* 16 (5): 1074–1081.

Obery, A.M., and W.G. Landis. 2002. A regional multiple stressor risk assessment of the Codorus Creek watershed applying the relative risk model. *Human and Ecological Risk Assessment* 8 (2): 405–428.

Pahkala, M., A. Laurila, and J. Merila. 2001. Carry-over effects of ultraviolet-B radiation on larval fitness in *Rana temporaria. Proceedings: Biological Sciences* 268 (1477): 1699–1706.

Parmesan, C., and G. Yohe. 2003. A globally coherent fingerprint of climate change impacts across natural systems. *Nature* 421 (6918): 37–42.

Pearson, R.G., and T.P. Dawson. 2003. Predicting the impacts of climate change on the distribution of species: Are bioclimate envelope models useful? *Global Ecology and Biogeography* 12 (5): 361–371.

Raimondo, S. Jr., C.L. McKenney, and M. G. Barron. 2006. Application of perturbation simulations in population risk assessment for different life history strategies and elasticity patterns. *Human and Ecological Risk Assessment* 12: 983–999.

Relyea, R., and J. Hoverman. 2006. Assessing the ecology in ecotoxicology: A review and synthesis in freshwater systems. *Ecology Letters* 9 (10): 1157–1171.

Rohr, J.R., and B.D. Palmer. 2005. Aquatic herbicide exposure increases salamander desiccation risk eight months later in a terrestrial environment. *Environmental Toxicology & Chemistry* 24 (5): 1253–1258.

Rohr, J.R., J.L. Kerby, and A. Sih. 2006. Community ecology as a framework for predicting contaminant effects. *Trends in Ecology & Evolution* 21 (11): 606–613.

Rowe, C.L., W.A. Hopkins, and V.R. Coffman. 2001. Failed recruitment of southern toads (Bufo terrestris) in a trace element-contaminated breeding habitat: Direct and indirect effects that may lead to a local population sink. *Archives of Environmental Contamination and Toxicology* 40 (3): 399–405.

Salice, Christopher J., and Thomas J. Miller. 2003. Population-level responses to long-term cadmium exposure in two strains for the freshwater gastropod, *Biomphalaria glabrata*: Results from a life-table response experiment. *Environmental Toxicology and Chemistry* 22 (3):678–688.

Salice, CJ, TJ Miller, and G Roesijadi. 2009. Demographic Responses to Multigeneration Cadmium Exposure in Two Strains of the Freshwater Gastropod, Biomphalaria glabrata. *Arch of Environ Contam Toxicol* 56 (4):785–795.

Stark, JD, and JE Banks. 2003. Population-level effects of pesticides and other toxicants on arthropods. *Annual Review of Entomology* 48 (1):505–519.

Suter, G.W. 2006. *Ecological Risk Assessment*. Boca Raton, FL: CRC Press.

Solomon, K., J. Giesy, and P. Jones. 2000. Probabilistic risk assessment of agrochemicals in the environment. *Crop Protection* 19 (8–10): 649–655.

Tannenbaum, LV. 2005. A critical assessment of the ecological risk assessment process: A review of misapplied concepts. *Integrated Environmental Assessment and Management* 1 (1):66–72.

Taylor, S.K., E.S. Williams, and K.W. Mills. 1999. Effects of malathion on disease susceptibility in Woodhouse's toads. *Journal of Wildlife Diseases* 35 (3): 536.

Thompson, K.M., and J.D. Graham. 1996. Going beyond the single number: Using probabilistic risk assessment to improve risk management. *Human and Ecological Risk Assessment: An International Journal* 2 (4): 1008–1034.

U.S. EPA. 2004. A Discussion with the FIFRA Scientific Advisory Panel Regarding the Terrestrial and Aquatic Level II Refined Risk Assessment Models (Version 2.0). *Scientific Advisory Panel,* 2004.

U.S. EPA. 2008. Framework for ecological risk assessment. EPA/630/R-92/001. Washington, DC.

Vyas, N.B. 1999. Factors influencing estimation of pesticide-related wildlife mortality. *Toxicology and Industrial Health* 15 (1): 186–191.

Walker, R., W. Landis, and P. Brown. 2001. Developing a regional ecological risk assessment: A case study of a Tasmanian agricultural catchment. *Human and Ecological Risk Assessment* 7 (2): 417–439.

Wiegers, J.K., H.M. Feder, L.S. Mortensen, D.G. Shaw, V.J. Wilson, and W.G. Landis. 1998. A regional multiple-stressor rank-based ecological risk assessment for the fjord of Port Valdez, Alaska. *Human and Ecological Risk Assessment: An International Journal* 4 (5): 1125–1173.

Zhou, X.N., G.J. Yang, K. Yang, X.H. Wang, Q.B. Hong, L.P. Sun, J.B. Malone, T.K. Kristensen, N.R. Bergquist, and J. Utzinger. 2008. Potential impact of climate change on schistosomiasis transmission in China. *American Journal of Tropical Medicine and Hygiene* 78 (2): 188.

Zolezzi, M., C. Cattaneo, and J.V. Tarazona. 2005. Probabilistic ecological risk assessment of 1,2,4-trichlorobenzene at a former industrial contaminated site. *Environmental Science Technology* 39 (9): 2920–2926.

10 Looking Forward
The Global Future of Wildlife Toxicology

Ronald J. Kendall, Todd A. Anderson, George P. Cobb, Stephen B. Cox, Lee Hannah, Thomas E. Lacher, Steven M. Presley, Christopher J. Salice, and Philip N. Smith

CONTENTS

10.1 EMERGING ISSUES

Traditionally, wildlife toxicology has focused on standardized testing to determine effects of chemical exposures. Indeed, in the early years, this was valuable for ranking toxicities of various chemicals, particularly pesticides to which wildlife may be exposed. Probably the most extensive testing involved the registration process for agricultural pesticides under strict environmental law. This involved standardized test protocols aimed at determining adverse effects in wildlife, and included not only metrics of lethality (LD_{50}, LC_{50}, etc.), but also data on reproductive toxicity. In the 1980s, we saw a rapid expansion of large field studies that would build on laboratory work to examine effects of a chemical on wildlife. Again, this was driven to a large extent by agricultural pesticide studies, where toxicity of chemicals in agroecosystems was evaluated for the purposes of either product registration or product re-registration. Although the evolution of the field of wildlife toxicology, extending from laboratory to field studies in various settings (waste sites, oil-polluted areas, agricultural fields treated with pesticides, etc.), has addressed a number of toxic issues, we are still challenged by the scope of evaluations necessary to properly evaluate the breadth of wildlife that are exposed to toxicants, which may be locally isolated or globally distributed. Such challenges require a new non-traditional perspective (Kendall 1982, 1992; Kendall and Smith 2003).

In the process of developing the book *Wildlife Toxicology: Emerging Contaminant and Biodiversity Issues*, we realized that assessing one chemical's effect on representative suites of wildlife species is not moving the field ahead quickly enough, especially when considering the current rate of global environmental degradation. Therefore, we encourage the consideration of non-traditional wildlife toxicology research techniques. In particular, consideration of toxicity issues at the interface of conservation biology and biodiversity deserves consideration. There should be more integration of ecologists and conservation biologists with wildlife toxicologists as we address increasingly complex research themes and strive to offer solutions.

An understanding of the impacts of multiple stressors, particularly in the context of shifting global environmental conditions and the subsequent impacts of those shifts, will be critical as the field of wildlife toxicology continues forward. Specifically, a suite of effects associated with climate change are likely to alter toxicant impacts on wildlife, and one must not only understand the behavior of individual toxicants and toxins, but must also consider the effects of multiple stressors on target organisms *and* the ecosystems they inhabit. For example, variation in macro- or micro-nutrient availability in certain ecoregions may alter wildlife's response to an introduced toxicant. Such variability also is likely to result in different species compositions within the region in which an organism resides. Therefore, exposure to the toxicant will not necessarily produce similar effects in ecosystems with vastly different macronutrient composition. Classic examples demonstrate that soil chemistry and water hardness significantly alter toxicant effects on wildlife. Both of these parameters can be significantly altered by other parameters such as pH (e.g., Al availability in soil, and Cu availability in water), which can be affected by acid precipitation or by increased CO_2 partitioning into water bodies.

Furthermore, as we have discussed in individual chapters, atmospheric CO_2 serves as a primary greenhouse gas, and temperature changes are likely to covary with atmospheric CO_2 content. Temperature changes are important factors in determining rainfall amounts and evaporative rates. As a result, the environment continues to be impacted by periodic (e.g., El Niño) and trending climate change including shifts in mean temperature and precipitation patterns (Walther et al. 2002) as well as surface water pH (Wooton et al. 2008). These subtle changes in mean temperature can have large impacts on environments experiencing extreme events and on wildlife and other species that occupy those environments. As new species are capable of thriving in an area previously outside their range, invasive species become an issue (Bradley et al. 2008) and may subsequently impact food webs. Substantial changes in food webs, protective cover, nutrient flows, and dominant species in a habitat, all of which will respond to toxicological impacts observed within that ecosystem, will likely follow as a consequence of the direct and indirect effects of climatic shifts.

Emerging diseases also should be considered an important class of stressors that interact with toxicants in the environment to severely impact wildlife species.

Emerging and resurgent diseases pose a significant threat to human and animal populations globally, and the threat has become increasingly alarming in recent years. This is especially important because individuals (or even populations) who are physiologically stressed by environmental contaminants may be compromised in their ability to immunologically counter infection. Furthermore, animals may be compromised by poor nutritional status, or may be more susceptible to infection by altered neurological and behavioral processes. When one considers the introduction of a new, previously unreported disease into a wildlife host population as a new stressor that may compromise the ability of wildlife to function and thrive, it is imperative to also consider the overall negative effects that the disease may cause in other fauna and flora of the ecosystem. Thus, the disruption of ecosystem function resulting from an emerging epizootic in one wildlife species may be exacerbated by acute or chronic exposures to environmental contaminants, both to the diseased species and indirectly to other animal and plant species.

A prime historic example of the immediate pronounced devastating effects on an ecosystem resulting from the emergence and establishment of a zoonotic infectious pathogen is the introduction of rinderpest (*Morbillivirus*) into Africa in 1889. In less than ten years, the leading front of the rinderpest panzootic raced across more than 5000 kilometers and reached the Cape of Good Hope by 1897. The devastation of the epizootic on the African buffalo (*Syncerus caffer*) population was unprecedented, killing more than 90% of the buffaloes in Kenya, and the direct and indirect effects on buffalo-dependent predator populations and parasites were equally destructive (e.g., local tsetse fly populations disappeared (Plowright 1982)). Current research paradigms and approaches for studying the influences of environmental contaminants on the physiological susceptibility and potential for exposure to infectious diseases, particularly emerging infectious diseases of wildlife, are lacking in their scope and should be more holistic to include the multitude of direct and indirect relationships that exist.

10.2 FUTURE STRATEGIES

Wildlife toxicology historically has operated under the traditional paradigm that the hallmark of sound toxicological science was a rigorous mechanistic understanding of exposure and effect. This paradigm was built on a strong foundation of deductive reasoning, and has been reinforced by technological advances that have increased our ability to observe molecular processes. To some extent, it was also flavored by reductionist thinking that assumed that solid mechanistic understanding of individual chemical stressors would allow prediction of actual risk in the environment. However, toxicological exposures occur in the context of a whole suite of natural and anthropogenic stressors. These stressors interact in complex and nonlinear ways that make prediction of effects extremely difficult.

In a sense, epidemiology has dealt with similar problems with respect to human health. Because human health is influenced by a whole host of variables, many of which interact in complex ways, epidemiologists have emphasized statistical evaluation of temporal and spatial patterns in disease dynamics. Although such a holistic approach does not negate the need for mechanistic studies, it does add a measure of understanding about the context in which human disease develops and points to potential cause-effect relationships. Thus, it is the interaction of reductionist and holistic approaches that advances human health sciences; it is logical to assume that wildlife toxicology would also benefit from such an interaction.

Initially, advances in genomics and other "omic" domains promised the ultimate reductionist understanding. Detailed understanding at the level of the genome was thought to facilitate prediction at unprecedented levels of accuracy. However, this has not proved to be the case. In addition to difficulties in dealing with such large amounts of data, increased recognition of the importance of nonlinear interactions and epigenetic effects has hampered efforts to base prediction solely on genomic data. One theme that has emerged, however, is that data mining techniques and multivariate analyses (ultimately based on correlative relationships) can effectively discern patterns from data that eventually lead to hypothesis generation and increased understanding. This does not negate the importance of hypothesis testing, strong inference, and the development of theory, but instead provides new tools to reanalyze past data and historical trends and use lessons learned to guide new research.

As a domain, wildlife toxicology in the future will see an expansion of investigators and techniques that, instead of attempting to reduce the estimation of risk to its constituent elements, will rely on advanced computation and statistical methods to examine correlates of effect in the context of the milieu of stressors that simultaneously affect organisms, populations, and communities. In addition, the application of advanced spatial analysis tools and scenario development, using GIS methodologies, will allow for the assessment of risk in a spatial context and the development of spatially explicit predictions. Concomitant with such approaches, environmental toxicologists will increasingly rely on novel frameworks for evaluating hypotheses, in light of data. Information theoretic and Bayesian approaches offer promise in this regard. However, growth in this area will not replace, or negate the need for, detailed study of mechanisms, modes of action, and so on. In fact, the embrace of both deductive and inductive means of conducting science will be a symptom of the further maturation of wildlife toxicology as a scientific discipline that effectively addresses questions critical to continued human and environmental health.

10.3 RISK ASSESSMENT

Mathematical models are a mainstay of ecological risk assessment (ERA) and ecotoxicology. The state of the science of ERA currently is the use of probabilistic risk assessment (PRA) models that move beyond some of the assumptions associated with more simplistic approaches for risk estimation (risk or hazard quotients, Chapter 9). There are several advantages of PRA, including (1) incorporation of greater ecological realism, (2) explicit consideration of variability and uncertainty, and (3) outputs that can be meaningful in a biological and/or ecological context. Despite the well-recognized value of PRA and the widespread availability of tools, it is only now being implemented formally by the EPA for pesticide risk assessments. Because PRA tools are widely available, and because they hold many advantages over traditionally used quotient estimates of hazard, it is reasonable to expect PRAs to become more prevalent in ERAs. Areas of improvement in PRAs should include greater consideration of spatial and temporal factors that impact risk, multiple sources of anthropogenic and natural stressors, and statistical approaches (e.g., Bayesian) that provide more robust and tractable estimates of variation and uncertainty. A strong contributor to improvements in PRA will likely come from education of students in applied ecology and environmental toxicology programs in the execution and implementation of mathematical approaches.

In addition to PRA tools, there is growing emphasis on the use of modeling techniques that explicitly address higher levels of biological organization such as populations, communities, and ecosystems (Barnthouse and Sorenson 2007; Landis 2004; Akçakaya 2001). Models used by ecologists and conservation biologists can be applied to issues in wildlife toxicology and to understand the impacts of multiple anthropogenic and natural stressors. A research priority for improving the ability to predict effects of stressors on wildlife populations must involve collaborations among scientists from historically disparate fields and an acceptance by decision makers, policy makers, and funding agencies that wildlife populations are not exposed to chemicals or invasive species or climate change in isolation. Improvements in tools such as stochastic population models, spatially and temporally specific meta-population models, and species distribution models should be a research priority among environmental scientists. In total, the use and development of more sophisticated mathematical models in environmental toxicology and ERA will also necessitate empirical research that moves beyond what has been traditionally required by environmental regulations. To some extent, this is ongoing by a number of researchers across the fields of ecology and ecotoxicology, and significant advances have been made regarding our understanding of complex interactions between stressors and receptors. Advances in ecotoxicology and risk assessment are tied strongly to communication of more sophisticated modeling approaches to decision makers, policy makers, and students.

10.4 TEMPORAL SCALES

While the focus of ecotoxicolgy and ecological risk assessment is fundamentally to protect wildlife populations from xenobiotic-induced stressors, it is essentially limited to analysis and data that are focused on a timeframe that is less than a generation of a receptor. In reality, however, many populations are likely to be exposed to stressors (including chemicals) over the course of generations. In the face of multi-generation exposure to contaminants (or other anthropogenic stressors) at levels that exert deleterious effects, populations either adapt or go extinct. That organisms adapt to toxicants has been well documented (Antonovics et al. 1971; Klerks and Levinton 1989; Forbes 1998). Insect resistance to pesticides (Georghiou 1972), plant resistance to heavy metals (Antonovics et al. 1971), and fish adapted to PAHs (Nacci et al. 1999) are several strong examples where wild populations of organisms have adapted to anthropogenic chemicals. Although it is well recognized that organisms can adapt (rapidly, in many cases) to anthropogenic chemicals, the ecological consequences of these adaptive responses are less well understood.

Generally, evolutionary theory supports the concept of tradeoffs among traits (Roff and Fairbairn 2007; Stearns 1992). In a toxicological context, an adaptive response to a toxicant may result in a tradeoff in fitness in non-contaminated environments or an inability to tolerate other stressors. There are examples in the literature of adaptation to xenobiotics resulting in costs of tolerance that manifest as both a reduction in fitness (reduced reproduction in metal adapted chironomics (Donker et al. 1993) or an inability to tolerate novel stressors (decreased temperature tolerance in metal-adapted snails (Salice and Roesijadi 2009)). However, the manifestation of costs is not guaranteed (Otto et al. 1986) and is likely dependent on the complexity of the genetic mechanism of resistance with a greater number of genes conferring greater costs. Despite the obvious importance of evolutionary responses to anthropogenic chemicals in understanding and predicting population-level responses, it has not been a research priority.

A glaring gap in our understanding of the evolutionary responses to anthropogenic chemicals is a complete lack of studies on the evolutionary responses to multiple, simultaneous anthropogenic stressors (see, for example, the Bickham case study in Chapter 5). To date, there are only a handful of studies that have investigated the impact of multiple stressors on the adaptive potential of organisms. Despite this, there are working hypotheses that are in need of rigorous scientific evaluation (Hoffmann and Merilä 1999). The basic question relates to whether, under conditions of environmental stress, populations have a decreased ability to adapt. Studies mostly on *Drosophila* have produced mixed results, with some populations showing an increased adaptive capacity in the face of stressors (De Moed et al. 1997) and some studies showing the opposite (Hoffmann and Schiffer 1998; Imasheva et al. 1999). While these research efforts provide the basis for hypothesis development, they do not provide enough insight to be useful for understanding the responses of wildlife populations to multiple, complex stressors. A research priority for the field of wildlife toxicology and applied ecology is an improved understanding of the evolutionary responses to single and multiple stressors to include thorough investigations into potential costs of stressor adaptation.

10.5 ENVIRONMENTAL POLICY AND SCIENCE

10.5.1 The Measurement of Environmental Baselines

The complexity of multiple stressors and interplay with global change suggests that traditional concepts of a reference baseline may need to be reexamined. The notion of "pristine" conditions is now recognized as outdated. Continual change has been the norm, even prior to human arrival, so how can management objectives or tests of contaminant effects find objective reference points?

Detecting change against a dynamic baseline requires sampling on multi-temporal and spatial scales. A better conception of baseline would be more of a background "trajectory of change" than a static condition. In contrast to past management objectives that sought a single reference point in time and space (such as pre-European contact in a particular protected area in North America), future management success must be measured against multiple points in time and space. An example of such monitoring comes from the Tropical Ecology Assessment and Monitoring network (TEAM), supported by the Gordon and Betty Moore Foundation (http://www.team-network.org/en/). TEAM is a network of biological research stations, arrayed across elevation, biome, and human land-use gradients. This system involves a network of partners dedicated to measuring landscape-level change and biodiversity metrics over long time periods. The National Ecological Observatory Network (NEON), supported by NSF, intends to provide similar temporal continuity, although the present network may not be dense enough to capture multiple environmental gradients (http://www.neoninc.org/). NEON, however, is exclusively a national program, while TEAM is becoming increasingly global.

While monitoring is often an unpopular investment, the complexity of current environmental issues gives it renewed relevance. Efficient allocation of resources to environmental problem-solving cannot be achieved unless the magnitude, relevance, and consequences of change are understood. Similarly, the efficacy of investments that are made cannot be judged with monitoring that is static in theory or practice, nor can effective policy be developed without an understanding of this inherent change.

10.5.2 Integration of Science and Policy

The evolution of monitoring is part of an overall evolution of the relationship between science and policy that is needed if environmental toxicology is to respond effectively to the next generation of problems. In the past, policy has responded to problems, and these policy responses have driven the science of solutions. For instance, environmental toxicology and risk assessment stemmed from environmental policies of the 1970s and 1980s that were first implemented in response to observed effects of single chemicals (such as DDT). A more proactive, less sequential relationship is needed between science and policy to effectively meet emerging threats and manage them before they become expensive problems.

Two steps can be taken to address these shortcomings. First, mechanisms that bring managers and researchers together for joint problem identification and solving can help derail the science-to-policy-to-science-to-management monorail. In the United States, much of the research capability of environmental management agencies has been disassembled. Reassembling this research capability can help managers solve problems before they reach levels that lead to public outcry and policy response.

For instance, the U.S. Park Service now has no research scientists to work collaboratively with managers in solving problems as complex as air pollution impacts on visibility or climate-change-driven range shifts in species. Management of contaminants is accompanied by a strong research program in the EPA, but remains isolated from other federal and local resource management agencies, and, therefore, from many of the most important synergetic problems of the day. The need for integrated problem-solving between researchers and managers and across agencies may be the strongest argument for a unified (e.g., Park Service, Forest Service, and the EPA) national environmental department. In many other countries, this research and management capability will have to be assembled for the first time.

Second, proactive global and national goals can provide a vital counterpart to reactive problem-solving. In the past, National Parks and National Forests provided positive models to counter the momentum of destructive exploitation. New models of positive integrated response are needed to counter the current momentum of multiple stressors and complex interactions. Integrated management of existing conservation areas such as the Greater Yellowstone Coalition is one example of this model. Integrating environmental performance into corporate goals may be another. Some international success in setting proactive goals gives some reason for hope.

10.5.3 Policy and Research Integration — A Global Example

International environmental policy, and especially conservation policy, has been driven by several international agreements and conventions over the past two decades. The Earth Summit in Rio de Janeiro in 1992 was a landmark event in the development of coordinated global conservation efforts. It marked the opening for signature of the Convention on Biological Diversity (CBD), now the major convention for global conservation policy and decision-making (http://www.cbd.int/). The CBD decision-making body is the Conference of the Parties (COP) that meets on a regular basis; there are currently 192 signatory nations. There are two non-parties: the Holy See has not signed the Convention, and the United States has signed but not ratified. The COP develops priorities and appoints committees and working groups. Scientific and technical input to the COP is provided by the Subsidiary Body on Scientific, Technical and Technological Advice (SBSTTA). Important documents produced by the CBD include the Global Biodiversity Outlook. The CBD, though the COP, also adopted the 2010 Target as a goal, which was also confirmed by the World Conference on Sustainable Development in Johannesburg in 2002. The CBD has subsequently developed indicators to help achieve this goal (Balmford et al. 2005).

The previous paragraph illustrates the desire at the international level to have coordinated science and policy programs to address global conservation needs. This level of coordination is often overlooked in the United States, given the country has never ratified the CBD. Difficulties persist, however, within the CBD given the use of consensus decision-making, much like the activities of the United Nations. Nevertheless, the Convention places the conservation of biodiversity within broader economic discussions, including the sustainable use of biodiversity and the sharing of economic benefits of commercial genetic resources and biotechnology derived from biodiversity. Costs of implementing conservation initiatives should therefore be balanced by the environmental, economic, and social benefits derived through the conservation of biodiversity and natural resources. Most importantly, the decisions made by the Convention are legally binding for the Parties.

The integration of conservation and economic benefit does raise issues about how prevailing and future economic conditions might affect the decision-making process within the CBD. Within the United States there is a continual assessment of the costs and benefits of all conservation initiatives, from biodiversity conservation to energy policy and climate change. The same occurs in the international arena. The current economic climate at the time of this writing (mid-2009 to 2010) has made the allocation of resources to resolve environmental problems a tough sell in many countries. Thus, even in the context of legally binding agreements such as the CBD, we might expect retrenchment of certain policies should economic conditions worsen.

We now must evaluate and address impacts on wildlife that involve the complex interactions of pesticides, industrial by-products, pharmaceuticals, personal care products, nanoparticles, disease, and climate change and the interplay of these stressors on a complex landscape of deforestation, agricultural development, and urban expansion. Add to this the need to work on transboundary problems, and frequently multinational problems, while dealing with the science policy and permitting processes of a number of governmental agencies in several countries simultaneously. All of this calls for a new approach that deals with uncertainty, spatial complexity, and policies that facilitate international collaboration. This will be essential for the development of global conservation priorities and the prevention of new extinctions (Brooks et al. 2006; Ricketts et al. 2005). This is the time for the development of a new vision of big science, closely integrated with national and international policy mechanisms. For the fields of wildlife toxicology and biodiversity conservation, this is where future efforts need to focus.

10.6 GLOBAL PERSPECTIVES

Chapter 5 in this volume presents a number of detailed suggestions on how wildlife toxicology and conservation biology might be better integrated to address issues of common concern. Table 5.1 in that chapter provides a number of specific recommendations. There are some bigger issues, however, and this book has put wildlife toxicology into a broader, global context. Twenty years ago the focus of wildlife toxicology was on single compound impacts on single species. This concept is no longer adequate to deal with the multi-stressor impacts on communities and, in cases such as amphibians, global biodiversity.

Data now exist that document the declining health of oceans and marine life, and depletion of fish stocks, which are indicative of the global nature of contaminant issues (Pew Oceans Commission 2003). The world's oceans and seas are unique in that they signal the cumulative global impact of humankind. They receive amalgamate runoff containing fertilizers, pesticides, and municipal and industrial effluents, and they assimilate volatile (persistent) pollutants from the atmosphere. Oceans are not static: they are moving systems that connect one another, distant lands, and all living estuarine, oceanic and many terrestrial organisms. As detailed by Smith et al. in Chapter 8, "Marine pollution is a major factor in the current and global marine ecosystem crisis (United Nations 2007; United Nations Educational Scientific and Cultural Organization 2002) . . . There is growing acknowledgement that marine ecosystem services are under threat, which will increase the economic burden on the already fragile global economy."

It is now inappropriate, even naïve, to consider environmental contamination (whether marine or terrestrial) as only a local, national, or even regional in scope. Environmental contamination issues may be localized, but are often transboundary, transnational, intercontinental, and global in nature. Indicative are the numerous studies of persistent organic pollutants in polar environmental matrices and organisms. Local efforts remain critically important for addressing contamination issues, but it has become clear that much broader perspectives and efforts are necessary to reverse current trends of global environmental degradation.

Until now, no entity (government, religion, convention, agreement) has demonstrated sufficient authority or the commitment to address global environmental issues affecting wildlife and biodiversity. Could this be because of a lack of unity of voice among the world's environmental scientists on these important issues? Is consensus on addressing global environmental contamination issues even possible? Until the dialogue is initiated among the world's leading environmental scientists, clarity in vision and a resulting path forward will not emerge. Environmental scientists must assume the responsibility for finding solutions to environmental issues that threaten all life. Thus we hope that this book will serve as a catalyst for such a dialogue within the scientific community.

Moreover, we propose that a global conference addressing environmental contamination and biodiversity issues take place in the near future. Participation of qualified environmental scientists from around the globe should be the requirement of the conference, not merely a goal. Topics in need of discussion include the status of global pollutants, biodiversity trends, wildlife populations, climate change, and anthropogenic factors contributing to effects on the former. In addition, a high priority must be placed on the discussion of novel statistical, computational, and cross-disciplinary approaches and their application to environmental problem-solving. Such a conference would permit prioritization of environmental issues to be addressed by international and national governing bodies.

10.7 SUMMARY

If you look at some of the current outlets for publication of wildlife toxicology studies, the general trend is heavily weighted toward traditional techniques and lines of inquiry. Although this information is valuable, it is neither robust nor aggressive enough to impact societal and political assessment of toxic chemical impacts on wildlife, especially when linked with changes in biodiversity and global climate change. We hope that *Wildlife Toxicology: Emerging Contaminant and Biodiversity Issues* will provide stimulus and ideas to actuate change in scientific thinking related to how we assess chemicals in the environment and their impacts on wildlife at the interface of the challenges to protect biodiversity.

REFERENCES

Akçakaya, H.R. 2001. Linking population-level risk assessment with landscape and habitat models. *The Science of the Total Environment* 274 (1–3): 283–291.

Antonovics, J., A.D. Bradshaw, and R.G. Turner. 1971. Heavy metal tolerance in plants. *Advances in Ecological Research* 7 (1): 85.

Balmford, A., L. Bennun, B.T. Brink, D. Cooper, I.M. Cote, P. Crane, A. Dobson, et al. 2005. Ecology: The Convention on Biological Diversity's 2010 target. *Science* 307 (5707): 212–13.

Barnthouse, L.W., M.T. Sorenson, and W. Munns, Jr. 2007. *Population-Level Ecological Risk Assessment*. Boca Raton, FL: CRC Press.

Bradley, B., M. Oppenheimer, and D. Wilcove. 2008. Climate change and plant invasions: Restoration opportunities ahead? *Global Change Biology* 15:1511–1521.

Brooks, T.M., R.A. Mittermeier, G. A. da Fonseca, J. Gerlach, M. Hoffmann, J. F. Lamoreux, C.G. Mittermeier, J.D. Pilgrim, and A.S. Rodrigues. 2006. Global biodiversity conservation priorities. *Science* 313 (5783): 58–61.

De Moed, G.H., G. De Jong, and W. Scharloo. 1997. Environmental effects on body size variation in *Drosophila melanogaster* and its cellular basis. *Genetics Research* 70 (01): 35–43.

Donker, M.H., C. Zonneveld, and N.M. Van Straalen. 1993. Early reproduction and increased reproductive allocation in metal-adapted populations of the terrestrial isopod *Porcellio scaber*. *Oecologia* 96:316–323.

Forbes, V. E., ed. 1998. *Genetics and Ecotoxicology*. Current Topics in Ecotoxicology and Environmental Chemistry Series, edited by G. Rand, P. Calow, and M. A. Lewis. Ann Arbor, MI: Braun-Brunfield.

Georghiou, G.P. 1972. The evolution of resistance to pesticides. *Annual Review of Ecology and Systematics* 3 (1): 133–168.

Hoffmann, A.A., and M. Schiffer. 1998. Changes in the heritability of five morphological traits under combined environmental stresses in *Drosophila melanogaster*. *Evolution* 52 (4): 1207–1212.

Hoffmann, A.A, and J. Merilä. 1999. Heritable variation and evolution under favourable and unfavourable conditions. *Trends in Ecology & Evolution* 14 (3): 96–101.

Imasheva, A.G., D.V. Bosenko, and O.A. Bubli. 1999. Variation in morphological traits of *Drosophila melanogaster* (fruit fly) under nutritional stress. *Heredity* 82 (2): 187–192.

Kendall, R. J. 1982. Wildlife toxicology. *Environmental Science & Technology* 16:448a–453a.

———. 1992. Farming with agrochemicals: The response of wildlife. *Environmental Science & Technology* 26 (2): 238–245.

Kendall, R.J., and P.N. Smith. 2003. Wildlife toxicology revisited. *Environmental Science & Technology* 37 (9): 178A–183A.

Klerks, P.L., and J.S. Levinton. 1989. Rapid evolution of metal resistance in a benthic oligochaete inhabiting a metal-polluted site. *Biological Bulletin* 176:135–141.

Landis, W.G., ed. 2004. *Regional Scale Ecological Risk Assessment: Using the Relative Risk Model*: Boca Raton, FL: CRC Press.

Nacci, D., L. Coiro, D. Champlin, S. Jayaraman, R. McKinney, T.R. Gleason, W.R. Munns, Jr., J.L. Specker, and K.R. Cooper. 1999. Adaptations of wild populations of the estuarine fish *Fundulus heteroclitus* to persistent environmental contaminants. *Marine Biology* 134 (1): 9–17.

Otto, E., J.E. Young, and G. Maroni. 1986. Structure and expression of a tandem duplication of the *Drosophila* metallothionein gene. *Proceedings of the National Academy of Sciences USA* 83 (16): 6025–6029.

Pew Oceans Commission. 2003. America's living oceans: Charting a course for sea change. Arlington, VA: Pew Oceans Commission.

Plowright, W. 1982. The effects of rinderpest and rinderpest control on wildlife in Africa. *Symp Zool Soc London* 50:1–28.

Ricketts, T. H., E. Dinerstein, T. Boucher, T.M. Brooks, S.H. Butchart, M. Hoffmann, J.F. Lamoreux, et al. 2005. Pinpointing and preventing imminent extinctions. *Proceedings of the National Academy of Sciences USA* 102 (51): 18497–501.

Roff, D.A., and D.J. Fairbairn. 2007. The evolution of trade-offs: Where are we? *Journal of Evolutionary Biology* 20 (2): 433.

Salice, C.J., and G. Roesijadi. 2009. Rapid adaptive responses and latent costs associated with multigenerational exposures to an anthropogenic stressor in two strains of a freshwater snail. *Evolution* In review.

Stearns, S.C. 1992. *The Evolution of Life Histories*. New York: Oxford University Press.

United Nations. *The United Nations Convention on the Law of the Sea: A Historical Perspective*. United Nations 2007. Available from http://www.un.org/Depts/los/convention_agreements/convention_overview_convention.htm, http://www.un.org/Depts/los/convention_agreements/convention_historical_perspective.htm.

United Nations Educational Scientific and Cultural Organization. 2002. The Final Design Plan for the Health of the Ocean (HOTO) Module of the Global Ocean Observing System (GOOS). Paris: GOOS.

Walther, G-R., E. Post, P. Convey, A. Menzel, C. Parmesan, T.J.C. Beebee, J.-M. Fromentin, O. Hoegh-Guldberg, and F. Bairlein. 2002. Ecological responses to recent climate change. *Nature* 416:389–395.

Wooton, J.T., C.A. Pfister, and J.D. Forester. 2008. Dynamic patterns and ecological impacts of declining ocean pH in a high-resolution multi-year dataset. *Proceedings of the National Academy of Sciences USA* 105:18848–18853.

Index

"f" indicates material in figures. "n" indicates material in notes. "t" indicates material in tables.

B

C

D

G

H

I

J

K

M

O

S

T

U

V